Remote Sensing
of the Lower Atmosphere

Remote Sensing
of the Lower Atmosphere

An Introduction

Graeme L. Stephens
Colorado State University

New York   Oxford
OXFORD UNIVERSITY PRESS
1994

Oxford University Press

Oxford   New York   Toronto
Delhi   Bombay   Calcutta   Madras   Karachi
Kuala Lumpur   Singapore   Hong Kong   Tokyo
Nairobi   Dar es Salaam   Cape Town
Melbourne   Auckland   Madrid

and associated companies in
Berlin   Ibadan

Published by Oxford University Press, Inc.,

198 Madison Avenue, New York, New York 10016-4314

Oxford is a registered trademark of Oxford University Press

Library of Congress Cataloging-in-Publication Data
Stephens, Graeme L.
Remote sensing of the lower atmospheric : an introduction
Includes bibliographical references and index.
1. Atmospheric physics - Remote sensing.
2. Atmosphere - Remote sensing.
3. Atmospheric radiation - Remote sensing.
4. Electromagnetic radiation - Remote sensing.
I. Title.
QC871.S86   1994   551.5'028 - dc20
ISBN 0-19-508188-9

2 4 6 8 9 7 5 3

Printed in the United States of America
on acid-free paper

Preface

With its growth, the atmospheric sciences have become increasingly complicated forcing the working scientist to specialize in one discipline or another. This especially applies to the subject of remote sensing. Yet remote sensing crosses over the imaginary disciplines of the science and is found in one form or other in most studies of the atmosphere. Data derived from remote sensing systems are incorporated into both the initialization and validation of forecast models, for example. They are also now used extensively in studies that seek to gain fundamental new knowledge about the workings of the atmosphere and the Earth's climate system. This book has grown out of the belief that it is now essential, for practicing meteorologists and for scientists who are to devote their careers to the study of the atmosphere, to familiarize themselves with general topics on remote sensing. While the book is intended for both graduate students and atmospheric scientists, it may also be useful to experts on the subject of remote sensing who seek to place their speciality in a broader perspective.

Interest in the remote sensing of the atmosphere has blossomed. The number of articles scattered in the literature representing various disciplines must run to thousands; there are special journals devoted to the subject and numerous books that deal with various topics of remote sensing are available to the interested scientist. The last time I counted, I found 27 of these books on my shelf alone. This is another book about remote sensing. Given the apparent abundance of books on the topic, an obvious question to ask is how does this book differ from others? I believe the difference is one of emphasis. This book focuses on understanding the basic interactions between radiation and the atmosphere rather than on cataloging numerous mathematical recipes. It is more conceptual and about a way of thinking of these interactions as much as it is a dialogue about remote sensing. Most remote sensing books are specialized in one way or another. I have books that address only the remote sensing associated with a specific technology (such as radar), or a specific platform (such as "satellite remote sensing"), or which focus on a specific class of properties such as surface properties. It was never my intention to present the newest methods of remote sensing in this book, nor was it my intention to emphasize the most recent developments of sensor systems, imaging systems, or platforms because this would surely date the book before ever reaching the reader. My aim is to describe the basic interactions between electromagnetic radiation and matter and to highlight these interactions using examples drawn from remote sensing. This book was written with the belief that an understanding of these interactions provides a solid foundation for understanding other topics of remote sensing along with the processes that govern the distribution of radiation in the Earth's atmosphere.

In writing this book I grew to appreciate the comments of professor Peter V. Hobbs who, in reference to his own writings (Hobbs, 1974), remarked on how one never finishes a book of this type but merely abandons it. I finally abandoned this book in 1992. I realized the futility in writing a comprehensive text on all the significant aspects of remote sensing from the beginning. Omission of certain topics was obviously necessary, as was the omission of various references to literature. I hope my colleagues will forgive me if my sense of priorities in selection does not match their own. I am guilty of committing a grave disservice to inversion theory, I have largely ignored the remote sensing of ocean properties, I have overlooked the remote sensing of energy fluxes, including the top of the atmosphere radiative budget, I have omitted the growing technique of Raman sensing, and I have refrained from highly technical discussions about the operation and performance of selected instrumentation.

This book, grown from two graduate courses taught in the Department of Atmospheric Sciences at Colorado State University, is intended both as an introductory text on atmospheric remote sensing and as a supplement to topics on atmospheric radiation. The topics discussed in the book are divided into two main sections. After the introductory chapter, the basic properties of radiation and how it interacts with matter are presented in Chapters 2–5. Chapters 6 and 7 build on these topics and, after a brief introduction to the subject of radiative transfer, focus on topics of passive remote sensing of various atmospheric parameters. The final chapter discusses the topic of active sensing and includes the subjects of radar and lidar sensing. The bibliographical notes at the end of each chapter list some historical papers and other important papers, relevant books, and review articles, as well as special papers that offer specific details about the topic in question. These references are a matter of personal judgment and, while the list is not extensive, should provide the interested reader with a starting point for his or her own literature search. The problems at the end of each chapter are more didactic in nature than is perhaps typical of practical problems encountered in remote sensing. Five more substantial projects, involving analyses of real data, are included in Appendix 2. The aim of these is to expose the interested student to some of the more practical problems one encounters in remote sensing. Details about how these data may be obtained are also provided in the Appendix.

Numerous colleagues contributed to this text in various ways. A number read selected chapters of the manuscript and suggested improvements in the presentation. One of my colleagues and a friend painstakingly read the entire manuscript in an effort to free it from errors and I am in your debt. My graduate students and students from different classes were also inflicted with the task of working through most of the problems and helped massage many of them into a meaningful form. A few of my graduate students also substantially contributed to the projects listed in the Appendix. I will not mention specific individuals because the list would be long and you know who you are.

This book took several years to organize and write. The time it took was time I never really had and it was time taken from my family. To my family I apologize and thank you for your understanding. I know that I can never give this time back.

Fort Collins, Colorado   G. L. S.
June 1993

Contents

5. Macroscopic Interactions—Particle Absorption and Scattering, 190

**6. Passive Sensing—Extinction and
   Scattering, 261**

7. Passive Sensing—Emission, 328

8. Active Sensing, 395

Appendix 2: Class Projects, 471

Remote Sensing
of the Lower Atmosphere

1
Introductory Surveys

This chapter is intended only as a brief survey of a number of properties of the atmosphere that will be needed later. The chapter also includes an abbreviated overview of satellite platforms since measurements from these platforms over the past two decades have advanced our ability to observe the Earth's atmosphere in important ways. Readers knowledgeable about these topics may find it convenient to skip this chapter and consult it only when reference to it is made in the subsequent text.

1.1   General Classifications

Broadly speaking, remote sensing is a way of obtaining information about properties of an object or volume without coming into physical contact with that object. While we may consider these to be indirect methods of measurement, most of our everyday measurements are indirect. Objects are weighed by observing how much they stretch a spring, or their temperature is deduced by the distance a column of mercury rises in a tube.

The variety of scientific disciplines that rely on remote sensing techniques is so broad that a single treatise could not possibly hope to cover all aspects of the subject. Remote sensing is by far the main observational tool of astrophysicists; it is used extensively by geologists in mineral exploration, by physical geographers, biologists and oceanographers, metallurgists sensing properties of blast furnaces, and by meteorologists and climatologists. While the more routine meteorological measurements of the upper air are currently performed using in situ devices on balloons, these traditional approaches are now being systematically complemented with sophisticated remote sensing methods. A study of the atmosphere by means of remote sensing involves the use of instruments on artificial satellites and orbiting probes, on aircraft, rockets, balloons as well as by instruments located at the ground.

The focus of this book is largely on remote sensing of the troposphere based on measurement of electromagnetic radiation. Within this viewpoint, it is possible to classify various methods of remote

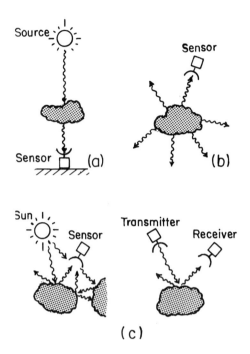

Figure 1.1 Three basic types of radiative properties used in electromagnetic remote sensing topics discussed in this book: (a) extinction, (b) emission, (c) scattering of either an incoherent or coherent source.

sensing in a number of different ways. For example, *passive* sensing relies on measuring natural levels of radiation, such as the radiation emitted from the Earth's surface or sunlight reflected by the atmosphere. On the other hand, *active* sensing makes use of electromagnetic energy transmitted from a specially chosen source to some "object" or "target," and then monitoring the interactions between this object and the radiation. The target may be a naturally occurring fluctuation in the atmosphere or a fixed target at a set distance from the transmitter.

Emphasis is placed on understanding the interactions between electromagnetic radiation and the atmosphere; these interactions provide a convenient way of categorizing the topics discussed in this book. The remote sensing topics are discussed in terms of
- extinction methods
- emission methods
- scattering methods

according to the different types of radiative processes (extinction, scattering, absorption–emission) that provide the basis for the approach under consideration. Figure 1.1 is a schematic view of these processes as they might be used in certain remote sensing applications. In the first case, the radiation from a given, known source is observed, and the amount of radiation attenuated or lost from the beam as it is transmitted to the sensor is used to determine the amount of matter along the path. This loss can either occur through absorption or scattering, or from a combination of both. A large number of remote sensing methods actually rely on this kind of approach (see Chapters 6 and 7). Radiation can also be emitted by the object and the second category deals with measurements of infrared and microwave emission which are used to obtain information about the thermal structure of the atmosphere and other properties (Chapter 7). The third category of remote sensing utilizes the properties of scattering.[1] In the example shown in Fig. 1.1c, the medium is illuminated either by a source of incoherent radiation, such as sunlight or infrared radiation emitted from the Earth's surface or by an artificial coherent source, like that transmitted by a laser source (as in the case of lidar – **l**ight **d**etection **a**nd **r**anging) or a source of microwave radiation (as in the case of radar – **ra**dio **d**etection **a**nd **r**anging). This radiation is then scattered by an object and received by a sensor optimally tuned to detect the signal.

1.2 The Nature of Inverse Problems

Problems of remote sensing fall into a category referred to as *inverse problems*. The nature of these inverse problems is portrayed in Fig. 1.2 in the form of an analogy and can be posed in the following way. How well can we infer the dragon from observations of its tracks? Clearly this line of inquiry is fruitless without some a priori knowledge of dragons and the types of tracks they leave. Therefore a central theme of the first part of this book is the understanding of how electromagnetic radiation interacts with matter since these are the tracks we exploit in remote sensing applications to infer information about the atmosphere.

---

[1] Scattering means elastic scattering, for which the frequency of light is unaltered. Another class of scattering is also relevant to remote sensing: Inelastic scattering results in a shift of the frequency of the incident light and the desired properties of the atmosphere are related to this frequency shift. Examples of these methods are Raman and fluorescent scattering and neither are dealt with in this book.

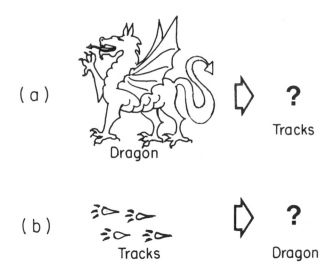

Figure 1.2 (a) The direct problem: Describe the tracks of a dragon.
(b) The inverse problem: Describe a dragon from its tracks (from
Bohren and Huffman, 1983.)

The theory of inverse problems is referred to as *retrieval or inversion theory*; atmospheric remote sensing is only one of a number of disciplines for which retrieval theory is an important topic. Detailed discussion of the general inversion problem and methods of solution by interactive, nonlinear and statistical approaches is considered beyond the scope of this book and the reader is referred to an introductory treatment of this subject provided by Twomey (1977) and others cited later in the notes.

Before leaving inversion theory, a simple example relevant to later topics will now be described. Suppose that an absorbing gas in the atmosphere emits radiation to a satellite in a way that depends on its temperature (which is true, as we see later in Chapter 2). Suppose also that the wavelength of the emission is related to a single level (this is not true, as we learn later) so that we obtain a relationship between wavelength and temperature as sketched in Fig. 1.3. Knowing how the intensity of radiation, the wavelength of the radiation, and the temperature of the emitting gas are related provides us with a way of estimating the temperature as a function of wavelength and thus altitude.

In reality, the actual situation is unfortunately more complex than the ideal one just described. The radiation at one wavelength does not come from a single height but is distributed over a broad layer of the atmosphere. This results in a blurring or a departure from a one–to–one correspondence between wavelength and height as supposed for the ideal case illustrated in Fig. 1.3b.

This example is typical of many of the inversion problems in remote sensing where a set of measurements (in this case emission as a function of wavelength) is influenced by all values of the unknown distribution (i.e., the vertical distribution of temperature throughout the layer that contributes to the emission). We represent this blurring effect in the following way. Let the sought–after distribution be $f(x)$ (which may be the temperature profile as a function of altitude z), and let $K_i(x)$ be the relative contribution curve for a wavelength λ_i. The interval between x and $x + \Delta x$ contributes to the measurement of the ith channel of a radiometer the amount $f(x)K_i(x)\Delta x$. The total measured radiation in this channel is

$$g_i = \int_a^b K_i(x)f(x)dx \tag{1.1}$$

where the limits of the integral depend on the details of the problem at hand. This equation is known as the *Fredholm integral equation of the first kind* because the limits of the integral are fixed and because $f(x)$ appears only in the integrand. The function $K_i(x)$ is known as the "kernel" or "kernel function."

We will see in later chapters how different remote sensing problems reduce to (1.1) or a form similar to (1.1) and its solution requires inversion to obtain the distribution $f(x)$. This is an unfortunate circumstance since the solutions to (1.1) suffer a number of difficulties, including non–existence, non–uniqueness, and instability. Non–existence is usually not an issue for most practical problems since we measure g_i and $f(x)$ exists through physical considerations. Non–uniqueness is related to the blurring effect discussed earlier. The practical consequence of this is that there exist several functions $f(x)$ which produce the same function g_i. This problem tends to be overcome to a certain extent by restricting the class of admissible solutions to physically realizable ones. In this way, a priori information is introduced into the retrieval scheme.

The major difficulty in solving (1.1), at least for most practical problems, concerns the problem of instability which arises, for example, from errors in the observations g_i. With a small error ϵ_i in

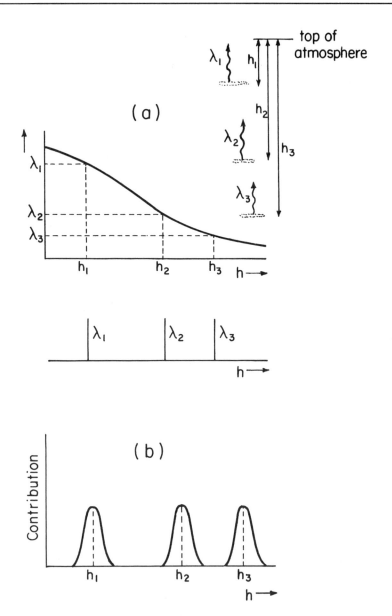

Figure 1.3 (a) Schematic diagram for a satellite–based atmospheric temperature profile measurement. The contribution to the radiation measured at the selected wavelength arises from a discrete level. (b) A version of the contribution function of Fig. 1.3a corresponding to a more practical case where the measured radiation originates from a range of different levels in unequal proportions (from Twomey, 1977).

g_i, (1.1) becomes

$$g_i + \epsilon_i = \int_a^b K_i(x)f(x)dx \tag{1.2}$$

where ϵ_i produces an arbitrarily large change in $f(x)$ and the ultimate success of any retrieval largely depends on the accuracy of the measurement g_i and on the shape of $K_i(x)$. A simple example that underlines this important point will now be presented.

Excursus: The Ill–Posed Nature of the Inversion Problem

Let us illustrate the nature of the instability using the problem of temperature retrieval. We can write the transfer equation of infrared radiation in the form of (1.1) (this equation and its interpretation is given in Chapter 7). The integral term of (7.32) can be expressed as

$$I_i = \int_{z_s}^{z_o} \mathcal{B}_i(z')\mathcal{W}_i(z', \infty)dz' \tag{1.3}$$

where I_i is the measured intensity of the radiation, $\mathcal{W}_i$ is the kernel function (for this particular problem we call this the weighting function, and a more complete discussion of this function is given in Chapter 7), and $\mathcal{B}_i$ is the distribution function containing temperature information. We use the subscript i to refer to the measurement of a particular satellite radiometer channel and assume there are $i = 1, \ldots, M$ spectral measurements.

Consider a simple example of the solution to (1.3). Suppose we have only two observations (i.e., $M = 2$), then we might discretize the integral in (1.3) as

$$\begin{aligned}
\mathcal{B}_1\mathcal{W}_{1,1} + \mathcal{B}_2\mathcal{W}_{1,2} &= I_1 \\
\mathcal{B}_1\mathcal{W}_{2,1} + \mathcal{B}_2\mathcal{W}_{2,2} &= I_2
\end{aligned} \tag{1.4}$$

where $\mathcal{W}_{i,j}$ is the weighting function of the jth layer for the ith channel. Let us suppose for the sake of illustration that the weighting functions have the numerical values $\mathcal{W}_{1,1} = 1, \mathcal{W}_{1,2} = 1, \mathcal{W}_{2,1} = 2$, and $\mathcal{W}_{2,2} = 2.000001$, and that the intensities are $I_1 = 2$ and $I_2 = 4.000001$. We then obtain

$$\mathcal{B}_1 = 1$$
$$\mathcal{B}_2 = 1$$

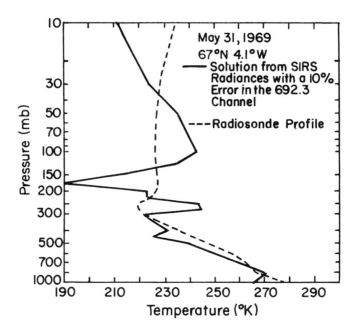

Figure 1.4 Comparison of radiosonde and a retrieved temperature profile when a spurious error of 10% is added to one channel (from Smith, 1972).

as a solution to (1.4). Suppose that there is a small uncertainty in one of the measured intensities, I_2, such that the value $I_2 = 4$ is recorded instead of the value 4.000001. Then the solution to (1.4) is

$$B_1 = 2$$
$$B_2 = 0$$

This is a dramatic change to the solution and nicely illustrates the problem of instability. In reality, matters become even worse as the number of measurements (i.e., as M) increases. Figure 1.4 provides an example of how random errors propagate in an actual temperature retrieval. The retrieved temperature profile obtained using infrared measurements from a radiometer when a spurious error of 10% is added to a single instrument channel is shown in this diagram. This point is explored further in one of the projects introduced in Appendix 2.

To summarize, it is important to remember for later discussion that the difficulties associated with inversion theory can be reduced if:

1. We make instruments as accurate (noise–free) as possible.
2. We select channels (i.e., wavelengths) so that the kernel functions are as sharp as possible.
3. We develop methods of solution that are stable (i.e., provide minimum distortion of the true solution) in the presence of unavoidable measurement noise.

1.3   The Chemical Composition of the Atmosphere

Table 1.1 lists the relative abundance of various species in the Earth's atmosphere. The concentrations are given as *mixing ratios* by volume, the unit commonly used by atmospheric scientists, and is identical to the chemists' *mole fraction*. In the table, the mixing ratios are given as fractions, but parts per million (p.p.m or p.p.m.v) are commonly used for minor constituents. The quoted mixing ratios are actually averages for the lower atmosphere. Figure 1.5 shows representative vertical profiles of the species listed in Table 1.1, for average, midlatitude conditions.

Concentrations of atmospheric gases are subject to chemical and photochemical alteration. Molecular oxygen, for example, decomposes into atoms above 90 km. Methane and nitrous oxide are unstable in the stratosphere. Relevant to remote sensing topics addressed later is the fact that both molecular oxygen and carbon dioxide are uniformly mixed below about 100 km.

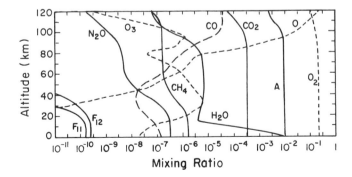

Figure 1.5 Vertical profiles of mixing ratios of selected species in the atmosphere (from Goody and Yung, 1989).

Table 1.1 The Composition of the Earth's Atmosphere

Molecule	Volume fraction+	Comments
N_2	0.7808	Photochemical dissociation high in the ionosphere; mixed at lower levels
O_2	0.2095	Photochemical dissociation above 95 km; mixed at lower levels
H_2O	<0.04	Highly variable; photodissociates above 80 km
A	9.34×10^{-3}	Mixed up to 110 km; diffusive separation above
CO_2	3.45×10^{-4}	Slightly variable; mixed up to 100 km; dissociated above
CH_4	1.6×10^{-6}	Mixed in troposphere; dissociated in mesosphere
N_2O	3.5×10^{-7}	Slightly variable at surface; dissociated in stratosphere and mesosphere
CO	7×10^{-8}	Variable photochemical and combustion product
O_3	$\sim 10^{-8}$	Highly variable; photochemical origin
$CFCl_3$ and CF_2Cl_2	$1-2 \times 10^{-10}$	Industrial origin; mixed in troposphere, dissociated instratosphere

+ Fraction of lower tropospheric air

Perhaps the most important gas in the atmosphere, from the point of view of its interaction with electromagnetic radiation, is water vapor. Water vapor is especially important in the troposphere because of its role in cloud formation and precipitation and in transporting significant amounts of energy in the form of latent heat and infrared radiation. Water vapor is also one of the most variable components of the atmosphere. In the tropics, water vapor may account for up to 4 % (by volume) of the atmosphere, while in polar regions or in dry desert air, the abundance may be only a fraction of a per

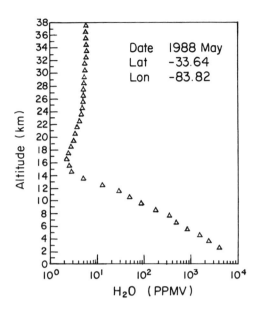

Figure 1.6 A vertical profile of water vapor extending from the middle troposphere into the stratosphere. These results are obtained from retrievals applied to SAGE measurements (McCormick, private communication).

cent. In the troposphere, the average relative humidity is close to 50% and the vapor pressure varies over a very wide range. Photochemical decomposition of water starts around 80 km, reducing the mixing ratio above this level. Figure 1.6 provides an example of the vertical profile of water vapor retrieved from measurements of direct attenuation of solar radiation using an instrument flown on a polar orbiting satellite. The profile shown is taken from data collected as part of the Stratospheric Aerosol Gas Experiment (SAGE II) which is described in further detail in later chapters of the book. The profile obtained indicates a vapor minimum in the lower stratosphere (the *hydropause*) where the concentrations of vapor are extremely low. It is still largely a mystery why concentrations of water vapor are so low at these levels, especially in the equatorial stratosphere.

Another important trace gas is ozone. Concentrations are variable and the mixing ratios are a few tenths of a part per million parts of air. If we compress the entire atmospheric column to 1 atmosphere of pressure, it would typically occupy a column only 3 mm tall (this column amount is also expressed in units called Dobson Units, which are defined later in Chapter 6 and is the topic of Problem 3.10). The vertical distribution of ozone differs from that of other atmospheric gases, having a maximum number density near 25 km (Fig. 1.5). Above 30 km, ozone is rapidly formed by photochemical reactions from oxygen so that an equilibrium is obtained during daylight hours. Ozone is created more slowly below this level such that ozone concentration depends on mixing and transport processes, and is highly variable. This variability is reflected in the seasonal and latitudinal changes of ozone. Although none is formed during the polar night, the maximum ozone amount occurs there because of the significant transports to these regions. Day–to–day ozone changes are related to the passage of weather systems. Satellite measurements demonstrate longer–term global changes in ozone amount (Chapter 6).

Carbon dioxide strongly influences the radiation field at all levels below 100 km. It is chemically unreactive and has its main sources and sinks in industrial and biological processes at the Earth's surface. In the planetary boundary layer, its concentration is variable but, at higher levels, its mixing ratio is essentially constant below the dissociation level of molecular oxygen; above this level carbon dioxide dissociates. The total amount of carbon dioxide in the atmosphere is slowly increasing with time (Fig. 1.7) because of industrial and agricultural activity.

1.4   Vertical Distribution of Pressure and Density

The vertical variability of pressure and density is much larger than either the horizontal or temporal variability of these quantities. Therefore, it is useful to define a "standard atmosphere" which represents the horizontal and time–averaged structure of the atmosphere as a function of height only. Such a standard view is shown in Fig. 1.8a. At any given level, up to about 100 km, the atmospheric pressure and density are nearly always within 30% of the corresponding "standard atmosphere" values. Within the lowest 100 km, the logarithm of the pressure drops off almost linearly with height;

$$\log\left[p(z)\right] \simeq \log\left[p(0)\right] - Bz \qquad (1.5)$$

where $p(z)$ is the pressure at height z above sea level, $p(0)$ is the pressure at sea level, and B is a constant which is related to the average slope of the pressure curve. Making use of the identity

$$\ln x = 2.3 \log x$$

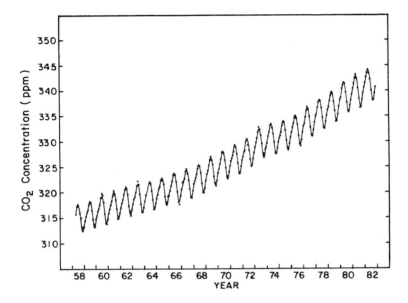

Figure 1.7 Atmospheric concentrations of CO_2 at the Mauna Loa observatory, Hawaii. Both long–term and seasonal changes are shown (after Bacastow et al., 1985).

we write

$$\ln \frac{p(z)}{p(0)} \simeq -\frac{z}{H} \tag{1.6}$$

where $H \equiv 1/(2.3B)$. It thus follows that

$$p(z) \simeq p(0) \exp(-z/H) \tag{1.7}$$

which states that pressure drops off by a factor e in passing upward through a layer of depth H. H is called the *scale height* of the atmosphere and has an approximate value of 7 km. Since the variation of density with z is similar to the variation of pressure with z, it follows that

$$\rho(z) \simeq \rho(0) \exp(-z/H) \tag{1.8}$$

It should be emphasized that the atmosphere is remarkably thin in comparison to the dimensions of the Earth. Half the mass of the atmosphere lies below the 500 mb level, which has a mean height of roughly 5.5 km above sea level (this is less than 0.001 of the Earth's radius), and about 99% of the mass of the atmosphere lies within the lowest 30 km.

1.5   The Thermal Structure of the Atmosphere

The vertical structure of atmospheric temperature is used to identify regions of the atmosphere. The vertical distribution of temperature for our example of the "standard atmosphere" is provided in Fig. 1.8b. It is customary to divide the atmosphere up to 100 km into four distinct layers: *troposphere, stratosphere, mesosphere,* and *thermosphere.* The levels separating these layers are referred to as the *tropopause, stratopause, mesopause,* and *thermopause,* respectively. The topics of this book largely focus on the remote sensing of the troposphere.

The actual thermal structure of the atmosphere is more complicated than shown in Fig. 1.8b. The lowest 1 or 2 km of the atmosphere differ from the remainder of the troposphere. Here interactions with the surface are strong and diurnal variations are large. This region has been intensively studied and is referred to as the planetary boundary layer (e.g., Garratt, 1992). At some latitudes, inversions exist in the lowest 2 or 3 km of the atmosphere. Above 3 km, however, there are some regular features such as a systematic decrease of temperature with height. This decrease is referred to as the *lapse rate* and has an approximate mean value of 6.5 K km^{-1} in the troposphere. A sudden change to isothermal or inversion conditions occurs at the tropopause. Multiple tropopauses also occur between latitudes 30 degrees and 50 degrees, where the high tropical tropopause overlaps the low arctic tropopause. In the stratosphere, temperatures are, curiously, lower in tropical regions than in the arctic. Above 30 km thermal data are more sparse.

Excursus: Microwave Measurements of Atmospheric Temperature

Study of global climate change requires adequate monitoring of global temperature. Present estimates of globally averaged surface temperatures, based on conventional measurements, for example, suffer from sampling difficulties and problems associated with other interpretative factors (e.g., Karl et al., 1988). By contrast, satellite observations seem particularly well suited to global monitoring, provided instruments can be made that are stable and readily calibrated. An example of the use of satellite observations in this type

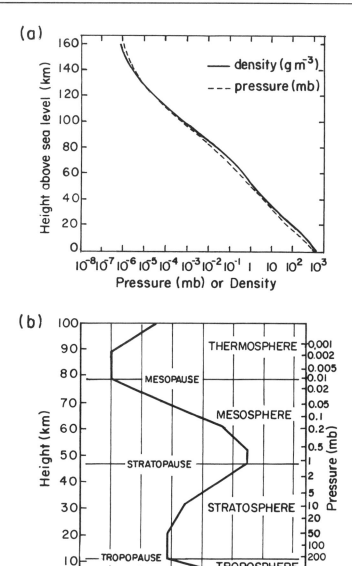

Figure 1.8 (a) Vertical profile of pressure in millibars (dashed) and density in grams per cubic meter (solid, adapted from Wallace and Hobbs, 1977). (b) The vertical profile of temperature of the U.S. Standard Atmosphere (Wallace and Hobbs, 1977).

of study is provided by the work of Spencer et al. (1990) which is based on the analysis of measurements obtained from the Microwave Sounding Unit (MSU). The MSU provides a weighted average of the atmospheric temperature determined by a vertical weighting function.[2] This function for channel 2 has a peak in the middle troposphere. Spencer et al. present a study of approximately six years of MSU data obtained from instruments flown on three satellites; TIROS-N, NOAA-6, and NOAA-7, for the period from November 17, 1978, to February 18, 1985. The zonally averaged channel 2 temperature for this approximate six year period is shown in Fig. 1.9a. The annual range of temperatures is largest at high latitudes, especially in the northern hemisphere (Fig. 1.9b). Due to a more efficient conversion of solar insolation to sensible heating by the land masses of the northern hemisphere, the average annual temperature of the northern hemisphere is about 1.6° C warmer than the southern hemisphere.

Figure 1.9c is a time series of the global mean channel 2 temperature anomalies. An identifiable feature in the anomaly time series is the atmospheric warming of a few tenths of a degree during the 1982–1983 El Nino–Southern Oscillation (ENSO) event which is associated with a warming of the equatorial eastern Pacific. However, just as significant are the cooler periods of 1979 and 1984 and the extended warm non–ENSO period of 1980.

1.6   The Particulate Composition of the Atmosphere

Particles suspended in a gas are called *aerosol*. In principle, atmospheric aerosol is a term that properly describes dust, haze particles, cloud water droplets, and ice crystals. Aerosols are so variable in concentration and, in the case of haze and dust particles, in chemical composition, that it is difficult to generalize their properties. Thus the properties to be described are meant only to be broadly representative of real aerosol. Hereafter noncloud particles are referred to as aerosol and cloud particles is used to represent either droplets, precipitation, or ice crystals.

---

[2] The shape of the weighting function is dictated by details of molecular absorption and the weighting functions of the MSU are governed by the absorption of molecular oxygen. The topic of weighting functions is extensively discussed in Chapter 7 and molecular absorption is a subject of Chapter 3.

MSU Channel 2
Troposphere

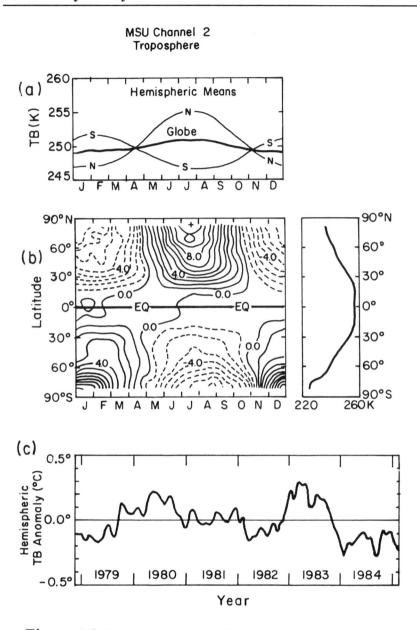

Figure 1.9 (a) Annual cycle in MSU channel 2 brightness temperatures for the globe and northern and southern hemispheres (top); (b) the zonally averaged difference between the brightness temperature and the annually averaged brightness temperature in 2.5 degree latitude bands (middle left); and (c) global anomalies (bottom panel) about the MSU channel 2 brightness temperature annual cycle which is shown in (a) for a six–year period from 1979 to 1984 (from Spencer et al., 1990).

1.6.1 Atmospheric Aerosol

Aerosol particles found in the atmosphere are produced both in nature and by people. We can distinguish the sources of these particles as either primary and secondary in nature. The former are mostly of natural origin, including meteorites (extraterrestrial or interplanetary dust), the world oceans (sea–salt particles), particles produced by weathering of arid and semi–arid areas, terrestrial material from volcanic debris, and particles derived from terrestrial biota. Secondary sources involve chemical conversion of atmospheric and anthropogenic trace gases (Gas–to–Particle–Conversion) into solid and liquid particles. This phase transition produces small particles with radii below about 0.1 μm and is the source of sulfate and nitrate particles in the atmosphere. Figure 1.10 is a convenient summary of different aerosol particles found in the atmosphere and identifies their sources and lifetimes and includes some information about their atmospheric effects.

The size of aerosol particles also varies considerably depending on the production mechanism. Junge (1955) introduced the following categories of aerosol based on their size:

- *Aitken particles* with dry radii $< 0.1 \mu m$
- *large particles* with dry radii $0.1 \le r \le 1.0 \mu m$
- *giant particles* with dry radii in excess of $1.0 \mu m$

The size distributions of aerosol are typically expressed via any of a number of different analytical formulas, including the one introduced later for cloud particles. The concentration of aerosol varies significantly with time and location. It depends on the proximity to sources, the strength of the sources, and on the activity of convection and turbulence among other factors that can act to remove them from the atmosphere. Observations confirm that the concentrations of aerosol decrease with increasing distance from the Earth's surface. This is expected from the atmospheric density profile and also because the surface is a major source of particles whereas removal mechanisms continually operate in the atmosphere. It is estimated that 80% of the total aerosol particle mass is contained within the lowest kilometer of the troposphere. The concentration of aerosol also decreases with increasing horizontal distance from the seashore toward open ocean. Land is a more prolific source of particles, and it is thought that 61% of the total aerosol is introduced in the northern hemisphere.

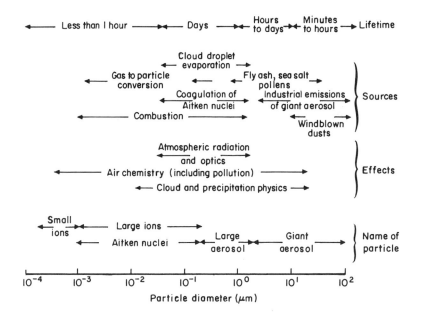

Figure 1.10 Names of atmospheric particles, together with effects, sources, and lifetimes. The lifetime of very small particles is short because they coagulate rapidly to form larger particles. Giant aerosols are also short–lived because they precipitate from the atmosphere (Wayne, 1985).

1.6.2 Cloud Microphysics

The microphysics of clouds is a topic extensively reviewed in a number of texts on the subject. Microphysical properties of clouds are taken here to mean the size and shape of the particles and their volume concentration. These properties vary considerably from cloud to cloud and whether the particles are water droplets, ice crystals, or large precipitation sized particles.

Water Droplets

Water droplets are typically smaller than 100 μm and are spherical. The distribution of cloud droplet sizes measured in different types of clouds under a variety of meteorological conditions commonly exhibit a characteristic shape. This distribution is expressed in terms of a concentration $n(r)dr$ (the number of droplets per volume existing in the radius range r to $r+dr$) and generally rises sharply as the radius increases to some maximum and then decreases gently toward larger sizes causing the distribution to be positively skewed with a long tail. This type of distribution can often be reasonably approximated by analytic functions such as a modified gamma distribution,

$$n(r)dr = \frac{N_o}{\Gamma(\alpha)r_n} \left(\frac{r}{r_n}\right)^{\alpha-1} \exp(-r/r_n) \, dr \qquad (1.9)$$

where N_o is the total number of droplets of all sizes in a given volume (typically numbers per cubic centimeter), Γ is the gamma function, r_n is a radius that characterizes the distribution and α is the variance of the distribution. This is a convenient form of the size distribution as it is straightforward to represent the following moments: cross–sectional area:

$$A = \int_0^\infty \pi r^2 n(r)dr = N_o \pi r_n^2 F(2) \qquad (1.10)$$

where $F(j) = \Gamma(\alpha + j)/\Gamma(\alpha)$, and
volume:

$$V = \frac{4}{3} \int_0^\infty \pi r^3 n(r)dr = \frac{4}{3}\pi N_o r_n^3 F(3) \qquad (1.11)$$

from which the cloud liquid water content ℓ follows $\ell = \rho_l V$, where ρ_l is the density of water.

Characteristics of the droplet distribution can be expressed numerically in terms of a number of parameters. Included in the list of such parameters are the mode radius[3] $r_d = (\alpha - 1)r_n$ corresponding to the maximum of the distribution, the mean radius $r_m = (\alpha+1)r_n$, which is the sum of all droplet radii divided by the total number of droplets, and the effective radius defined as $r_e = \frac{V}{A} = (\alpha + 3)r_n$. Table 1.2 presents values of $r_m, \ell,$ and N_o derived from size distributions measured in a variety of clouds as well as estimates of α and r_e derived from these measured quantities. The droplet concentration N_o generally varies considerably depending on whether the cloud forms in nuclei rich continental air where concentrations can exceed 1000 cm^{-3} or in maritime air where concentrations are often

---

[3] In these definitions α is taken as an integer quantity.

Table 1.2 Characteristics of Selected Dropsize Distributions

Cloud type	N_o (cm^{-3})	r_m (μm)	r_{max} (μm)	r_e^+ (μm)	ℓ (gm^{-3})
Stratus (ocean)	50	10	15	17	0.1–0.5
Stratus (land)	300–400	6	15	10	0.1–0.5
Fair weather cumulus	300–400	4	15	6.7	0.3
Maritime cumulus	50	15	20	25	0.5
Cumulonimbus	70	20	100	33	2.5
Cumulus Congestus	60	24	40–80	40	2.0
Altostratus	200–400	5	15	8	0.6

$^+$ Assumes the size distribution (1.9) with $\alpha = 2$. Adapted from Mason (1971).

less than 100 cm^{-3}. The liquid water content, however, does not necessarily follow the droplet concentration.

Raindrops

Raindrops are not spherical like the smaller cloud drops (Fig. 1.11) and their shape depends on their size. Drops smaller than about 1 mm in diameter are slightly deformed and resemble oblate spheroids, but this shape is further distorted as the size of the drop increases. The aspect ratio of the drop (i.e., the width–to–length ratio) decreases as the drop size increases. We will see in later chapters how this property is used in radar remote sensing of rainfall.

Perhaps the most common type of distribution used to represent raindrops is the Marshall–Palmer (MP, Marshall and Palmer, 1948) distribution

$$n(r) = n_o \exp(-2\Lambda r) \tag{1.12}$$

where Λ is related to the rainfall rate $\Re$ via $\Lambda = 4.1\Re^{-0.21}$ mm^{-1} and the parameter $n_o = 8 \times 10^3$ m^{-3}mm^{-1} is taken to be constant. The MP distribution does not represent all types of rainfall and variants of this distribution are often used. For example, n_o may vary from rain type to rain type and this parameter is sometimes expressed

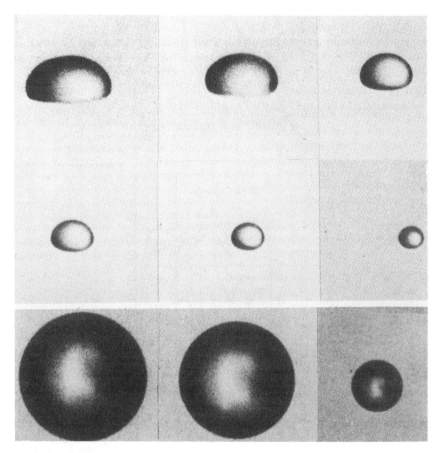

Figure 1.11 The shape of cloud and raindrops as determined from wind tunnel experiments. The drops in the third row are expanded in size to show sphericity (Pruppacher and Beard, 1970).

as a function of the rain rate. The MP distribution is also used to represent the distribution of hail.

Ice Crystals

There is a great variety of growth forms and thus shapes of ice crystals. The relation between the form of ice crystal and the temperature T and supersaturation S_i with respect to ice of the environment in which they form has been studied extensively. Figure 1.12 illustrates the different crystal growth forms as a function of T and S_i. Crystal habits change from plates to columns as the temperature changes and sector plates, dendrites, and needles form as the super-

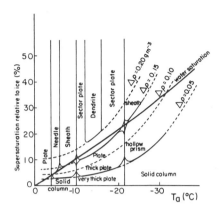

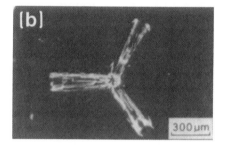 

Figure 1.12 (a) Habit changes as a function of temperature and supersaturation with respect to ice (after Kobayashi, 1961). (b) Replicas of a combination of bullets forming a bullet rosette (from Ueyeda and Kikuchi, 1979).

saturation is increased. In addition to the simple polyhedral forms, irregular crystals or combinations of simple shapes readily appear in nature (Fig. 1.12b). The size distribution of ice crystals measured in cirrus clouds also demonstrates a temperature dependence (Heymsfield and Platt, 1984).

1.7 Satellite Platforms

Observation of the weather was one of the earliest uses of Earth—orbiting satellites. A large number of satellites have been launched since the 1960s for that purpose when the United States launched the first in an experimental series of satellites, TIROS 1 (Television and Infrared Observation Satellite), on April 1, 1960. From the observations made with the imaging cameras carried by that satellite, the value of a large–scale view of clouds in the atmosphere, particularly over remote oceans, was appreciated. Since then, the use of Earth–orbiting satellites has become the principal means of observing the entire Earth and its atmosphere, whether for routine surveillance of the Earth or as a research tool for tackling basic problems in understanding weather and climate. Because of the important role satellite observations play in studying the atmosphere, a brief out-line will now be given of the general characteristics of satellites and their orbits.

Excursus: Mechanics of a Satellite Orbit

It is reasonable to suppose for this discussion that satellites move in a circular orbit around the Earth. These circular orbits are predicted by matching the centripetal acceleration to the gravitational force. The balance of forces is

$$mv^2/R = Gm_Em/R^2$$

where v is the tangential velocity, m is the mass of the satellite, G is the universal gravitational constant ($G = 6.673 \times 10^{-11}$ Nm^{-2} kg^{-2}), m_E is the mass of Earth, and R is the orbit radius. From the proceeding identity, we obtain

$$v = \sqrt{Gm_E/R}$$

and from the definition of the period of the orbit,

$$P = 2\pi R/v$$

it follows that

$$P = 2\pi\sqrt{R^3/(Gm_E)} \qquad (1.13)$$

As a typical example, NOAA polar orbiting satellites are approxi-mately 850 km above the Earth. Assuming that the radius of the Earth is 6378 km, then R=7228 km and $P = 102$ minutes.

Equation (1.13) predicts that the higher the orbit, the longer its period of rotation. If the satellite is moved far enough out from Earth, then the period of this orbit can be selected to match the ro-tation rate of Earth. A satellite in that orbit, at the chosen distance from Earth in the equatorial plane, moving in a counterclockwise sense, moves around the Earth at exactly the same speed as the Earth rotates. The satellite then stays fixed in the same place above the Earth in a geostationary orbit. The altitude of a geostationary orbit can be estimated as follows. The angular velocity of the satellite is

$$\Omega_{sat} = 2\pi/P$$

and it follows from (1.13) that

$$R^3 = Gm_E/\Omega_{sat}^2 \tag{1.14}$$

With Ω_{sat} chosen to match the angular velocity of Earth, $\Omega_E = 7.29221 \times 10^{-5}$ s^{-1}, then R=42165 km . Thus the altitude of geostationary satellites is approximately 36000 km, meaning that an instrument on a polar orbiting satellite that sees an area 1 km^2 on the surface has a much larger field of view (approximately 1600 km^2) when flown on a geostationary platform. This is a particularly troublesome issue for microwave instruments

1.7.1 Satellite Orbits

Satellites used to observe the atmosphere fly principally in one of two orbits: polar orbits or geostationary orbits. Polar orbiting satellites overfly higher latitudes even though the satellite might not pass directly over either pole. The altitude of polar orbiting satellites is normally lower than 2000 km, and these satellites are sometimes termed low Earth orbit (LEO) satellites. Satellites in this orbit view a swath of the Earth below with a periodicity that is related to the altitude of orbit as predicted by (1.13). The higher the orbit, the longer is its period. Typical periods are less than about two hours for orbits below about 1600 km. A satellite coverage capabilities depend on its orbit height and the sensor's field of view. The density of the grid traced out by the ground track of a satellite is determined by the time required for the satellite to return to the same point. The longer the time allowed, the tighter the grid. Landsat, for example, with narrow field of view instruments, requires approximately 16 days to achieve full coverage, whereas polar orbiting meteorological satellites provide full coverage twice a day.

An important characteristic of the polar orbit is the angle of inclination which determines the poleward extent of the orbit (Fig. 1.13a). This angle is defined as the angle measured in the counterclockwise direction between the equatorial plane and the plane of the orbit. Only orbits with an inclination greater than 90 degrees cover the whole globe. The *nodes* of the orbit correspond to the point of the orbit where the satellite crosses the equator. If the satellite passes this point moving in a northerly direction as in Fig 1.13a, then the node is referred to as an *ascending* node. Southward passage across the equator distinguishes the *descending* node.

While satellite orbits defined by inclination angles greater than 0 are loosely referred to as polar orbits, it is perhaps appropriate to call low–inclination orbits "tropical" orbits. An example of such an orbit is that of the proposed tropical rainfall measurement mission (TRMM) satellite and the ground track of this satellite is shown in Fig. 1.14a. The inclination of the orbit is 35 degrees, which offers a better coverage of the tropics than do higher inclination orbits. The track shown represents the path of the point on the Earth's surface, the subsatellite point, lying directly below the satellite on the surface along the line between the satellite and the center of Earth. A total of 16 orbits for each 24 hour period are planned which will provide space–time coverage of the tropics that is not available from present polar orbiting weather satellites.

Orbits are frequently designed so the satellite crosses the equator at the same local time every day (Fig. 1.13b). This is important for instruments that require sunlight. Such an orbit is referred to as a *sun–synchronous* orbit and the satellite orbit is always the same in relation to the sun. This does not mean that the satellite orbit remains fixed in space, as it must move to compensate precisely for the Earth's rotation about the sun (1 degree per day). Figure 1.13b shows how the orbit moves or precesses to provide this compensation. The mechanism for providing this precession of the orbit is achieved using the effects of the non–uniformity of the Earth's gravitational field that arises from the fact that the Earth is not a perfect sphere. Orbital mechanics predicts that the drift in the orbit is proportional to $\cos i$ and inversely proportional to orbit altitude. Orbits defined by $i < 90°$ precess to the west in a clockwise direction (a prograde orbit), whereas orbits with $i > 90°$ drift to the east in a retrograde orbit. By adjusting both the orbital altitude and inclination angle, the precession can be chosen to compensate for the Earth's rotation exactly.

Sun–synchronous orbits are used by many satellites to observe the Earth and its atmosphere. The NOAA polar orbiting weather satellites are sun–synchronous with an orbit altitude of approximately 850 km and $i \approx 99°$.

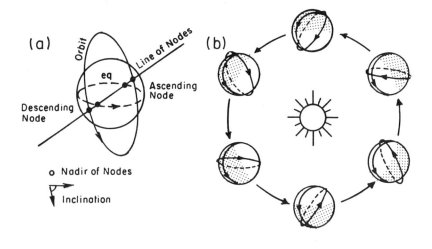

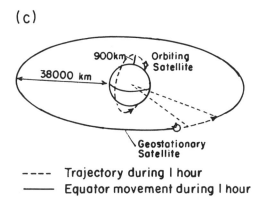

Figure 1.13 (a) The polar orbit of an artificial satellite in relation to the Earth's surface. The inclination angle is i and the angular position of the satellite relative to the center of Earth is ϕ. The direction of the Earth's rotation is shown by the arrow at the north pole. (b) Sun–synchronous orbit; the angle between the orbit plane and the Earth–sun distance is fixed; (c) geostationary orbit.

(a)

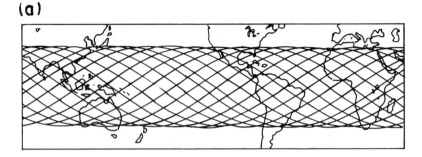

(b)

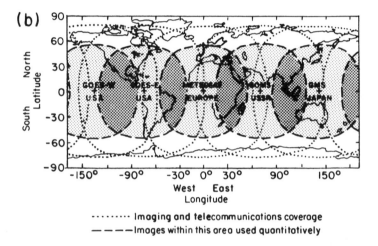

········· Imaging and telecommunications coverage

————— Images within this area used quantitatively

Figure 1.14 (a) The proposed track of the TRMM satellite inclined at 35 degrees. (b) Coverage of the globe provided by geostationary satellites in orbit in the 1980s.

The geostationary orbit (GEO) provides the capability of detailed observations of the time evolution of atmospheric phenomena. If the orbit lies in the equatorial plane, as shown in Fig. 1.13c, the satellite will stay above the same point on the equator and thus an instrument can stare at the Earth. Several meteorological satellites, as well as telecommunication satellites, fly in geostationary orbits. The present constellation of geostationary satellites acts as a coordinated international system (Fig. 1.14b), stationed around the equator, giving complete coverage to about 60 degrees latitude. This coverage of the Earth by GEO satellites is an important ingredient of global climate research programs like the International Satellite Cloud Climatology Project (ISCCP) which is discussed later in Chapter 7.

1.7.2   Selected Historical Highlights

The first truly operational weather satellite was ESSA–1 in 1966. In 1970, NOAA–1 carried instruments that were improvements on the TIROS–1 payload and included infrared imaging instruments which made night–time coverage possible. In 1972, NOAA–2 carried an infrared radiometer for operational sounding of the vertical temperature structure of the atmosphere. Further improvements and the addition of the MSU were included on TIROS–N, launched in 1978, as the first of the current series of operational TIROS/NOAA satellites. The principal instruments flown on these satellites include the Advanced Very High Resolution Radiometer (AVHRR) and the TIROS Operational Vertical Sounder (TOVS). The AVHRR is a visible and infrared imager with a horizontal resolution of about 1 km in the visible and about 4 km in the infrared. This instrument is principally used for the study of clouds and monitoring sea–surface temperature (Chapter 7). It has also been applied to a wide range of meteorological, climatological, and environmental problems, some of which are discussed in Chapters 4 and 6. TOVS is a group of three radiometers, infrared and microwave instruments that are sensitive to the vertical temperature structure of the troposphere and stratosphere as well as sensitive to atmospheric moisture. Global monitoring provided by these radiometers is not only important for weather prediction but is becoming increasingly important in the study of global climate (refer to Fig. 1.9 and related discussion). As a complement to NOAA's weather satellites, the United States launched a parallel series of polar orbiting military weather satellites as part of the Defense Military Satellite Program (DMSP) which includes both visible and infrared imaging radiometers and an advanced microwave imager (the SSM/I). The DMSP satellites are the major source of input for the U.S. Air Force Global Weather Central which provides worldwide meteorological and space environmental support to the United States and other national defense agencies.

The first geostationary weather satellite was the United States GOES–1, launched in 1975. This was followed by Europe's Meteosat and the Japanese GMS or "Himawari" in 1977, and later by the Indian geostationary series, INSAT, in 1988. In the same year, the European Space Agency (ESA) launched the first in a series of operational geostationary satellites, the Meteosat Operational Programme–1 (MOP–1). Each of these satellites carries visible and

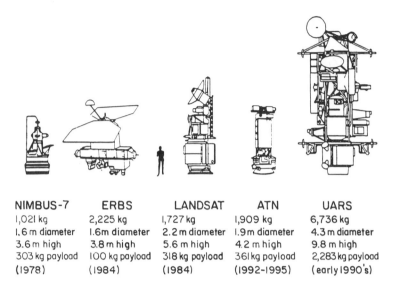

NIMBUS-7	ERBS	LANDSAT	ATN	UARS
1,021 kg	2,225 kg	1,727 kg	1,909 kg	6,736 kg
1.6 m diameter	1.6m diameter	2.2 m diameter	1.9 m diameter	4.3 m diameter
3.6 m high	3.8 m high	5.6 m high	4.2 m high	9.8 m high
303 kg payload	100 kg payload	318 kg payload	361 kg payload	2,283 kg payload
(1978)	(1984)	(1984)	(1992-1995)	(early 1990's)

Figure 1.15 Comparison of the size and mass of some recent and forthcoming Earth satellites (from Taylor and Eyre, 1989).

infrared imaging radiometers, and a vertical sounding component (VAS) is flown on the GOES series.

The atmospheric research community has been served for most of the decades of the 1970s and 1980s by a series of Nimbus satellites which were first introduced by NASA in 1964. The last of the Nimbus research satellites is the Nimbus 7 which was launched in October 1978. In 1984, NASA's Earth Radiation Budget Satellite (ERBS) was launched primarily for measurement of the Earth's Radiation Budget. Another important experiment that is also carried out on this satellite is the SAGE II (discussed further in Chapter 6). A long gap has followed since 1984, but several new research satellites are proposed for the 1990s. A large and sophisticated Upper Atmosphere Research Satellite (UARS) launched in 1991 and a second ERBS, carrying the TRMM payload, is to be launched in the mid–1990s.

An important series of satellite missions began in 1972 with the U.S. Earth Resources Technology Satellite (ERTS) which was renamed Landsat–1. This satellite focuses on measurement of land surfaces with a high spatial resolution; these measurements are used in land surface inventory, geological and mineralogical exploration, crop and forest assessment, and cartography. Landsat became operational with the launch of Landsat–4 in 1982. In 1986, France launched its new Satellite Pour l'Observation de la Terre (SPOT), the first in a series of land–sensing satellites to complement Landsat observations.

The remote sensing of oceans from space has been developed since the 1960s. Today, sea surface temperature, ocean currents, surface winds, wave heights and distribution, surface topography, ocean color, and plankton concentration are all sensed from satellites. To a large extent, these measurements derive from NASA's 1978 Seasat mission and from the highly successful Nimbus–7 mission. Unfortunately, Seasat became non–operational after just three months into the mission. Despite this early failure, the data continue to be analyzed. Another important satellite is the GEOSAT satellite launched by the U.S. Navy in March 1985 which provides important altimeter data for monitoring sea level. The Marine Observation Satellite (MOS–1) was launched by the Japanese in 1987 into a polar orbit and observes reflected visible and emitted infrared radiation in a number of channels and also carries a microwave radiometer for observing clouds, precipitation, and sea–ice, among other parameters.

1.8   Notes and Comments

1.2. A general discussion of the retrieval of atmospheric temperature and composition is given in the book by Houghton et al. (1984). This is referred to as remote sounding and the topic of sounding is discussed in Chapter 7. The book by Twomey (1977) is an excellent introductory text on the subject of inversion theory as is the Menke's text (1989).

1.3, 1.4, and 1.5. The properties of the atmosphere are sketchily described in this section. More details can be found in a number of basic references, such as Wallace and Hobbs (1977) and Piexoto and Oort (1991).

1.6. There are a number of basic texts on the subject of cloud microphysics. Mason (1971) and Pruppacher and Klett (1980) give a broad treatment of the topic and the book of Hobbs (1974) describes the microphysics of ice. Aerosol properties are also discussed in Pruppacher and Klett (1980) and in d'Alemeida et al. (1991).

1.7. Other details about the mechanics of satellite orbits as well as a summary of other launches can be found in the books of Kidder and Vonder Haar (1993) and Harries (1987). A historical account of satellite applications to the study of the atmosphere and oceans is contained in Baker (1990).

Not all satellites orbit in polar or geostationary orbits. Since much of the former Soviet Union lies in high latitudes, the use of geostationary satellites orbiting in the equatorial plane is useless for their communication needs. Soviet communication satellites were placed in a highly elliptical orbit (called the Molynia orbit, Kidder and Vonder Haar, 1993).

Satellites do not remain in their orbits indefinitely. Ultimately, loss of energy through the action of atmospheric drag causes the satellite to loose speed and fall back to Earth, or burn up in the atmosphere. This action therefore provides an upper limit to the useful lifetime of the satellite. We can estimate this lifetime by considering the reduction on the radius of the satellite δr in a circular orbit

$$\delta r \approx 4\pi d A \rho_{air} R^2 / m$$

where d is a drag coefficient, A is the satellite's cross–sectional area (normal to its motion), and ρ_{air} is the atmospheric density at satellite altitude. The uncertainty in this expression hinges on the uncertainty to which the drag coefficient is known and this is probably only accurate to within a factor of 2. For example, consider the Landsat 5 satellite. The mass of this satellite is about 1700 kg, its area A is about 10 m^2 and its orbital altitude is 700 km where the atmospheric density is about 10^{-13} kg m^{-3}. Thus the satellite descends about 0.4 m per orbit, or about 5 m per day and we infer that the maximum useful time of Landsat 5 in orbit (not the operation of instruments on Landsat) is in excess of 100 years. By contrast, the low orbit of the TRMM satellite limits its usefulness to only a few years.

The general concept of the Tropical Rainfall Measurement Mission (TRMM), the satellite orbit and its implications for sampling the tropical atmosphere is given in Simpson et al. (1988).

2
The Nature of Electromagnetic Radiation

For the topics considered in this book, it is electromagnetic radiation, in the form of a wave, that communicates information from the atmosphere to the observer. The light that we detect by our eye is such a wave. This chapter deals with the fundamental properties of such waves: the frequency at which they oscillate, the way in which they propagate, and the manner by which they are created. These properties form an important basis for understanding the way that electromagnetic waves interact with matter, and, therefore, the way we utilize this radiation for remote sensing.

Electromagnetic waves are generated by oscillating (or, more generally, time varying) electric charges which, in turn, generate an oscillating electric field. A characteristic of an oscillating electric field is that it produces an accompanying oscillating magnetic field that further produces an oscillating electric field. Therefore these fields, initiated by the oscillating charge, proceed outward from the original charge, each creating the other. A visualization of such a propagating wave is given in Fig. 2.1. It was James Clerk Maxwell, who, more than a century ago, provided us with the theoretical synthesis of this phenomenon. While detailed mathematical accounts of this work are omitted here, they can be found in most standard texts on electromagnetics.

Excursus: The Electric Dipole

Oscillating charges produce electromagnetic radiation. A particular charge distribution of basic importance to our understanding of radiation and its interaction with matter is the *electric dipole*. The electric dipole consists of two charges: a positive and a negative charge of the same magnitude q, separated by a distance s (Fig. 2.2a). From this concept emerges the definition of the *dipole moment*

$$\vec{p} = q\vec{s} \tag{2.1}$$

which is a vector of magnitude qs directed from the negative to the positive charge. Oscillation of this dipole produces an electromagnetic wave which has properties that are discussed more fully below.

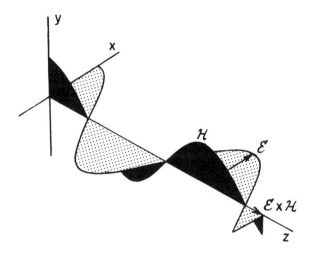

Figure 2.1 A schematic view of a time harmonic electromagnetic wave propagating along the z axis. The oscillating electric $\mathcal{E}$ and magnetic $\mathcal{H}$ fields are shown. Note that the oscillations are in the x–y plane and perpendicular to the direction of propagation.

The concept of an oscillating electric charge is fundamental to our understanding of particle scattering. There are, however, other arrangements of charges that also oscillate and thus contribute to the electromagnetic field emerging from matter. These charge distributions can be considered as *multipoles* and are defined as follows. A single point charge is a *monopole*. A dipole is obtained by displacing a monople through a small distance (s in Fig. 2.2a) and replacing the original monopole by another, but of the opposite sign. Likewise, a *quadrupole* is obtained by displacing a dipole a small distance and replacing this dipole with one of opposite sign (Fig. 2.2b). This idea can be continued to build up higher order multipoles. In this way, the charge distribution in a particle can be considered as a composition of various orders of multipoles. We shall see later how the oscillation of these multipoles contribute to the radiation scattered by a particle.

2.1   The Electromagnetic Spectrum

Electromagnetic theory predicts that the electromagnetic wave travels at a unique speed c, which is the speed of light. The wavelength of the wave depends upon how rapidly, or the *frequency* at which, the charge oscillates. We shall denote this frequency (number of

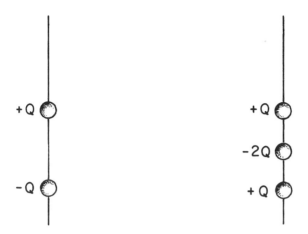

Figure 2.2 (a) The two charges $\pm q$ form a dipole. The dipole moment is a vector which has a magnitude of qs directed toward the positive charge. (b) Charges $+q$ and $-2q$ are arranged along a line to form an axial quadrupole.

oscillations per second) by ν, and it is related to c by

$$\nu = \frac{c}{\lambda} \tag{2.2}$$

where λ is the wavelength of the wave. For example, red light with a wavelength of 0.7 micrometers (μm) corresponds to a frequency of 4.3×10^{14} oscillations per second while violet light, at 0.4 μm, corresponds to 7.5×10^{14} oscillations per second. An alternate way of describing the frequency of radiation is in terms of *wavenumber*

$$\tilde{\nu} = \frac{1}{\lambda} \tag{2.3}$$

which is a count of the number of wave crests or troughs in a given unit of length. For example, red light has 14,286 wave crests in a centimeter whereas 25,000 crests can be counted in a centimeter of violet light. Wavenumber is the measure often used by spectroscopists and others involved in experimental measurements of the interaction of radiation with matter.

Even before Maxwell, the spectrum of electromagnetic radiation (that is the range of wavelengths or frequencies of the radiation) was extended beyond the visible (i.e., beyond those wavelengths detectable by the human eye). In fact, we now know that the visible portion of the spectrum, from 0.4 to 0.7 μm, is just a tiny part of a much broader spectrum of electromagnetic radiation. The various portions of the spectrum and the terminology commonly used to refer to these regions are given in Fig. 2.3.

The remote sensing techniques addressed in this book are generally concerned with the portion of the spectrum from the ultraviolet to the microwave and radiowavelength regions. A sense of the importance of the various spectral regions to present methods of atmospheric remote sensing is also provided in Fig. 2.3, which shows the considerable range and variety of sensors flown on the Nimbus 7 experimental satellite.

2.2 Wave Propagation

2.2.1 Mathematical Description

Figure 2.1 provides a snapshot of an electromagnetic wave and illustrates how the electric and magnetic fields move back and forth as the wave travels along its path. An obvious characteristic of this propagation is that these fields repeat themselves at set distances along the path; this distance is the wavelength of the radiation. In the most general sense, suppose we have a displacement $\mathcal{E}$ (for our purposes, this is the magnitude of the electric field) which is specified as a function $\mathcal{E} = f(x)$. The displacements at the point $x \pm \lambda$ are the same as at x. We conclude that a mathematical expression of the form

$$\mathcal{E} = f(x \pm ct) \tag{2.4}$$

adequately describes the physical situation that repeats as it travels or propagates along the $\pm x$ directions. While $\mathcal{E}$ is taken here to represent the electric field, $\mathcal{E}$ may in fact represent a great diversity of physical quantities, such as a deformation in the Earth's crust, the pressure of a gas, and many others.

An especially important example of repetitive wavelike motion is the harmonic wave which is described by the formula

$$\mathcal{E} = \mathcal{E}_o \cos k(x - ct) \tag{2.5a}$$

The quantity $\mathcal{E}_o$ has a special meaning. It is the amplitude of the wave and, as we shall see later, the energy carried by the wave is

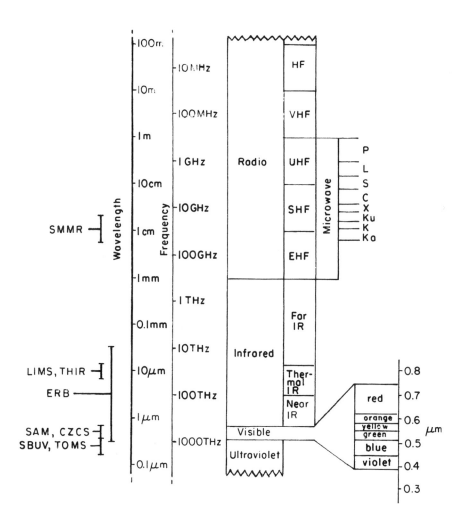

Figure 2.3 The electromagnetic spectrum. The diagram shows those parts of the electromagnetic spectrum which are important in remote sensing, together with the conventional names of the various regions of the spectrum. The letters (P, L, S, etc.) used to denote parts of the microwave spectrum are in common use in remote sensing, being standard nomenclature among radar engineers in the United States. Various terminologies are in use for the subdivisions of the infrared (IR) part of the spectrum. That adopted here defines the thermal band as lying between 3 and 15 μm, since this region contains most of the power emitted by black bodies at terrestrial temperatures. Also shown are wavelength regions of sensors on the Nimbus 7 satellite.

related to the square of this amplitude. The quantity $k(= 2\pi\tilde{\nu})$ is also referred to as wavenumber, but this should not prove to be a source of confusion as $\tilde{\nu}$ and k are used in different contexts; k generally applies to wave propagation, whereas $\tilde{\nu}$ is used, as in the previous section, to discriminate regions of the electromagnetic spectrum. Equation (2.5a) can also be written in the form

$$\mathcal{E} = \mathcal{E}_o \cos(kx - \omega t) \tag{2.5b}$$

where $\omega = kc = 2\pi c/\lambda$ is the angular frequency of the wave and, according to (2.2), $\omega = 2\pi\nu$.

The argument of the cosine function in (2.5a) also has a particular meaning. It is represented by the function ϕ

$$\phi = k(x - ct) \tag{2.6}$$

and is referred to as the *phase* of the wave. If, for instance, the plane wave example of Fig. 2.4 is considered, then the plane that contains the wave crests is also a plane over which the phase is the same at all points. This situation is most easily visualized by the analogy to a plane surface wave such as the one shown in Fig. 2.4.

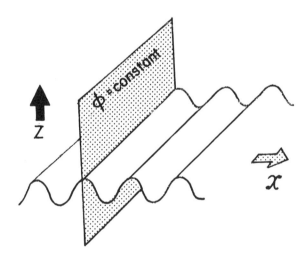

Figure 2.4 An undulating surface as an analogy to a propagating plane wave. All points along lines of equal displacement (such as along the ridges of the surface) correspond to lines of equal phase and are referred to as wavefronts. A light "ray" is simply the line drawn normal to these fronts. This ray can be characterized by the unit vector $\vec{n}$ along the direction of the ray.

The mathematical form of the harmonic wave can be written in a more general way by introducing complex variables and by noting, through the use of Maclaurin's theorem, that

$$e^{\pm i\phi} = \cos\phi \pm i\sin\phi$$

from which it follows that (2.5a) becomes

$$\mathcal{E}(x,t) = \mathcal{E}_o e^{ik(x-ct)} \tag{2.7}$$

and it is taken for granted that the real part of this expression represents the wave. The general representation of the harmonic wave requires that the displacement $\mathcal{E}$ be specified at $x = 0$ and $t = 0$. We specify this initial displacement (initial since it is defined at $t = 0$) in terms of a constant phase ϕ_o at $x = 0$ and $t = 0$. For this general case,

$$\phi = \phi_o + k(x - ct) \tag{2.8}$$

and

$$\mathcal{E}(x,t) = \mathcal{E}_o e^{i\phi} \tag{2.9}$$

Simple algebraic manipulations show that the square of the wave amplitude, given as

$$\mid \mathcal{E}(x,t) \mid^2 = \mid \mathcal{E}_o \mid^2 \tag{2.10}$$

is the same for all x and t since $\mathcal{E}_o$ is a constant. The energy transferred by the wave, related to $\mid \mathcal{E}_o \mid^2$, does not vary along its path of propagation and is independent of our definition of ϕ_o. It is only the interaction of the wave with matter that alters the energy of a propagating wave. It is ultimately this energy modulation that is exploited in remote sensing.

Excursus: The Intensity and Irradiance of Electromagnetic Radiation

An electromagnetic wave, traveling through space at the speed of light, carries electromagnetic energy which is detected by sensors that respond to this energy. Energy flows in the direction in which the wave advances and this direction of propagation is defined by the vector cross–product $\vec{\mathcal{E}} \times \vec{\mathcal{H}}$. The energy per unit area per unit time flowing perpendicular into a surface in free space is given by the *Poynting vector* $\vec{S}$, where

$$\vec{S} = c^2\epsilon_o\vec{\mathcal{E}} \times \vec{\mathcal{H}}$$

where c is the speed of light and ϵ_o is the vacuum permittivity. Energy per unit time is power, so the SI units of $\vec{S}$ are Wm^{-2}. At the frequencies of interest to the topics of this book, the fields $\vec{\mathcal{E}}, \vec{\mathcal{H}}$, and $\vec{S}$ oscillate at rapid rates and it remains impractical to measure an instantaneous value of $\vec{S}$ directly. We measure its average magnitude $<S>$ over some time interval that is a characteristic of the detector. This time averaged quantity is referred to as the *radiant flux density*.

Strictly speaking, the flux density emerging from the surface is known as the *exitance* and the flux density incident on the surface is called the *irradiance*. To avoid unnecessary complications with nomenclature, we refer to the flux density onto or from a surface as either irradiance or flux and use the symbol F to represent this quantity.

When the flow of light is nonparallel and when the detector collects the light confined to a range of directions, specified by a small element of solid angle $d\Omega$, then the quantity sensed is the *intensity*, defined as $<S>/d\Omega$ and has units of Wm^{-2}ster^{-1}. This is a quantity that is used throughout this book and we will denote it by the symbol I.

We can consider a more direct relationship between the energy carried by an electromagnetic wave and the amplitudes of the electric and magnetic fields by considering the simple case of a plane wave of the form

$$\vec{\mathcal{E}} = \vec{\mathcal{E}}_o \cos(kx - \omega t)$$

The magnetic field also has the form $\vec{\mathcal{H}} = \vec{\mathcal{H}}_o \cos(kx - \omega t)$ and therefore

$$S = c^2 \epsilon_o \vec{\mathcal{E}} \times \vec{\mathcal{H}} = c^2 \epsilon_o \vec{\mathcal{E}}_o \times \vec{\mathcal{H}}_o \cos^2(kx - \omega t)$$

Hence

$$<S> = c^2 \epsilon_o \mid \vec{\mathcal{E}}_o \times \vec{\mathcal{H}}_o \mid < \cos^2(kx - \omega t) >$$

and the time average is calculated for an interval of length T according to

$$< \cos^2(kx - \omega t) > = \frac{1}{T} \int_t^{t+T} \cos^2(kx - \omega t')dt'$$

$$= \frac{1}{2} - \frac{1}{4\omega T}[\sin(2kx - 2\omega(t + T)) - \sin 2(kx - \omega t)]$$

When $T \gg t$, $\omega T \gg 1$ and $< \cos^2(kx - \omega t) > \to 1/2$. Since $\mathcal{E}_o = c\mathcal{H}_o$,

$$F = <S> \approx \frac{c\epsilon_o}{2} \mathcal{E}_o^2$$

or

$$F \approx c\epsilon_o < \mathcal{E}^2 >$$

where $< \mathcal{E}^2 > = \mathcal{E}_o/2$.

2.2.2 Waves in Three Dimensions

Although our simple relation $\mathcal{E} = f(x - ct)$ represents a wave motion propagating along the x axis, it does not mean that a wave is actually concentrated *on* the axis. If we consider the physical disturbance extended over all space at a specified time t, then the function $\mathcal{E} = f(x - ct)$ takes the same value at all points having the same x. In three–dimensional space, $x = constant$ represents a plane perpendicular to the x axis such as demonstrated in Fig. 2.5a. Thus $\mathcal{E} = f(x - ct)$ represents a propagating plane wave in three dimensions.

Suppose that instead of propagating along the x axis, this plane wave propagates along a general direction characterized by the unit vector $\vec{n}$. If $\vec{r}$ is the position vector of a point on the wave front, then $\vec{n} \cdot \vec{r}$ is the distance measured from an origin along the direction of propagation. Thus we write

$$\mathcal{E} = f(\vec{n} \cdot \vec{r} - ct) \qquad (2.11)$$

for the general wave equation. It is convenient to introduce the vector $\vec{k}$ for $k\vec{n}$ in which case the phase of a harmonic wave is

$$\phi = \phi_o + \vec{k} \cdot \vec{r} - \omega t \qquad (2.12)$$

A particular type of wave propagation that is important for later considerations is the spherical wave shown in Fig. 2.5c. We can think of this wave as a surface propagating out from a point source like that of a single oscillating dipole. As the wave propagates outward, the wave surface becomes progressively larger (increasing as r^2). Since the energy associated with the flow of photons is the same through each wave front, the total energy flowing out from the point source through each spherical surface is proportional to $\mathcal{E}_o^2$. The wave displacement at a point far from the source is approximated by

$$\mathcal{E} = \frac{\mathcal{E}_o}{kr} e^{i\phi}$$

where ϕ is again defined according to (2.8). Thus we can conclude that *for a spherical wave, the energy flow per unit area confined to*

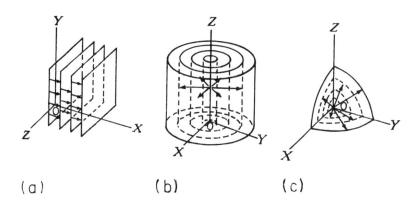

(a)   (b)   (c)

Figure 2.5 Examples of three–dimensional waves: (a) a plane wave, (b) a cylindrical wave, and (c) a spherical wave.

a particular direction) decreases as the inverse of the square of the distance from the source.

2.2.3   Doppler Effects

When the source of an electromagnetic wave and an observer are in relative motion with respect to the material medium in which the waves propagate, the frequency of the waves observed is different from the frequency of the emitting source. This is the so–called Doppler effect named after the German–born, Austrian physicist C. J. Doppler, who first noticed the effect in sound waves. The Doppler effect for elastic waves consists of matter in motion and is well described in standard texts on physics. However, the Doppler effect for electromagnetic waves, as exploited in the remote measurement of atmospheric motion using techniques discussed in Chapter 8, requires a treatment different from the more mechanical discussion of this effect. The reasons for this are that electromagnetic waves do not involve matter in motion, and therefore the velocity of the moving source relative to the medium does not enter into discussion. Second, the velocity of propagation is c, the speed of light, which is the same for all observers regardless of their relative motion.

Consider the example of one observer at O' moving relative to a source at O with velocity v along the line joining O and O'. The Doppler effect for electromagnetic waves is derived from the principle of relativity which requires that the phase of the wave, $kx - \omega t$, remains invariant when passing from one inertial system (at rest) to another (in motion with a velocity v). Therefore,

$$kx - \omega t = k'x' - \omega't' \qquad (2.12)$$

and with arguments that use Lorentz transformations, the frequency observed at O' is shifted relative to the frequency ν at O according to

$$\nu' = \nu \frac{1 + v/c}{\sqrt{1 - v^2/c^2}} \qquad (2.13)$$

For velocities typical of atmospheric motion, $v^2/c^2 << 1$, and

$$\nu' \approx \nu(1 + v/c) \qquad (2.14)$$

It is customary to express the Doppler effect in terms of the frequency shift defined relative to the fixed observer;

$$\Delta\nu_D = (\nu' - \nu) = v/\lambda \qquad (2.15)$$

where $\Delta\nu_D$ is referred to as the *Doppler shift*. This shift is defined such that $\Delta\nu_D > 0$ when O' is moving toward the observer and negative as O' recedes from the target as in the example of the red shift of white star light of receding galaxies.

When the relative motion is not along the line joining O and O', it follows that

$$\nu' \approx \nu(1 + v\cos\theta/c) \qquad (2.16)$$

and

$$\Delta\nu_D = v\cos\theta/\lambda \qquad (2.17)$$

where θ is the angle between the direction of motion and the line connecting the source and observer.

Our applications deal with the frequency shift of an electromagnetic wave scattered by a moving target with the source and the observer stationary relative to each other. To treat this rigorously, the relativistic arguments introduced to derive (2.13) are needed, but the same answer may be obtained using a non-relativistic treatment given a consistent but nonphysical assumption about the motion of the "medium" transmitting the wave motion. Consider the geometry illustrated in Fig. 2.6. Light of frequency ν from a source at O is scattered by an object at P and observed at O'. The angles that the direction of motion make with OP and PO' are θ_1 and θ_2, respectively. The frequency observed at P according to (2.16) is

$$\nu' = \nu(1 + v\cos\theta_1/c) \tag{2.18}$$

If we now consider a hypothetical source at the particle P emitting at ν' to an observer at O' moving relative to P with a velocity $v\cos\theta_2$, then the frequency of the light received at O' is

$$\nu'' = \nu'(1 + v\cos\theta_2/c) \tag{2.19}$$

Combining (2.18) and (2.19) produces

$$\Delta\nu_D = \nu'' - \nu = \frac{\nu v}{c}(\cos\theta_1 + \cos\theta_2) \tag{2.20}$$

for velocities much smaller than c. For a system with a collocated source and receiver (one we refer to as a *monostatic* system), $\theta_1 = \theta_2 = \theta$, and it follows that

$$\Delta\nu = 2v\cos\theta/\lambda \tag{2.21}$$

where $v\cos\theta$ in this sense is the radial velocity of the target relative to the observer. Equation (2.21) provides the basis for wind measurements using the Doppler shift associated with the motion of certain targets (such as aerosol) as they are advected by the wind.

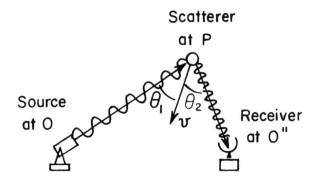

Figure 2.6 Diagram for the calculation of the Doppler shift on scattering by a moving object at P for a bistatic system.

2.3 Polarization

A subtle feature of electromagnetic radiation, and one discovered some two centuries before Maxwell by the peculiar way certain materials reflect light, is the property of *polarization.* This feature played an important role in shaping the electromagnetic theory as it developed during the nineteenth century. Although polarization is not an obvious property of an electromagnetic wave — for example, human vision is not very sensitive to polarized light — it is nevertheless an important property, especially for remote sensing. Two electromagnetic waves, identical in all respects except for their polarization state, can interact differently with matter, and it is the very nature of these differences that are exploited in certain remote sensing methods.

The amount of light transmitted through two sheets of a material arranged as in Fig. 2.7 depends on the rotation of one sheet with respect to the other. For one case, light is transmitted straight through the two sheets although its character is altered on transmission. As one of the sheets is rotated with respect to the other, the intensity of the light decreases until an orientation is reached when no light is transmitted. This type of experiment suggests that an electromagnetic wave has properties in directions other than along the line of propagation and that the wave in some sense is three dimensional. Polarization is a property of this dimensionality.

The previous experiment illustrates the property of polarization. We note in reference to Fig. 2.1 that as the electromagnetic wave propagates along the z axis, the direction of the oscillations of the fields may vary about the direction of propagation. At one point in space the electric vector might be directed along the y axis, while at another point along the beam this vector might have turned to point along the x axis. For the example shown in Fig. 2.7, the polarizing crystal allows only oscillations which are preferentially aligned along a specific direction (the vertical in this example), and the light is said to be linearly polarized along the vertical direction. Although the property of polarization seems to be a subtle property of electromagnetic radiation, it is a property that demonstrates a basic characteristic of electromagnetic waves. It is one that emphasizes the difference between these waves and other types of waves. Electromagnetic waves are *transverse waves* for which the oscillations are perpendicular to the path of propagation.

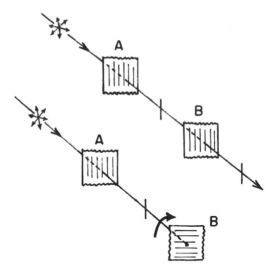

Figure 2.7 A polarizing crystal A allows only that part of an incident beam that is polarized along a particular direction defined by the properties of the crystal. Polarized light enters the second crystal B which passes the light depending on the relative alignment of A with respect to B. If B is rotated 90 degrees, then none of the polarized beam is transmitted. When used in this way, B is referred to as an analyzer.

2.3.1   Mathematical Description

We are now left to devise some way of describing the state of polarized radiation. It is customary to use the behavior of the electric field for this description since the magnetic field is perpendicular to it and a description of one defines the other. The electric field may oscillate in many different ways; it could point in a single direction after transmission through the first of the polarizing crystals as in our proceeding experiment (an example of linear polarization), or it may oscillate in such a way that the superpositions of all directions of oscillations trace a circular pattern, in which case the radiation is said to be *circularly polarized*. There are many more possibilities, but with all of these we can consider that the electric field at any point is simply a superposition of two waves linearly polarized at right angles to each other. Such a decomposition is more than just a mathematical device since, according to our experiment described in relation to Fig. 2.7, polarizers can be used to isolate these perpendicular components and therefore provide a way of analyzing the state of polarized radiation.

We choose to write the electric fields as a superposition of two linearly polarized waves, namely

$$\vec{\mathcal{E}} = \mathcal{E}_\ell \hat{\ell} + \mathcal{E}_r \hat{r} \tag{2.22}$$

where $\mathcal{E}_\ell$ and $\mathcal{E}_r$ are the components of the vibrations along two orthogonal directions shown in Fig. 2.8. The direction vectors $\hat{\ell}$ and $\hat{r}$ are defined so that both are perpendicular to the direction of propagation with the $\hat{\ell}$ component lying parallel to the *plane of reference* and $\hat{r}$ perpendicular to this plane in the sense that $\hat{r} \times \hat{\ell}$ is along the direction of propagation. We will refer to this representation of the $\mathcal{E}$ field as the "linear basis". We could choose other bases to represent the electric field, and an example is discussed later.

When dealing with problems of scattering, it is customary to specify the plane of reference so that it includes both the direction of the incident and the direction of the scattered waves.

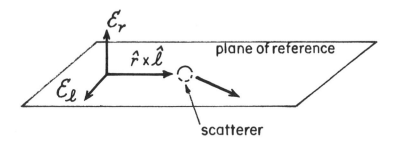

Figure 2.8 A schematic illustration of the relation between $\hat{\ell}, \hat{r}$, and the direction of propagation of the electromagnetic wave and the plane of reference. The electric field is considered to be a superposition of $\mathcal{E}$ fields along the $\hat{\ell}, \hat{r}$ directions. The symbols ℓ and r are taken from the last letters of the words parallel and perpendicular, respectively.

According to (2.5b) we can express each of the orthogonal components in the form

$$\mathcal{E}_\ell = \mathcal{E}_{o,\ell} e^{i(\phi+\phi_\ell)}$$

$$\mathcal{E}_r = \mathcal{E}_{o,r} e^{i(\phi+\phi_r)} \tag{2.23}$$

where $\phi = kx - \omega t$ and ϕ_r and ϕ_ℓ are the constant phase shifts of the perpendicular and parallel vibrations relative to some origin point. The precise specification of this origin is not important to the mathematical description of polarization as only the difference of the phases of the two orthogonal waves matters. If $\phi_r = \phi_\ell$, then (2.23) defines a straight line and the radiation is said to be *linearly polarized*. We can mathematically represent this linearly polarized radiation by eliminating the harmonic factor in (2.23) to give

$$\mathcal{E}_r = (\mathcal{E}_{o,r}/\mathcal{E}_{o,\ell})\mathcal{E}_\ell \tag{2.24}$$

where the slope of the line is $\mathcal{E}_{o,\ell}/\mathcal{E}_{o,r}$. Figures 2.9a and b illustrate examples of the electric field vector projected on the plane normal to the direction of propagation for two cases of linear polarization. For the more general case when the phase constants differ by a fixed amount, then the polarization is no longer linear. Examples of this situation are presented in Figs. 2.9c to f for phase differences of either 90 or 45 degrees. When both the phases and the amplitudes of the fields differ in a fixed way, the electromagnetic wave is said to be *elliptically polarized* as in the cases shown by Figs. 2.9d to f.

In summary, we are able to describe the state of polarization of an electromagnetic wave in terms of two linearly polarized waves vibrating at right angles to each other with some fixed phase difference. We both visualize and describe this state of polarization as a projection on two perpendicular axes of the electric vector rotating with a circular frequency ω around the direction of propagation. Figure 2.10 provides a clear visualization of this idea for the case of two equal amplitude waves out of phase by 90 degrees from each other. The position of the electric vector as the wave propagates along the z axis is also drawn at eight different locations along the path. In this case, the tip of the vector rotates to the right as the beam propagates along. As seen by an observer looking toward the light, the electric field rotates counterclockwise.

In defining the complete state of polarization only three pieces of information are required. These are the amplitudes of each wave and the phase difference between them. A fourth parameter, the

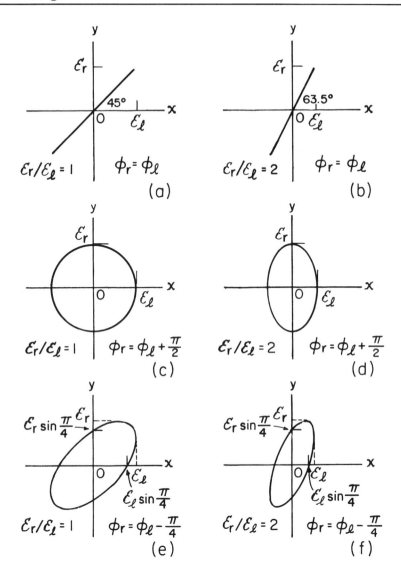

Figure 2.9 Simple harmonic motions in two dimensions: (a) The amplitudes along the r and ℓ directions are the same, as are their phase constants. (b) rs amplitude is twice ℓs but their phase constants are the same. (c) Their amplitudes are equal, but ℓ leads r in phase by 90 degrees. (d) Same as (c), but r's amplitude is twice ℓs. (e) Equal amplitudes, but ℓ lags r in phase by 45 degrees. (f) Same as (e) except rs amplitude is twice ℓs.

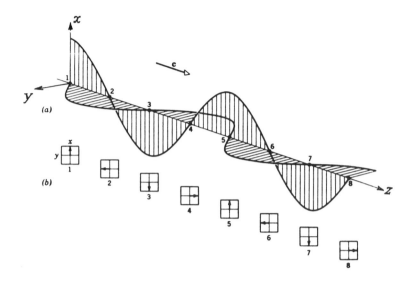

Figure 2.10 (a) Two linearly polarized waves of equal amplitude at right angles to each other moving along the z axis. For the case drawn, they differ in phase by 90 degrees; where one wave has a maximum value, the other is zero. (b) Views of the resultant amplitude of the approaching wave as seen by observers located at the positions shown on the z axis. The electric vector moves to the right as seen lookin along the z axis. The electromagnetic wave is said to be right–handed circularly polarized.

sign of the phase difference, defines the sense by which the vector rotates and is referred to as the *handedness* of the polarization. The polarization is said to be right–handed if the $\mathcal{E}$ field at an instant in time traces out a right–hand screw or spiral and left–handed if the $\mathcal{E}$ traces out a left–handed screw.

2.3.2 Examples of Polarized Radiation

Electromagnetic waves in the radio and microwave range are generated by surging a charge up and down a wire to create an oscillating dipole in the form of a transmitting antenna as shown in Fig 2.11. The field transmitted by such a dipole tends to be linearly polarized in a direction parallel to the dipole axis. When this polarized wave in turn falls on a receiving antenna, the alternating electric field of the transmitted wave causes the charges in the receiving antenna to

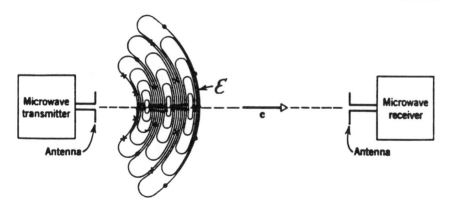

Figure 2.11 The $\mathcal{E}$ vectors of the linearly polarized transmitted electromagnetic wave are parallel to the axis of the receiving antenna so that a wave will be detected. If the antenna is rotated through 90 degrees about the direction of propagation, no signal is detected. The electric field radiates out from the transmitter in a manner indicated by the contours of the $\mathcal{E}$ field.

surge back and forth producing a reading on the detector. If the receiving antenna is turned through 90 degrees around the direction of propagation, the detector reading drops to 0. In principle, two transmitters aligned perpendicular to each other, one with the current out of phase by 90 degrees with the other, transmit an electric field that is circularly polarized.

Common sources of visible and infrared light differ from radio and microwave sources in that the elementary radiators are the atoms and the molecules of matter which tend to act independently of one another. The light that propagates in a given direction consists of many independent wave–trains that vibrate in a random fashion relative to each other. The electric fields are thus randomly distributed in the plane normal to the direction of propagation; consequently the electric vector forms no well–defined pattern as the field oscillates. Despite this irregularity, we are still able to describe this random case in terms of two waves vibrating at right angles to each other. In this case the wave may be viewed as two linearly polarized waves with a random phase difference. Thus the fluctuations of one of these superimposed fields is independent of the other, and if the average amplitudes of the two are equal, then the light is said to be *unpolarized*. Sunlight, radiation emitted from the Earth, and light radiated from incandescent sources are examples of unpolarized radiation sources.

Radiation is said to be *partially polarized* when its state of polarization falls between the extremes of total polarization and complete unpolarization. Such radiation can be thought to be composed of varying contributions of two beams, one completely polarized and the other unpolarized. As we shall see, the degree to which a partially polarized beam is polarized can then be assessed in terms of the ratio of the intensity of this polarized component to the total intensity. This ratio is referred to as the *degree of polarization* and varies from 0 for completely unpolarized radiation to unity for a fully polarized beam.

2.4   Stokes' Parameters

The state of polarization is completely specified by the four parameters described earlier (two amplitude quantities, the magnitude, and the sign of the phase difference). While it is not difficult to devise ways of measuring amplitude quantities (since these are just related to intensities), it is an entirely different matter to measure phase differences. We now turn to an alternate description of the polarization of light which is tied directly to observable intensity quantities. These intensity quantities are the four *Stokes parameters*, I, Q, U, V, named after G. N. Stokes, who systematically studied polarized light fields in 1852. It is quite remarkable that a description of the myriad forms of polarized light reduce to the specification of just four parameters. Methods to measure these four parameters are described, and some attempt is made to connect these parameters to the more heuristic description of polarization given earlier.

Consider the simple, hypothetical instrument shown in Fig. 2.12. Light is collected by the instrument and passed through two special plates, a spectral filter, and ultimately down to the detector. The relative placement of the wave plate W and the polarizer P (used here as an analyzer) is important, although where the filter is relative to W and P is immaterial provided the properties of W and P are spectrally flat. We will suppose for the purpose of this discussion that W and P are ideal and that there is no attenuation of radiation when transmitted through either plate. In order to understand the workings of this instrument it is appropriate first to review some special properties of the materials used to construct W and P.

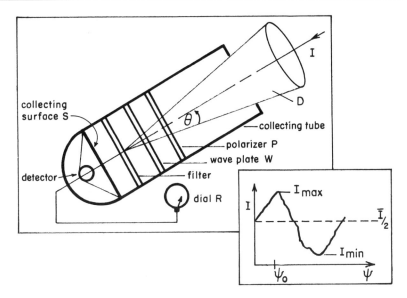

Figure 2.12 Schematic details of a radiometer fitted with a polarizer P and wave plate W which is used to measure polarized radiation and to obtain the four Stokes parameters. A typical output of the instrument is shown as the polarizer is rotated through ψ.

When a transverse wave propagates through certain crytalline solids, the oscillations excited in the material[1] depend on the direction of polarization of the wave as well as its direction of propagation. The molecules of these solids tend to be oriented and unable to rotate about their equilibrium positions within the crystal lattice. For certain types of material, the ability of molecules to oscillate is not necessarily the same in all directions. This material is then said to be *anisotropic*. When an electromagnetic wave penetrates a slab of anisotropic material, no matter what the initial state of polarization, the electromagnetic wave behaves as two waves that are polarized at right angles to each other and propagate at different phase velocities. The material is then said to be *birefringent*.[2] Since the two

---

[1] The basic nature of these oscillations is the subject of much of Chapters 4 and 5.

[2] Many substances that are normally isotropic become anistropic and birefringent when subjected to mechanical stresses. This fact is useful in engineering design studies in that strains in gears, in bridge structures, and so on, can be studied quantitatively by stressing plastic models and examining the optical anisotropy that results.

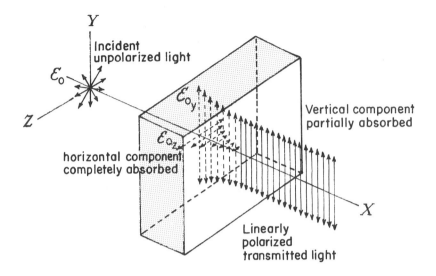

Figure 2.13 An illustration of dichroism.

waves travel through the slab of crystal at different speeds, there
will be a phase shift between them as they emerge from the crystal. If the slab thickness is chosen so that this phase difference is
precisely 90 degrees (the retardation angle), then the slab is called
a *quarter–wave plate*, and, according to our discussion in relation to
Fig. 2.10, linearly polarized light will become circularly polarized as
it emerges from the crystal. It is this type of crystal, in the form of
a quarter–wave plate, that we use in our simple instrument for W.

Certain anisotropic substances may also absorb more in one
direction than in another. An electromagnetic wave that propagates
through a sufficiently thick piece of this material becomes gradually
polarized in only one direction. This situation is called *dichroism*
and is shown in Fig. 2.13. Dichroic materials offer a simple and
inexpensive way of producing and analyzing polarized light. We use
this material for the polarizing plate P.

The *optical axis* is a characteristic direction in the crystal that
allows us to establish the orientation of the polarization of light on
transmission through the crystal. The orientation of the optical axis
of the polarizer in the instrument is important in the experiments
described next.

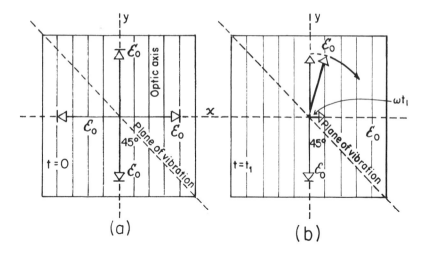

Figure 2.14 (a) Plane polarized light falls from behind on a quarter–wave plate oriented so that light emerging from the page is circularly polarized. (b) In this case, the electric vector $\mathcal{E}$ rotates clockwise as seen by an observer facing the light.

Excursus: A Circular Basis for the Description of Polarization

There is an alternate mathematical way to describe polarization other than the linear basis described earlier. Here we discuss the circular basis for polarization. Let us first contemplate our experiment using the radiometer of Fig. 2.12 where W is a quarter–wave plate. Consider light entering this instrument and suppose that the optical axis of our wave plate is oriented at 45 degrees to the direction of polarization of the beam that emerges from the polarizer (see Fig. 2.14). The wave that emerges from the wave plate is circularly polarized and we will determine in what direction the electric vector rotates assuming the direction of propagation is out of the page for the geometric arrangement shown.

The component of wave vibrations parallel to the optical axis of the wave plate (we denote this direction as the ℓ axis) as the wave emerges from W is

$$\mathcal{E}_\ell = \mathcal{E}_o \sin 45° e^{i\omega t} = \frac{1}{\sqrt{2}} \mathcal{E}_o \cos(\omega t) \qquad (2.25a)$$

where we assume for convenience that $x = 0, \phi_o = 0$ in (2.8). The wave component perpendicular to the optical axis is

$$\mathcal{E}_r = \mathcal{E}_o \cos 45° e^{i\omega t - \pi/2} = \frac{1}{\sqrt{2}} \mathcal{E}_o \sin(\omega t) \qquad (2.25b)$$

where the $-\pi/2$ phase shift represents the action of the quarter–wave plate. To decide the direction of rotation, we locate the tip of the rotating electric vector at two instants of time, say $t = 0$ and, at a short time later, $t = t_1$. At $t = 0$,

$$\mathcal{E}_\ell = \frac{1}{\sqrt{2}} \mathcal{E}_o \text{ and } \mathcal{E}_r = 0.$$

At $t = t_1$, these coordinates become approximately

$$\mathcal{E}_\ell = \frac{1}{\sqrt{2}} \mathcal{E}_o \cos \omega t_1 \approx \frac{1}{\sqrt{2}} \mathcal{E}_o (1 - \omega t_1)$$

$$\mathcal{E}_r = \frac{1}{\sqrt{2}} \mathcal{E}_o \sin \omega t_1 \approx \frac{1}{\sqrt{2}} \mathcal{E}_o (\omega t_1)$$

Thus the vector representing the emerging circularly polarized light is rotating clockwise when the observer faces the light source and the emerging light is therefore *left–circularly polarized.*

One important result of this analysis is that it demonstrates how a circularly polarized can be written as a combination of linearly polarized waves. For example, we write for left–hand circular polarization,

$$\mathcal{E}_{LH} = \frac{1}{\sqrt{2}} (\mathcal{E}_\ell - i\mathcal{E}_r) \qquad (2.26a)$$

and similarly

$$\mathcal{E}_{RH} = \frac{1}{\sqrt{2}} (\mathcal{E}_\ell + i\mathcal{E}_r) \qquad (2.26b)$$

for right–hand circular polarization. It is also relevant to note how a linearly polarized wave can be represented as the sum of right– and left–hand polarized waves (e.g., $\mathcal{E}_\ell = [\mathcal{E}_{RH} + \mathcal{E}_{LH}]/\sqrt{2}$). We can write the relationship between the circular and linear polarization basis as

$$\begin{pmatrix} \mathcal{E}_{RH} \\ \mathcal{E}_{LH} \end{pmatrix} = \frac{1}{\sqrt{2}} \begin{pmatrix} 1 & i \\ 1 & -i \end{pmatrix} \begin{pmatrix} \mathcal{E}_{\ell} \\ \mathcal{E}_{r} \end{pmatrix}. \tag{2.27}$$

2.4.1 Measurement of I,Q,U,V

We will now perform two experiments with our radiometer. In the first experiment, the instrument operates without the wave plate, and the polarizer is initially aligned with the optical axis pointing along a specified reference direction (say along the vertical direction). The angle of the polarizer's optical axis measured from this reference direction is denoted by ψ. This angle is then systematically varied from zero radians along the reference direction in a clockwise direction (when looking along the direction of the traveling beam and into the instrument) to π radians. A typical example of the intensities recorded by this instrument as P rotates in this way is also shown in Fig. 2.12. If $I(\psi)$ is the intensity measured by the instrument for the given angle ψ, then we obtain

$$I(\psi) = \frac{1}{2}[\bar{I} + \Delta I \cos 2(\psi - \psi_o)] \tag{2.28}$$

where it simply follows that

$$\bar{I} = I_{max} + I_{min} \tag{2.29a}$$

$$\Delta I = I_{max} - I_{min} \tag{2.29b}$$

where I_{max} and I_{min} are the maximum and minimum readings, respectively, and ψ_o is the angle corresponding to I_{max}. Introducing

$$Q = \Delta I \cos 2\psi_o \tag{2.30a}$$

$$U = \Delta I \sin 2\psi_o \tag{2.30b}$$

causes equation (2.28) to become

$$I(\psi) = \frac{1}{2}[\bar{I} + Q \cos 2\psi + U \sin 2\psi] \tag{2.31}$$

The second experiment repeats the procedure of the first experiment except now a wave plate of a fixed retardation ϵ is used. In this case, the intensity is

$$I(\psi, \epsilon) = \frac{1}{2}[\bar{I} + Q \cos 2\psi + (U \cos \epsilon - V \sin \epsilon) \sin 2\psi] \tag{2.32}$$

where $I(\psi, \epsilon)$ is the intensity measured for the polarizer angle ψ and the wave plate retardation ϵ.

Only four direct readings from our instrument are needed to obtain the Stokes parameters, I, Q, U, and V. These four intensity measurements are $I(0,0), I(\pi/2,0), I(\pi/4,0)$, and $I(\pi/4, \pi/2)$; the first three are obtained with the instrument configuration used in our first experiment, and the fourth measurement is carried out with the configuration of our second experiment with a quarter–wave plate for W.

The connection between the four intensity measurements and the four Stokes parameters is easily established by means of (2.32). Substituting the appropriate values of ϕ and ϵ in (2.32) produces

$$I(0,0) = \frac{1}{2}[I + Q]$$

$$I(\pi/2,0) = \frac{1}{2}[I - Q]$$

$$I(\pi/4,0) = \frac{1}{2}[I + U]$$

$$I(\pi/4, \pi/2) = \frac{1}{2}[I - V]. \tag{2.33}$$

The relationship between the four Stokes parameters and the four intensity quantities can be written in the following mathematical way:

$$\mathbf{I} = \mathbf{P}\mathbf{I}_{obs} \tag{2.34}$$

where

$$\mathbf{P} = \begin{pmatrix} 1 & 1 & 0 & 0 \\ 1 & -1 & 0 & 0 \\ -1 & -1 & 2 & 0 \\ 1 & 1 & 0 & -2 \end{pmatrix} \tag{2.35}$$

and where the (observable) intensity vector and the Stokes vector are

$$\mathbf{I}_{obs} = \begin{pmatrix} I(0,0) \\ I(\pi/2,0) \\ I(\pi/4,0) \\ I(\pi/4, \pi/2) \end{pmatrix} \quad \text{and} \quad \mathbf{I} = \begin{pmatrix} I \\ Q \\ U \\ V \end{pmatrix} \tag{2.36}$$

Thus $\mathbf{I}_{obs}$ and $\mathbf{I}$ are equivalent descriptions of polarized light fields in the sense that one allows the deduction of the other. There is a difference, however, that is important. Our hueristic discussions of polarization tend to focus on the Stokes vector whereas $\mathbf{I}_{obs}$ is more suited to discussion of actual measurements and measurement systems. For these discussions, we adopt the short–hand notation I_ℓ for $I(0,0)$ to represent the intensity of the component parallel to the reference axis of the linear polarizer. In a similar way, we use I_r for $I(\pi/2,0)$ for the intensity perpendicular to the reference direction. Further discussion of these components is given in Section 5.8.

Table 2.1 provides examples of $\mathbf{I}_{obs}$ and $\mathbf{I}$ for commonly occurring polarization states. We note from studying this table how the parameter V specifies the extent of circular polarization with the sign of V determining the handedness of this state. By contrast, Q and U characterize the extent to which the light field is linearly polarized; Q refers to parallel and perpendicular polarization, and U refers to polarization along the ±45 degree direction. If the light is unpolarized, then $Q = U = V = 0$, which naturally leads to a definition of the *degree of polarization* as $\sqrt{Q^2 + U^2 + V^2}/I$. Furthermore, the ratios $\sqrt{Q^2 + U^2}/I$ and V/I are referred to as the degrees of linear and circular polarization, respectively.

Only three of the Stokes parameters are actually independent for a single electromagnetic wave and are related by the identity

$$I^2 = Q^2 + U^2 + V^2 \tag{2.37}$$

In general, however, the light field that flows into our instrument is not a single continuous monochromatic wave; rather, it consists of many "simple waves" that flow in rapid succession, and the measurable intensities refer to a superposition of many millions of these simple waves with independent phases. Because the definition of Stokes parameters involves only intensities, it follows that the Stokes parameters of such a collection of incoherent beams (i.e., beams exhibiting no fixed relationship among the phases) are simply additive. Thus we understand that by Stokes parameters we mean the sums

$$I = \sum I_i, Q = \sum Q_i, U = \sum U_i, V = \sum V_i \tag{2.38}$$

where the index i denotes each independent wave. It may be shown (but omitted for brevity) that under these circumstances

$$I^2 \geq Q^2 + U^2 + V^2 \tag{2.39}$$

Table 2.1 Radiance and Stokes Vectors of Common States.

Verbal Description	Observable Intensity Vector	Stokes Vector [I,Q,U,V]
Vertically or parallel polarized intensity	$[1,0,\frac{1}{2},\frac{1}{2}]$	[1,1,0,0]
Horizontally or perpendicular polarized intensity	$[0,1,\frac{1}{2},\frac{1}{2}]$	[1,-1,0,0]
Linearly polarized intensity at +45°	$[\frac{1}{2},\frac{1}{2},1,\frac{1}{2}]$	[1,0,1,0]
Linearly polarized intensity at -45°	$[\frac{1}{2},\frac{1}{2},0,\frac{1}{2}]$	[1,0,-1,0]
Right circularly polarized intensity	$[\frac{1}{2},\frac{1}{2},\frac{1}{2},0]$	[1,0,0,1]
Left circularly polarized intensity	$[\frac{1}{2},\frac{1}{2},\frac{1}{2},1]$	[1,0,0,-1]
Unpolarized intensity	$[\frac{1}{2},\frac{1}{2},\frac{1}{2},\frac{1}{2}]$	[1,0,0,0]

where the equality holds for completely polarized light and the inequality holds for partially polarized light.

Excursus: Example Calculation of I,Q,U,V

We will now analyze our method for measuring the four Stokes parameters by presenting a calculation of the polarization state illustrated in Fig. 2.9e. As in problem 2.2b, we express the orthogonal components of the electric field as

$$\mathcal{E}_r = \mathcal{E}_o \cos(kx - ct)$$

$$\mathcal{E}_\ell = \mathcal{E}_o \cos(kx - ct + \frac{\pi}{4})$$

We choose our origin point as $x = 0$ and $t = 0$ for convenience. Let us write these components as

$$\mathcal{E}_r = e^{i0}$$

$$\mathcal{E}_\ell = e^{i\pi/4}$$

where we also assume $\mathcal{E}_o = 1$. The first intensity measurement $I(0,0)$ is

$$I(0,0) = \mid \mathcal{E}_\ell \mid^2 = \mid \frac{1}{\sqrt{2}}(1+i) \mid^2 = 1$$

and the second intensity measurement is

$$I(\pi/2, 0) = \mid \mathcal{E}_r \mid^2 = 1$$

It follows from (2.35) that $I = I(0,0) + I(\pi/2, 0) = 2$ and $Q = I(0,0) - I(\pi/2, 0) = 0$. We derive the U component from the third measurement with the polarizer aligned at 45 degrees with respect to the parallel direction. Thus the projection of the two orthogonal components onto this direction yields the amplitude

$$\mathcal{E}_{\pi/4} = \sin\frac{\pi}{4}e^{i0} + \cos\frac{\pi}{4}e^{i\pi/4} = \frac{1}{\sqrt{2}}(1 + e^{i\pi/4})$$

Therefore

$$I(\pi/4, 0) = \frac{1}{2}(2 + \sqrt{2})$$

and

$$U = 2I(\pi/4, 0) - I = \sqrt{2}$$

The fourth intensity measurement is obtained with the instrument configured as shown in Fig. 2.12. The action of the wave plate in this configuration is to retard the $\mathcal{E}_r$ component relative to $\mathcal{E}_\ell$. Thus on passing through the wave plate, the orthogonal components are

$$\mathcal{E}_r = e^{-i\pi/2}$$

$$\mathcal{E}_\ell = e^{i\pi/4}$$

The projections of these at 45 degrees are

$$\mathcal{E}_{\pi/4} = \sin\frac{\pi}{4}e^{-i\pi/2} + \cos\frac{\pi}{4}e^{i\pi/4}$$

Therefore

$$I(\pi/4, \pi/2) = \frac{1}{2}(2 - \sqrt{2})$$

and

$$V = -2I(\pi/4, \pi/2) + I = \sqrt{2}$$

From these calculations we obtain $\mathbf{I} = [2, 0, \sqrt{2}, \sqrt{2}]$ for elliptically polarized light. We also establish that the degree of polarization

$$\sqrt{Q^2 + U^2 + V^2}/I = 1$$

as expected.

Excursus: Mueller Matrices

A beam of arbitrarily polarized radiation is represented by the column vector $\mathbf{I}$. As this beam encounters an optical element (such as the polarizer or wave plate) the state of polarization changes. These changes are represented in terms of *Mueller matrices*. These matrices provide a convenient way of relating the incident and transmitted Stokes vectors. The usefulness of the Mueller matrices appears when assessing the effects of a series of optical elements on an incident beam. In this case, the combined effect of all elements is merely obtained from the product of their associated Mueller matrices. We will now consider the Mueller matrices of the elements of our radiometer since these matrices are also relevant to later discussions. The Mueller matrix for our ideal linear polarizer (e.g., Bohren and Huffman 1985) is

$$\frac{1}{2} \begin{pmatrix} 1 & \cos 2\psi & \sin 2\psi & 0 \\ \cos 2\psi & \cos^2 2\psi & \cos 2\psi \sin 2\psi & 0 \\ \sin 2\psi & \cos 2\psi \sin 2\psi & \sin^2 2\psi & 0 \\ 0 & 0 & 0 & 0 \end{pmatrix},$$

where ψ is the angle of the polarizer relative to the given reference direction. The Mueller matrix of our ideal polarizer for parallel transmission (i.e., for $\psi = 0$) is

$$\mathbf{M}_\ell = \frac{1}{2} \begin{pmatrix} 1 & 1 & 0 & 0 \\ 1 & 1 & 0 & 0 \\ 0 & 0 & 0 & 0 \\ 0 & 0 & 0 & 0 \end{pmatrix} \tag{2.40}$$

The Mueller matrix of the wave–plate W (i.e., of an ideal linear retarder) is

$$\begin{pmatrix} 1 & 0 & 0 & 0 \\ 0 & C^2 + S^2 \cos \epsilon & SC(1 - \cos \epsilon) & -S \sin \epsilon \\ 0 & SC(1 - \cos \epsilon) & C^2 + S^2 \cos \epsilon & C \sin \epsilon \\ 0 & S \sin \epsilon & -C \sin \epsilon & \cos \epsilon \end{pmatrix}$$

where $C = \cos 2\psi, S = \sin 2\psi$. The Mueller matrix of W configured to measure $I(\pi/4, \pi/2)$ thus follows as

$$\mathbf{M} = \begin{pmatrix} 1 & 0 & 0 & 0 \\ 0 & 0 & 0 & -1 \\ 0 & 0 & 1 & 0 \\ 0 & 1 & 0 & 0 \end{pmatrix}$$

with $\epsilon = \pi/2$ and $\psi = \pi/4$. The combined effect of the linear polarizer and wave-plate is

$$\frac{1}{2} \begin{pmatrix} 1 & 0 & 0 & 0 \\ 0 & 0 & 0 & -1 \\ 0 & 0 & 1 & 0 \\ 0 & 1 & 0 & 0 \end{pmatrix} \begin{pmatrix} 1 & 1 & 0 & 0 \\ 1 & 1 & 0 & 0 \\ 0 & 0 & 0 & 0 \\ 0 & 0 & 0 & 0 \end{pmatrix} = \frac{1}{2} \begin{pmatrix} 1 & 1 & 0 & 0 \\ 0 & 0 & 0 & 0 \\ 0 & 0 & 0 & 0 \\ 1 & 1 & 0 & 0 \end{pmatrix} \quad (2.41)$$

If light of arbitrary polarization, specified by the four parameters I, Q, U, V, flows through this particular configuration of elements (P first and W next), the transmitted light emerges as 100% right circularly polarized and of the form

$$\mathbf{I} = \frac{1}{2} \begin{pmatrix} I + Q \\ 0 \\ 0 \\ I + Q \end{pmatrix}$$

Since matrix multiplication is not commutative, we deduce that the order of the elements is important to the final outcome.

2.5 Creation of an Electromagnetic Wave

There are three stages in the life of a photon: creation, propagation, and destruction. Both the theory of Maxwell and the discussion so far focus only on the propagation of the electromagnetic wave through space. Moreover, Maxwell's theory is formulated so that it is independent of the properties of matter. The creation and destruction of a photon, by contrast, occurs as a direct result of its interaction with matter. The final section of this chapter focuses on how electromagnetic waves are created through a process referred to as *emission*. As we shall see throughout this book, the relationship between the emitted radiation and the constituents of the atmosphere that emit this radiation is exploited in many remote sensing methods. For instance, emission properties are used to derive atmospheric temperature, to derive concentrations of certain trace gases, and in the estimation of precipitation among other important atmospheric properties. Thus, it is important in the context of remote sensing to understand how and why objects emit radiation.

2.5.1   Equilibrium Radiation and Kirchhoff's Law

The generation of electromagnetic waves occurs as a general result of accelerating electric charges. In general, any object is composed of a vast number of molecules. Even without the aid of external excitation, these molecules oscillate over a continuous range of frequencies and therefore emit radiation of all frequencies. However, this radiation is not emitted equally at all frequencies rather it is distributed according to the *emission spectrum* which, as we shall see, depends strongly on the temperature of the object.

The nature of the emission spectrum and its relationship to the temperature of the body loomed as a major challenge to physicists late in the nineteenth century. In fact, the relationship could not be accounted for by using the principles of classical physics, and its description marked one of the major turning points in the history of science. The hypothetical concept of a blackbody, (i.e., a body whose surface absorbs all radiation incident upon it) emerged in attempting to formulate the description of the emission spectrum. It also follows that any two blackbodies at the same temperature emit precisely the same radiation and that a blackbody emits more radiation than any other type of object at the same temperature.

It is more appropriate to view blackbody radiation as *equilibrium radiation*. One can appreciate this by considering an isolated cavity with walls opaque to all radiation. The cavity walls constantly emit, absorb, and reflect radiation until a state of equilibrium is reached (i.e., until the temperature of the cavity walls no longer change with time). This equilibrium radiation fills the cavity uniformly and is just the same as the radiation emitted by a hypothetical blackbody at the same temperature of the cavity. To understand why this is so, imagine that a blackbody is placed in the cavity. This body absorbs all of the equilibrium radiation incident on its surface and, since the cavity is in a state of equilibrium (it cannot cool or warm), the radiation emitted by the object must be precisely equal to that absorbed by it. This radiation happens to be the radiation that fills the cavity. Therefore, under the conditions of equilibrium, the ability of a body to radiate is closely related to its ability to absorb radiation. The mathematical formulation of this statement is known as *Kirchhoff's Law*, which can be written as

$$E_\lambda = a_\lambda B_\lambda (T) \qquad (2.42)$$

where E_λ is the emitted radiation and $B_\lambda (T)$ is the radiation (expressed in intensity units) of the hypothetical blackbody. The proportionality constant is the absorption coefficient which varies between 0 and 1. If $a_\lambda = 0$, then (2.42) states that a body neither absorbs radiation at the given wavelength nor emits radiation at the same wavelength. For $a_\lambda = 1$, the emitted radiation is the blackbody radiation. As we shall see in Chapter 3, the absorption coefficient contains information about the type of matter that emits radiation and is therefore an important parameter in emission–based sensing methods. The wavelength dependence of this coefficient varies dramatically according to the nature of the matter emitting the radiation and the portion of the electromagnetic spectrum under consideration.

It is through the statement of Kirchhoff's Law that the whole point of blackbody radiation becomes apparent. All blackbodies at some temperature behave identically and the radiation emitted by such bodies at a given λ depends only on the temperature of the body. Thus the emission of radiation at some chosen wavelength is solely determined by the characteristics of the emitting matter (through a_λ) and temperature (through B_λ).

2.5.2 Planck's Blackbody Function and Related Laws

The theoretical question regarding the form of the wavelength distribution of the intensity of this cavity radiation and how this radiation in turn depends on the temperature of the walls of the cavity occupied the attention of many of the world's leading physicists during the 1890s. It was Max Planck who provided us with the theoretical description of the blackbody radiation however in doing so he was forced to make an assumption that proved to be one of the most daring departures from the philosophies of physics to that time. The assumption is that each oscillator in the walls of the cavity possesses only one of a discrete set of energies rather than the more conventional view that energy assumes any value above or equal to 0. The discrete energy level is given as

$$E = nh\nu \qquad (2.43)$$

where n is an integer, and is referred to as *the quantum number* that defines the permitted number of discrete units of energy of the oscillator. The fundamental unit of energy turned out to be proportional to the frequency of the oscillator ν where the proportionality constant h is known as *Planck's constant*. It is these discrete packets, or *quanta*, of energy that are emitted by the oscillators in the cavity walls after the oscillator undergoes a transition from one quantized energy state to another. On the basis of these arguments, Planck was able to demonstrate that the relationship,

$$B_\lambda(T) = \frac{2hc^2}{\lambda^5 (e^{hc/k_B \lambda T} - 1)} \qquad (2.44)$$

adequately describes blackbody radiation where k_B is Boltzmann's constant and T is the absolute temperature of the cavity walls.

Graphic examples of the function B_λ, known as *Planck's function*, are given in Fig. 2.15 for three different temperatures. The three examples given demonstrate an obvious relationship. For example, consider an ordinary electrical element on a stove. On the highest and thus hottest setting the element glows brightest with a reddish hue (Fig. 2.16a). When the electricity is turned off and the element is allowed to cool, the color of the element fades until its luminosity vanishes; however it still radiates, a fact evident when a hand is placed above the cooling element. This simple experiment helps illustrate that the hotter the object the shorter the wavelength of the maximum intensity. This observation can be mathematically stated by an expression known as *Wien's displacement law* which establishes a connection between the wavelength of maximum emission (λ_{max}) and the temperature of the radiator. This law is simply derived from

$$\frac{\partial B_\lambda}{\partial \lambda} = 0$$

from which it follows that

$$T\lambda_{max} = 2898 \quad (\mu m.K) \qquad (2.45)$$

Wien's displacement law is indicated on Fig. 2.15 as the solid, diagonal line joining the maxima of the three Planck functions. For example, at 6000 K, the maximum emission is in the blue region of the visible spectrum ($\lambda_{max} = 0.48$ μm) and substantial amounts of radiation are emitted across the entire visible spectrum. This distribution is very similar to the wavelength distribution of radiation

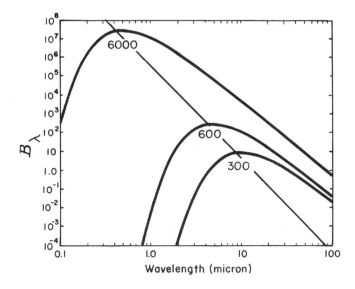

Figure 2.15 Planck's blackbody curve at the three temperatures shown. The units of this function are given in $Wm^{-2}ster^{-1}$. The diagonal line intersecting the curves at their maxima depicts Wien's displacement law.

emitted by the sun. At temperatures more typical of the element of the stove, this maximum is shifted to the longer wavelengths so that more red light is emitted than blue light, whereas at 300 K, the maximum intensity occurs at far infrared wavelengths around 10 μm. Even at this temperature, however, radiation is still emitted at all wavelengths including visible light, although the emission of visible light is too dim to see with the naked eye (Fig. 2.16b).

Another obvious characteristic of blackbody radiation is the hotter the object, the greater the total amount of radiation emitted from a given surface area. This is just a statement of *Stefan–Boltzmann's law*, which derives from integration of B_λ over the entire wavelength domain,

$$B(T) = \int_0^\infty B_\lambda(T)d\lambda = \frac{\sigma}{\pi}T^4, \qquad (2.46)$$

where $\sigma = 5.67 \times 10^{-8}$ $Wm^{-2}K^{-4}$ is the *Stefan–Boltzmann constant*. The radiation emitted by a 6000 K blackbody, for instance, is 160000 times that emitted by a 300 K blackbody.

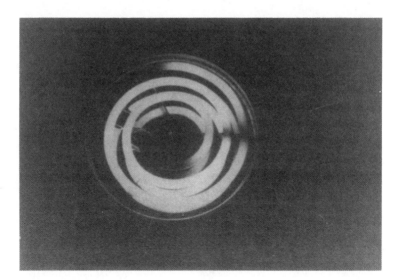

Figure 2.16 The element of a stove with the heating control on the highest setting (upper panel). The bottom panel, like that for the upper panel, was taken in a darkened room but with the control turned down to the point where the element was no longer visible; the exposure time was 12 hours (from Bohren, 1987).

It is often convenient to use the Planck function defined in terms of wavenumber rather than wavelength. The relationship between different forms of the Planck function is obtained from the simple requirement that the energy integrated over the same spectral domain be equivalent. Thus,

$$B_\lambda(T)d\lambda = -B_{\tilde{\nu}}(T)d\tilde{\nu}$$

and, using (2.44) along with (2.2), it follows that

$$B_{\tilde{\nu}}(T) = \frac{2hc^2\tilde{\nu}^3}{(e^{hc\tilde{\nu}/k_BT} - 1)} \tag{2.47}$$

There follows from either (2.44) or (2.47) two important limits of the Planck function. The first of these limits is *Wien's distribution* which applies as $\lambda \to 0$. In wavenumbers, this limit is

$$B_{\tilde{\nu}} = 2hc^2\tilde{\nu}^3 e^{-hc\tilde{\nu}/k_BT} \tag{2.48}$$

The long wavelength limit $\lambda \to \infty$ is known as the *Rayleigh–Jeans distribution* and in terms of wavenumbers is expressed by

$$B_{\tilde{\nu}} = 2k_BTc\tilde{\nu}^2 \tag{2.49}.$$

This longwave limit has a direct application to passive microwave remote sensing problems. At these wavelengths, the emission by the Earth's atmosphere is directly proportional to temperature. Thus, intensity is synonymous with temperature at these wavelengths.

From Table 2.2 we infer certain consequences of the long– and short–wavelength limits of the blackbody function and Wien's displacement law as they apply to atmospheric remote sensing. The relevant characteristics of the emission are tabulated at three wavelengths in the longwave spectrum: one in the shortwave limit at 4.3 μm, one at 15 μm, and the third in the longwave limit at 5 mm for the two temperatures indicated. The two shortest wavelengths lie in an atmospheric carbon dioxide absorption band and the third lies in a molecular oxygen absorption band. These wavelengths are suitable for temperature sounding (Chapter 7). The emission (in relative units) and the sensitivity of the emission to a small change in temperature (which can be thought of as related to the dB_λ/dT) are listed. Based purely on the availability of energy, it seems that the 15μm region is superior given equivalent detector sensitivities

Table 2.2 Properties of black body emission.

Wavelength	200 K	300K
Energy in relative radiance units		
4.3 μm	1.25	200
15 μm	5000	15000
5 mm	1	1
Temperature sensitivity*		
4.3 μm	1	20
15 μm	10	6
5 mm	4	1
Cloud transmission in %		
4.3 μm	6	1
15 μm	1	1
5 mm	96	99.98

* Relative to detector noise of 0.002 or 0.7 K.
From Smith (1972).

for each region. If a high temperature sensitivity is desirable, then the 4.3μm region is superior for detecting warmer temperatures but is still inferior to the 15μm region at colder temperatures, and the 5mm wavelength is the least sensitive of all three spectral regions. However, the optimum choice of wavelength is complicated by factors other than instrument sensitivity and available energy–the transmission characteristics of the atmosphere is one factor. This aspect is illustrated in the table in terms of typical transmission properties of clouds for the wavelengths under consideration (expressed as a percentage). We see that clouds in this example complicate matters further by strongly attenuating the shorter wavelengths, yet they are essentially transparent to millimeter wave radiation.

Excursus: Brightness Temperature

We refer to the intensity expressed in units of temperature as the *brightness temperature*. This is the temperature that is required to match the measured intensity to the Planck blackbody function at the given wavenumber or wavelength. For microwave radiation this is simply obtained from (2.49). At other wavelengths, the brightness temperature is obtained from either (2.44) or (2.47). For example, it follows from (2.47) that the brightness temperature T_b is

$$T_b = \frac{hc\tilde{\nu}}{k_B} \left[\ln \left(\frac{2hc^2\tilde{\nu}^3}{I_m} + 1 \right) \right]^{-1} \quad (2.50)$$

where I_m is the intensity measured at the characteristic wavenumber $\tilde{\nu}$ of the radiometer. The brightness temperature provides some indication of the temperature (and thus altitude in the atmosphere) at which the radiation is emitted. Some caution is needed here as the brightness temperature determined at one wavelength is generally different from that determined at another wavelength. Thus T_b cannot be used to determine the total emission by the atmosphere. It is only the emission of a pure blackbody that is defined by a single temperature.

2.5.3   Blackbodies and Cavity Radiometers

The radiation emitted from cavities is a subject that is of more than just of heuristic interest as cavities are often an important element in the design of radiometers. In many instrument applications, it is important to provide a source of blackbody radiation. A particular configuration that closely approximates a blackbody is the conical cavity which directs any small amount of radiation reflected from the sides of the cavity further into the cavity (Fig. 2.17). If the internal surfaces of the cavity are coated with special highly emissive paint, then the absorption by the cavity is almost complete, and the radiation leaving the cavity is very nearly blackbody radiation emitted at the temperature of the cavity walls. This radiation can be accurately estimated from measurements of the cavity temperature.

Cavities of this type are used in two important but different ways. Cavities are used to provide a well–defined source of radiation for calibration of radiometers. In fact, many well–designed radiometers actually include a cavity as a means of internal calibration. In the second application, the highly efficient absorption properties of the cavity are exploited as a way of measuring the radiation that flows into the cavity. Cavity radiometers operating in this way have been flown on spacecraft to measure the radiation output of the sun.[3]

---

[3] This output is referred to as the solar constant which is defined as that radiation received on a horizontal surface at the top of the atmosphere when the Earth–sun distance is one astronomical unit.

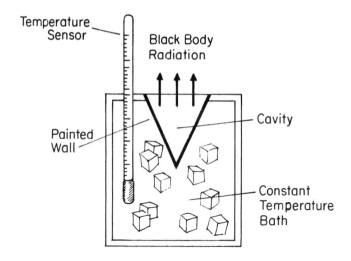

Figure 2.17 A schematic illustration of the components of a simple calibration black body showing a conical cavity with a painted internal surface, a constant temperature sink which can be produced in a variety of different ways and a temperature measuring device.

An example of a cavity radiometer is shown in Fig. 2.18 which is a self–calibrating instrument aboard the ERBE satellite. The cavity design is slightly different from the one sketched in Fig. 2.17. It is composed of an inverted cone within a cylinder, the interior of which is coated with a specularly reflecting paint. The absorptivity of this particular cavity was determined to be 0.999 for the entire solar spectrum. Calibration is performed by turning the instrument to view space. The basic sensing of the instrument is the thermopile attached to the cavity, which effectively measures the temperature of the cavity after it absorbs the radiation. Other elements of the radiometer are the front baffle, which prevents any stray light from entering the instrument, and a constant temperature shield that alleviates any unwanted thermal influences on the cavity.

An important example of cavity radiometers on satellites are the radiometers flown on the Solar Maximum Mission (SMM), Nimbus 7, and ERBE satellites. Measurements obtained from these radiometers are used to study the variability of the solar irradiance. Measurements from SMM and Nimbus 7, collected over the past

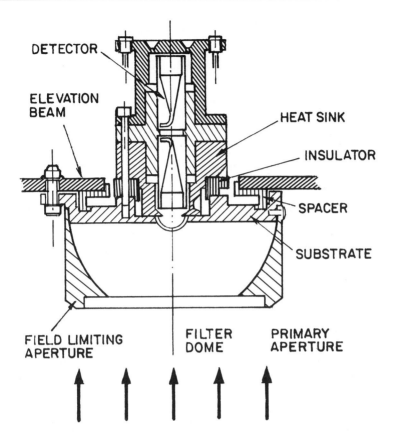

DETECTOR

ELEVATION
BEAM

HEAT SINK

INSULATOR

SPACER

SUBSTRATE

FIELD LIMITING
APERTURE

FILTER
DOME

PRIMARY
APERTURE

Figure 2.18 Expanded view of the cavity radiometer flown on the ERBE satellite to measure solar output. Shown are the major components of the instrument described further in the text (Barkstrom and Smith, 1986).

10 years, show a long–term trend associated with the 11–year solar activity cycle with superimposed higher frequency variabilities (Fig. 2.19). The relationship between the sun's luminosity and the sunspot number is, at first glance, counterintuitive; the luminosity decreases with a decrease in the sunspot number. Solar luminosity is most affected by areas of bright faculae that are associated with sunspot activity and these bright areas outweigh the darker areas of the sunspots. Both sunspots and areas of faculae increase as solar activity increases.

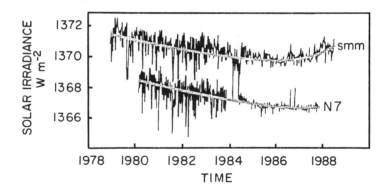

Figure 2.19 The flickering of the sun is recorded by cavity radiometers flown on two satellites, Nimbus 7 and the SMM. On average (shaded line), the sun is brightest at time of maximum sunspot activity (adapted from Foukal, 1990).

2.6   Notes and Comments

2.1. Naming the regions of the spectrum, and the colors of visible light in particular, is somewhat arbitrary since there is really a continuum of "color". Newton distinguished seven colors in visible light: red, orange, yellow, green, blue, indigo, and violet.

2.2. Drain (1975) develops the Doppler shift given by (2.22) using a number of different arguments.

2.3. Polarization is described in elementary texts on electromagnetic radiation. A detailed account of polarization and its measurement is given by Kliger et al. (1990). They provide a brief historical account of polarization and discuss the pitfalls encountered in describing circular polarization. These arise from the definition of right versus left circular polarization and the definition of the sense of rotation relative to an observer. The convention used in this book is widely used in optics.

2.5. The history of the theory of black body radiation, culminating in Planck's famous radiation formula, is indeed fascinating. A brief account of this history can be found in Baggott (1992). In 1896 Wien developed a successful model of black body radiation at high frequencies, but his model failed at infrared frequencies. The model developed by Rayleigh in 1905, followed by a correction to that model by Jeans, led to the Rayleigh–Jeans law which was successful in matching observations at the lower frequencies (this is also sometimes referred to as the Rayleigh–Jeans catastrophe in the ultraviolet because of the model's failure at these higher frequencies). Planck's formula was able to fit the data at all frequencies although many physicists at that time were critical of the formula as merely providing an empirical fit to experimental data. Planck's concern at that time was to establish a theoretical basis for the formula which he was able to do through the use of thermodynamics.

Two blackbodies at the same temperature emit the same radiation. Proof of this lies in the second law of thermodynamics. In the case of two black surfaces A and B at the same temperature, suppose A radiates more energy than the other. Imagine placing these surfaces next to each other and allow each to absorb the radiation from the other. Thus B must absorb more radiation than it emits, receiving more energy and becoming hotter. A, correspondingly becomes cooler. Thus the second law of thermodynamics is violated and our assumption that A radiates more than B is false.

Why is the conical cavity a good blackbody? It is simply that light falling upon the hole enters the cavity and has little chance of emerging from it. Even if the absorption of the surface is not close to unity, almost all light is absorbed before any escapes the cavity.

In 1837, when Louis–Philippe was yet the King, the French physicist Claude Pouillet first measured and then named the solar constant. The following December, while the sun at Cape Town was near the zenith, Sir John Herschel tried his hand at the same measurement—with a crude apparatus—a thermometer encased in a small tin box filled with a measured amount of water, which, with the help of a black umbrella, was alternately shaded and then exposed to sunlight. The measured heating of the water provided a numerical definition of the most fundamental of climatic parameters: the precise amount of solar energy that falls upon the Earth.

Victorian interest in the solar constant evolved from the time of Herschel and Pouillet and focused not so much on its absolute value as on the possibility of predictable change, and the consequences of such changes for weather and climate. Specifically at issue in 1881 was whether the solar constant would increase, or decrease, with the coming and going of spots and other signs of activity on the solar surface. We have since been able to answer this question from the measurements collected on satellites (Foukal, 1990).

Before satellites, to answer this direct and simple question Balfour Stewart, a Scottish meteorologist, developed a new and more accurate actinometer, and in the 1880s sent it off to India, where, he reasoned, sunny skies would bring the nagging matter more quickly to an end. But little was achieved. After a few years, when British interest faded, the quest was taken up in America by Samuel Langley and later Charles Greeley Abbot, who for half a century doggedly pursued the answer, without success, from mountaintops around the world.

At the start of the present decade, after more than 140 years of effort, climatology still had no clear answer as to how the solar constant varied, or indeed whether Pouillet's choice of words to describe this quantity was right, after all. We now have a much better understanding of how this quantity varies with solar activity thanks largely to high quality satellite observations. An excellent overview of solar variability and a discussion of the satellite measurements of solar irradiance is given by Foukal (1990).

2.7   Problems

The following are physical constants useful for a number of problems given throughout this book: speed of light $c=2.99792\times10^8$ ms^{-1}, vacuum permittivity $\epsilon_o = 8.854 \times 10^{-12}$ m^{-3}kg^{-1}s^{2}Co^2 where Co is used for coulomb, the Stefan–Boltzmann constant $\sigma = 5.67051\times10^{-8}$ Wm^{-2}K^{-4}, Planck constant $h = 6.62608\times10^{-34}$ J s, Boltzmann's constant $k_B = 1.38066 \times 10^{-23}$ J K^{-1}.

2.1. Briefly explain or interpret the following:
    a. Could a sound wave in the air be circularly polarized?
    b. Unpolarized light falls on two polarizing sheets so oriented that no light is transmitted. If a third polarizing sheet is placed between them and arranged so that light is transmitted, can this transmitted light be polarized?
    c. Devise a way to identify the direction of the optical axis of a quarter wave plate.
    d. If we look into a cavity whose walls are maintained at a constant temperature no details of the interior are visible.
    e. Where and why in the atmosphere does Kirchhoff's law fail?
    f. The solar constant is not constant.
    g. The colors of stars are related to their temperatures whereas the colors of planets are not.

2.2. Describe the state of polarization represented by the following:
    a. $\mathcal{E}_x = \mathcal{E}_o \sin(kx - wt), \quad \mathcal{E}_y = \mathcal{E}_o \cos(kx - wt)$
    b. $\mathcal{E}_x = \mathcal{E}_o \cos(kx - wt), \quad \mathcal{E}_y = \mathcal{E}_o \cos(kx - wt + \frac{\pi}{4})$
    c. $\mathcal{E}_x = \mathcal{E}_o \sin(kx - wt), \quad \mathcal{E}_y = \mathcal{E}_o \sin(kx - wt)$

2.3. Electromagnetic radiation from the sun falls on the top of the Earth's atmosphere at the rate of 1.37×10^3 Wm^{-2}. Assuming this to be plane wave radiation, estimate the magnitude of the electric and magnetic field amplitudes of the wave. The units of the electric field are m kg s^{-2}Co^{-1} and the units of the magnetic field are kg s^{-1}Co^{-1} which is also known as a telsa (T).

2.4. Assume that a 100 W lamp of 80% efficiency radiates all its energy isotropically. Compute the amplitude of both the electric and magnetic fields 2 m from the lamp.

2.5. A plane, sinusoidal, linearly polarized electromagnetic wave of wavelength $\lambda = 5.0 \times 10^{-7}$m travels in a vacuum along the x axis. The average flux of the wave per unit area is 0.1 Wm^{-2} and the plane of vibration of the electric field is parallel to the y axis. Write the equations describing the electric and magnetic fields of the wave.

2.6. Certain characteristic wavelengths in the light from a galaxy in the constellation Virgo are observed to be increased in wavelength, compared with terrestrial sources, by about 0.4%. What is the radial speed of this galaxy with respect to Earth? Is it approaching or receding?

2.7. The difference in wavelength between an incident microwave beam and one reflected from an approaching or receding automobile is used to determine its speed on the highway. (a) Show that if v is the speed of the car and ν the frequency of the incident beam, the change in frequency is approximately $2v\nu/c$, where c is the speed of the electromagnetic radiation. (b) For microwaves of frequency 2450 megacycles/sec what is the change of frequency per mile/hour of speed?

2.8. Calculate the Doppler shift (in Hertz) of radar and laser beams backscattered by particles with radial speeds of 0.01, 0.1, 1.0, and 10 ms^{-1}. Choose the wavelengths of 0.69 μm, 1.04 μm, and 10.6 μm for the laser and 3 cm, 5 cm, and 10 cm for the radar. Compare these frequency shifts to the frequency of the carrier beam.

2.9. Determine the Stokes parameters for the states of polarization represented by (b), (d), and (e) in Fig. 2.10.

2.10. Determine the Stokes parameters of the $\mathcal{E}$ fields defined in Problem 2.2.

2.11. An infrared scanning radiometer aboard a meteorological satellite measures the outgoing radiation emitted from the Earth's surface at a wavelength of 10μm. Assuming a transparent atmosphere, what is the temperature of the surface if the observed intensity is 0.98×10^4 ergs cm^{-2} μm^{-1} sr^{-1}?

2.12. What is the ratio of the spectral radiances of black bodies at 300 K and 6000 K at (a) 1 GHz, (b) 1000 GHz, (c) 1 μm, and (d) 0.1 μm?

2.13. Show that, for a black body, the wavelength at which $B_{\tilde{\nu}}$ is maximum is about 1.76 times greater than the wavelength at which B_λ is maximum at the same temperature.

2.14. Derive the Rayleigh–Jeans distribution (2.49) from (2.47).

2.15. Find the wavelength at which the incoming solar irradiance at the top of the Earth's atmosphere is equal to the outgoing terrestrial irradiance. Assume the sun and Earth to be emitting as blackbodies at 6000 K and 255 K, respectively. The radius of the Earth and the Earth–sun distance are given in Prblem 6.3.

3

Microscopic Interactions — Atomic and Molecular Absorption

Our ability to interpret measurements and subsequently infer information about the atmosphere from them could not be possible without some knowledge of the way that electromagnetic radiation interacts with matter and, in turn, not possible without some understanding of the fundamental properties of matter itself. It is certainly beyond the scope of this book to provide a thorough discussion of such a basic topic. The focus of this and the next two chapters concerns the properties of matter that are most relevant in its interaction with radiation. The specific topics discussed in this chapter apply to the microscopic scale (i.e., how radiation interacts with matter on the atomic and molecular level) and primarily concentrate on the absorption of radiation by gases in the Earth's atmosphere.

3.1   The Atomic Absorption Spectrum

3.1.1   The Bright Line Spectrum

The concept of a blackbody was introduced in Chapter 2. It was shown how the emission by a solid body produces a spectrum that is continuous in wavelength and, temperature aside, independent of the nature of matter. However, this is hardly the complete story. Almost two centuries before Planck's description of blackbody radiation, it was noted how vaporization of certain volatile metals produces distinct colors in flames and how these colors clearly relate to the type of metals being vaporized. In 1752 Thomas Melvill studied the color of the flame after its light was passed through a prism and discovered that the spectrum was not continuous like that of the sun or of the radiation emitted from cavities; rather, it existed as a series of distinct bright lines, the spectral locations of which varied according to the particular substance placed in the flame. This work, along with an instrument developed in the early part of the nineteenth century (the *spectroscope*) to observe the spectrum, heralded the modern science of *spectroscopy*. Analyzed in the spectroscope, the spectra of

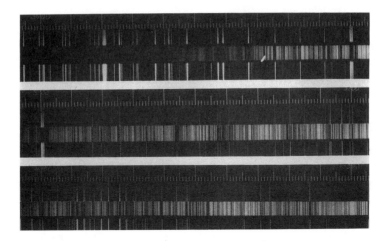

Figure 3.1 The bright line spectrum of the vaporized element iron is shown with the spectrum of the sun. The wavelength region is from 300 to 330 nm, in the ultraviolet. The solar spectrum is in the center of each strip, and the emission spectrum of iron lies above and below it. The bright lines of iron occur at the same wavelengths as some of the dark lines in the solar spectrum (Sobel, 1987).

flames (and electric discharges in gases) produce a line spectrum unique to the particular chemical element being vaporized.

Explanation of the line spectra phenomenon posed a challenge to theorists perhaps as significant as that posed by blackbody radiation. In some ways the challenge was even greater because, unlike the continuous emission from hot solid objects, the line spectrum was directly dependent on the nature of the object being observed. What transpired in the earlier part of this century with the experiments of Rutherford and the insights of Bohr was the development of modern quantum theory and a revolution that shook the foundations of both physics and scientific philosophy thereafter.

Bohr provided us with an early explanation of line spectra like that shown in Fig. 3.1. He proposed a planetary model of the atoms in which an electron moved around the nucleus in discrete orbits, with the position of the orbit relative to the nucleus representing discrete or quantized energy levels. As long as the electron remains in a fixed energy state it can emit no radiation since its energy does not change. The electron, however, can spontaneously fall from one level to a lower level losing energy in the form of radiation. A photon

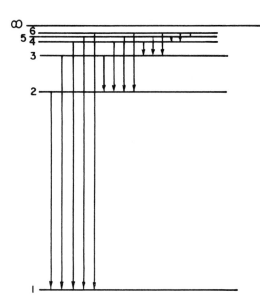

Figure 3.2 Energy levels of a hydrogen atom according to the Bohr theory. The first six levels are shown and drawn to scale in the sense that distance between levels is proportional to energy. Level 1 is the atom's ground state. Light is emitted whenever an atom makes a transition from a higher state to a lower one and the frequency of light is proportional to the energy difference. Higher levels are closer together. The line labeled by (∞) is the energy the electron would have to be barely able to escape the nucleus.

of light is therefore created with a frequency that is determined by the change in energy state of the atom,

$$h\nu = E_2 - E_1 = \Delta E \tag{3.1}$$

where E_2 and E_1 are the upper and lower energy states. Thus isolated atoms emit light at only certain definite frequencies as observed in the line spectra of Fig. 3.1. Moreover, the energy loss occurs by transitions to several intermediate levels such as that shown schematically in Fig. 3.2 for the simple hydrogen atom. The result is a spectrum of several lines, many of which are separated by only small frequency shifts.

To understand how these lines occur, we need to reconsider the Bohr model of an atom. Despite the success of the Bohr theory, it could not explain a number of fundamental issues underlying the quantum theory of Planck and Einstein. For example, why are states of oscillators and atoms quantized? How can this quantum or photon possess both particle and wavelike properties? We learn from quantum theory that the position of an electron cannot be precisely located at any particular point in space. As a result, the position of an electron must be treated as some sort of probability of existing at a particular point. It is here that the concept of a wave emerges as some kind of distribution in space, whereas a particle is thought of as a fixed point. Since it is fundamentally impossible to observe the position of the electron, the Bohr concept of an electronic orbit has no physical meaning. [1] The new theory replaces an orbit with a probability distribution, the highest probability being concentrated in the region of the Bohr orbit.

Our inherent inability to describe the motions of electrons in a classical way is stated in terms of the *uncertainty principle* of Heisenberg

$$\Delta E \approx \frac{\hbar}{\Delta t} \tag{3.2}$$

where $\hbar = h/2\pi$. This formula expresses the impossibility of determining both the energy states of an atom or molecule and the lifetime of these states with unlimited accuracy. According to (3.1), this limitation also means that the spectral lines are not sharply defined but are smeared over a finite frequency interval,

$$\Delta \nu = \frac{\Delta E}{h} \tag{3.3}$$

which prescribes the minimum possible line width. This broadening about the position of the line is referred to as *natural broadening*.

3.1.2   The Absorption Line Spectrum

One way to excite an atom to a higher energy level is to use light as a source of energy. Let the frequency of the light be such that the quantum energy $h\nu$ precisely equals the difference in energy between the ground state and some higher state of the atom. The quantum is absorbed and an electron jumps to a higher state; this is the reverse process of atomic emission. If the original light source has

---

[1]   There is another inconsistency with the circularly orbiting electron model of Bohr. This motion requires an accelerating electron which, by definition, must radiate and fall to a lower orbit.

a continuous distribution of frequencies, like the radiation from the sun, then only those frequencies that correspond to transitions of that atom are absorbed. A spectrum of dark lines is produced on viewing the light source through a cell containing a gas of atoms and the positions of these lines correspond to the absorbed frequencies. We refer to such a spectrum as the *line absorption spectrum*.

The key to our understanding of the underlying physics of absorption lines of an atom came when it was discovered that the dark absorption lines occur at the same frequencies as the bright line spectra of a glowing gas composed of the same atoms (as shown in the example of Fig. 3.1). In Bohr's quantized atom, we found a simple explanation of both bright line spectra and absorption spectra. The energy levels are characteristic of each element. In an excited gas, such as in a discharge tube, the atoms are in higher levels and fall to lower levels, emitting light. On illumination, the same kind of atoms absorb light undergoing transitions from lower to higher levels.

Absorption and emission spectra are of profound importance to remote sensing. These absorption lines are a form of spectral fingerprint which can be used to identify the composition of the atmosphere of distant stars and planets. For instance, we can determine that the sun contains vapors of sodium, calcium, and iron, among other elements. Using the same techniques to identify molecules, we learn that the atmosphere of Mars is predominantly composed of carbon dioxide and further that the atmosphere of Titan has significant concentrations of methane.

3.2 Molecular Absorption Spectra

The absorption spectrum of a molecule is substantially more complex than is that of an atom. Both transitions between the energy states of the atoms that make up the molecule and transitions between energy states associated with movements of the atoms themselves are possible.

Since the energy required to produce a transition from a lower to a higher state is inversely proportional to the wavelength of the photon, the types of mechanisms that induce absorption also depend on the wavelength of the absorbed photon. These mechanisms must induce either a magnetic or an electric effect that can be influenced by electromagnetic radiation. Mechanisms responding fastest occur at the shortest wavelengths whereas the more sluggish mechanisms produce absorption at longer wavelengths. We can use this wave-

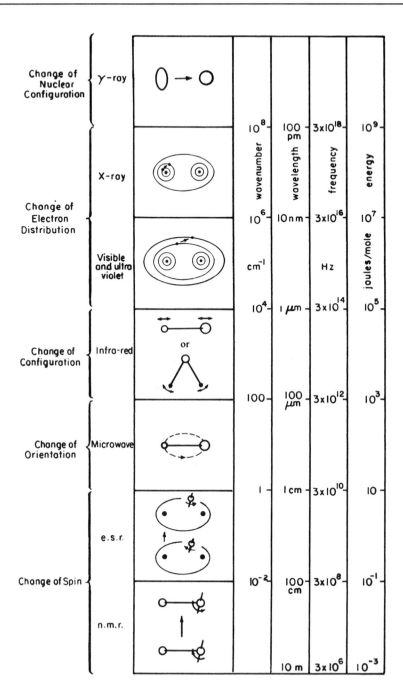

Figure 3.3 The electromagnetic spectrum and the possible types of interactions between photons and a molecule or atom (Banwell, 1983).

length dependence as a convenient classification of the absorption mechanisms as shown in Fig. 3.3 although the dividing boundaries are by no means precise.

• In the radio frequency regime, the absorption is associated with the nucleons and electrons which we consider to be tiny charged particles that spin producing tiny magnetic dipoles. The reversal of this dipole due to spin reversal interacts with the magnetic field at frequencies in the range 3×10^6 to 3×10^{10} Hz.

• In the visible and ultraviolet region excitation of valence electrons results in moving electric charges in the molecule. Changes in the electric dipole give rise to a spectrum by its interaction with the oscillating electric field of radiation. These electronic transitions occur within the individual atoms of molecules and dominate the visible and ultraviolet portions of the electromagnetic spectrum. At even shorter wavelengths, photons can actually disrupt the absorbing molecule by *photodissociation* or even produce *photo–ionization* of individual atoms.

• Absorption by molecules in the mid– and near infrared occur by vibration (although a mixture of vibrations and rotations are usually induced at these frequencies). Induction of vibrations requires more energy than rotations and thus takes place at higher frequencies corresponding to infrared wavelengths between about 0.7 μm and about 20 μm.

• In the microwave and far infrared, the molecule undergoes a rotation like that depicted in Fig. 3.4a and the component of the dipole in a given direction fluctuates in a regular fashion as shown in the lower part of Fig. 3.4a. These fluctuations are more sluggish than are the fluctuations associated with vibrations or the fluctuations associated with electronic transitions. Rotational lines generally occur in bands at the longer infrared wavelengths beyond about 20 μm extending into the microwave spectral region where individual rotational lines can be resolved.

As a consequence of the vibrational–rotational transitions, absorption lines are spread into bands containing many lines (as illustrated in Fig. 3.4b) that are used, either individually or as a group, to fingerprint molecules in the same way that atomic spectral lines fingerprint atoms. It is the vibrational–rotational absorption spectrum of molecules that is largely of interest to topics discussed in this book.

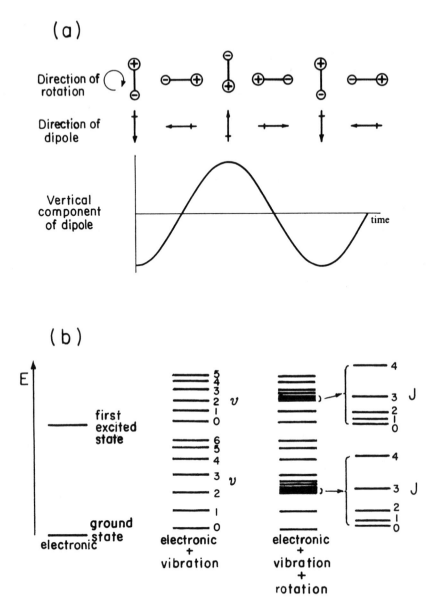

Figure 3.4 (a) The rotation of a simple diatomic molecule showing the fluctuation in the dipole moment measured in a particular direction (Banwell, 1983). (b) Molecular absorption spectra actually consist of closely spaced lines due to rotational and vibrational transitions. J and υ refer to the quantum numbers associated with the rotational and vibrational transitions, respectively. This diagram shows how these transitions are superimposed on electronic states.

3.2.1 Molecular Bonding and the Molecular Dipole Moment

To appreciate the nature of molecular absorption, it is useful to reflect on the characteristics of the forces that bond atoms to form molecules and to reflect on the properties of molecules themselves which give rise to the absorption spectra we observe. The *covalent bond* is the prevalent type of bond that forms radiatively active molecules of the Earth's atmosphere. The bond is considered to be the result of the electrostatic forces that arise from the sharing of electrons between the atoms that compose the molecules. In circulating among the atoms, the electrons spend more of their time between atoms than on the outside of the molecule. This lopsided circulation produces a net attractive electrostatic force bonding atoms into molecules.

The basic importance of the oscillating electric dipole to our view of the way radiation interacts with matter was emphasized in Chapter 2 as well as in relation to Fig. 3.3. For the interactions discussed here, a photon can only be absorbed to produce vibrational and rotational movements in the molecule if the charge around a molecule is separated in some way to produce a dipole moment (Section 2.1). How can an electrically neutral molecule then possess such a dipole moment and hence be radiatively active? To understand how this is so, we consider the molecule as a charge distribution which can be described by a probability function in the same way that atoms are considered as a distribution of charge. If the geometric center of this distribution happens to coincide with the center of mass, as it will for a symmetric homonuclear molecule like N_2, then the electric dipole moment is 0. Such molecules are said to be *homopolar* because they lack a permanent dipole moment. Therefore, most *homonuclear* molecules such as N_2 and H_2, when isolated from each other, are also homopolar and are radiatively inactive to infrared radiation[2].

---

[2] Under conditions of high pressure, such as those that occur in the atmospheres of Jupiter and Saturn, the charge distribution of homonuclear molecules is distorted sufficiently by frequent collisions with other molecules that a dipole moment is produced. The resulting spectrum produces a pressure–induced absorption band. Since the collision time between molecules is very brief, the absorption lines are correspondingly broad. This general characteristic is evident in the emission spectra between 300 and 800 cm^{-1}, which will be presented in Fig. 3.27 for Saturn. This feature is unimportant for the topics of this book.

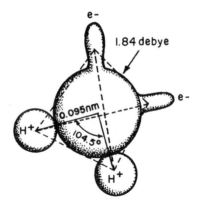

Figure 3.5 Schematic of the atomic configuration and electronic orbitals of the water molecule, with an oxygen atom located at the center and hydrogen atoms at an angle of 105 degrees. Dipolar character comes from protons at H+ positions and unshared electrons at e− locations; the direction of the dipole moment is along the symmetry axis. The atoms and the electronic orbitals have tetrahedral symmetry.

Molecules that possess a permanent electric dipole moment are said to be *heteropolar*. This dipole moment arises from the particular bonding configuration of the molecule. An important example of a heteropolar molecule is the H_2O molecule. We see by studying the molecule how a dipole moment is established through the geometric arrangement of atoms in the molecule. The specific structure of the water molecule, highlighted in Fig. 3.5, shows how the bonding of hydrogen atoms around the oxygen atom help produce a permanent dipole moment from the charge separation of the protons at the $H+$ positions and from the unshared electrons as shown in the diagram.

Molecular oxygen is a special case of a homonuclear molecule that is radiatively active at those wavelengths which we normally associate with rotational transitions. Rotational transitions arise by virtue of the permanent magnetic dipole of the molecule resulting from the presence of an unshared electron in the outer orbit of each oxygen atom. As we shall see, this magnetic dipole gives rise to absorption bands in the microwave region that are exploited by the MSU for temperature sounding of the atmosphere.

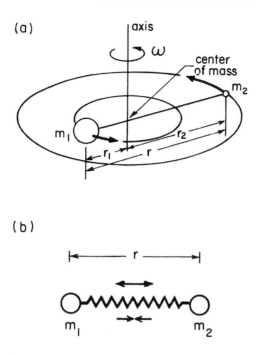

Figure 3.6 (a) A simple analog to a diatomic molecule that rotates about its center of mass. (b) A vibrating diatomic molecule can be thought of as two masses held together by a spring. Vibrations occur when the spring is distorted from its equilibrium position.

3.2.2 Vibration and Rotation Spectra of Simple Diatomic Molecules: Illustrative Examples

Simple mechanical analogs to a rotating–vibrating diatomic molecule are now described to provide an elementary understanding of these vibrational–rotational absorption spectra.

Rotating Molecules

First we invoke a simple mechanical model of a two body rigid rotator, as shown in Fig. 3.6a, as an analog of an absorbing diatomic molecule. An elementary treatment of rigid body rotation tells us that the energy of such a system rotating with an angular velocity ω is

$$E = \frac{1}{2}\mathcal{I}\omega^2 = \frac{\mathcal{L}^2}{2\mathcal{I}} \tag{3.4}$$

where $\mathcal{L}$ is the angular momentum and $\mathcal{I}$ is the moment of inertia,

$$\mathcal{I} = m_1 r_1^2 + m_2 r_2^2 \tag{3.6}$$

With the definition of the center of mass as

$$m_1 r_1 = m_2 r_2 \tag{3.7}$$

then $\mathcal{I}$ follows as

$$\mathcal{I} = \frac{m_1 m_2}{m_1 + m_2} (r_1 + r_2)^2 = m' r^2 \tag{3.8}$$

where m' is the reduced mass of the molecule and r is the distance between the two atoms. Schroedinger's equation shows that the angular momentum allowed to the rigid diatomic molecule has discrete values given by $\mathcal{L} = \sqrt{J(J+1)}\hbar$ where J is the *rotational quantum number* $(J = 0, 1, 2, ...)$. When this expression is used for the angular momentum factor in (3.4), we find that the energy of the Jth rotational state of the molecule is

$$E_J = \frac{J(J+1)\hbar^2}{2\mathcal{I}} \tag{3.9}$$

and the change in energy associated with a transition from this state to $(J+1)$ is

$$\Delta E = E_{J+1} - E_J = h\nu \tag{3.10}$$

It follows from a combination of these expressions that the spectral position of an absorption line of a diatomic molecule is defined by

$$\nu = \frac{\hbar}{2\pi m' r^2}(J+1) = 2B(J+1) \tag{3.11}$$

where $B = \hbar/4\pi\mathcal{I}$ is referred to as the rotational constant. Since this is a factor that contains the moment of inertia of each molecule, it is a factor that is specific to that molecule. Equation (3.11) predicts absorption lines as shown in Fig. 3.7a that are equally spaced by a frequency interval $2B$. From (3.9) and (3.11), the moment of inertia of a molecule can be deduced from its rotational spectra and given the masses of its constituent atoms, the interatomic separation r can be calculated.

For more complex molecules, their rotational energy may be generally specified in terms of three principal moments of inertia and three sets of rotational quantum numbers. Molecules can be divided into four types of rotators based on their geometric structure and resulting moments of inertia:

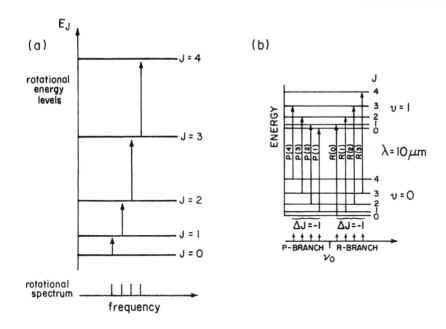

Figure 3.7 (a) Energy levels and the resulting spectrum of molecular rotation. (b) Vibrational–rotational transitions between the lowest vibrational and rotational levels.

• **Linear molecules** (CO_2, N_2O, C_2H_2, all diatomic molecules): For these molecules, the moment of inertia along the internuclear axis is zero and the other two principal moments of inertia about axes perpendicular to the internuclear axis are equal. Rotational energy for these molecules is characterized by a single rotational constant and absorption spectra resemble the simple spectra predicted by our rigid rotator model shown earlier.

• **Symmetric top molecules** (NH_3, CH_3Cl, CF_3Cl): These are non-linear molecules possessing two moments of inertia that are equal. Thus two rotational constants define the rotational energy states of the molecule.

• **Spherical symmetric top molecules**: This special case is of interest because one of the important trace gases in the atmosphere, namely methane (CH_4), is a classic example of a spherical top rotator. A spherical top is one in which all three principal moments of inertia are equal.

• **Asymmetric top molecules** (H_2O, O_3): These molecules possess three different moments of inertia and are thus characterized by three rotational constants and three sets of rotational quantum numbers. The rotational absorption spectra are therefore quite complex.

Vibrating Molecules

When sufficiently excited, polar molecules vibrate as well as rotate. Vibrational transitions in diatomic molecules can also be studied via the simple mechanical model shown in Fig. 3.6b. In this case, two atoms can be thought of as being held together by a spring that permits vibrations along the line connecting the two atoms. In a nonvibrating state, the atoms are separated by some distance r, which is determined by the tension in the spring. A distortion of the "spring" occurs by absorption to displace one atom with respect to another and the force,

$$F = -k\,(r' - r) \tag{3.12}$$

acts to restore the molecule to its original undisturbed state. This sets up an oscillation at a frequency

$$\nu' = \frac{1}{2\pi}\sqrt{\frac{k}{m'}} \tag{3.13}$$

which follows from the classic theory of a harmonic oscillator where k is the spring constant and m' is the reduced mass of the system. Now quantum theory predicts that the frequency of a harmonic oscillator is quantized according to

$$\nu = \left(v + \frac{1}{2}\right)\frac{1}{2\pi}\sqrt{\frac{k}{m'}} = \left(v + \frac{1}{2}\right)\nu' \tag{3.14}$$

where v is the *vibrational quantum number*. The energies corresponding to these frequencies are

$$E_v = h\nu = (v + 1/2)h\nu' \tag{3.15}$$

The energy required for a vibrational transition is larger than that required for a rotational transition. Vibrations, however, are typically accompanied by rotations so the rotating molecule is not exactly like a rigid rotator. We learn from quantum mechanics, however, that only certain types of vibrations and rotations are permitted together. These are defined by selection rules which, for the diatomic molecule (or a longitudinal polyatomic molecule like the carbon dioxide molecule), the transition $\Delta v = \pm 1$ occurs simultaneously with a $\Delta J = \pm 1$ transition. This selection rule produces pairs of transitions of the form shown in Fig. 3.7b. As a rule, each vibrational transition frequency is split up into a series of spectral lines with mutual separations that approximately correspond to the respective rotational constant. In Fig. 3.7b, the vibrational transition from $v = 0$ to $v = 1$ is shown. Two branches of rotation lines result for this vibrational transition: one for $\Delta J = +1$, which is referred to as the *R branch*, and the other for $\Delta J = -1$, the *P branch*.

3.2.3   Absorption by Two Triatomic Gases

Because of the dependence of both rotations and vibrations on m' we can conclude that it is possible for molecules with the same approximate total mass to possess very different absorption spectra (Problem 3.5). These spectra are determined not only by the mass of the molecule but also by the distribution of mass within the molecule.

In additional to the effects of the distribution of mass, the rotational and vibrational absorption spectra of polyatomic molecules are much more complex than are the spectra of diatomic molecules because of the higher degrees of freedom of both vibrational and rotational motions. The absorption spectra of the CO_2 molecule is highly relevant to atmospheric remote sensing. The CO_2 molecule vibrates in four different modes, two of which are energetically equivalent. These modes are referred to as *the symmetric stretch mode,* the *asymmetric stretch mode,* and the *bending mode* which has two equivalent modes of vibration. The dipole moment of the symmetric stretch mode is plainly 0 throughout the whole motion (Fig. 3.8a) and this vibration is radiatively inactive. The asymmetric stretch produces a periodic alteration of the dipole moment and this mode is "infrared active" as is the bending mode (Figs. 3.8b and c). The bending mode actually permits $\Delta v = \pm 1, \Delta J = 0$ transitions. These transitions then produce a large absorption peak centered on the fundamental frequency of the oscillator. The absorption is strong at these frequencies is referred to as the Q branch and is especially strong because all the $\Delta v = 1$ transitions accumulate for all available J-levels.

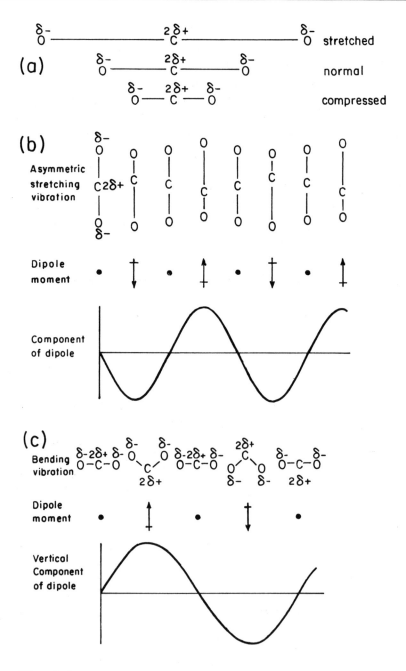

Fig. 3.8 (a) The symmetric stretching of a vibrating CO_2 molecule. (b) The asymmetric stretching of the CO_2 molecule showing the fluctuating dipole moment. (c) The bending motion of the carbon dioxide molecule and its associated dipole fluctuation (Banwell, 1983).

(d)

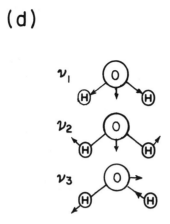

Fig. 3.8 (Cont.) (d) Vibrational modes of a water–vapor molecule.

Another molecule of considerable importance to the study of the Earth's atmosphere is the H_2O molecule. Since this molecule is not linearly arranged like the CO_2 molecule, geometrically different modes of vibration and rotation occur. Figure 3.8d indicates the three modes of vibration of the water molecule. Superimposed on these three modes are the rotational modes around three axes of rotation. The spectra arising from the multiplicity of vibration–rotation transitions are accordingly complex producing irregular absorption spectra composed of bands of thousands of lines.

Excursus: The Infrared Carbon Dioxide Laser

The infrared CO_2 gas laser exploits vibration–rotation transitions of the CO_2 molecule. This particular laser is successfully used in laser studies of the atmosphere (Chapter 8). Lasers (light **a**mplification by **s**timulated **e**mission of **r**adiation) possess four extraordinary properties compared to ordinary light: coherence, monochromaticity, directionality, and intensity. That laser light is emitted in an extremely narrow range of directions is a consequence of the geometry of the laser tube. The property of monochromaticity is a consequence of quantum mechanics and the special property of coherence is a property of *stimulated emission*. Stimulated emission is an idea that dates back to Einstein in 1917. Bohr's theory of the atom explains two types of transitions; excitation from a lower state to a higher one by absorption of a photon and the reverse process which results in emission. Einstein argued for a third type of transition, stimulated emission, in which an atom or molecule in an upper state is forced to emit radiation and decay to a lower state by the presence in its vicinity of a photon of the same frequency. The original photon, together with that from the stimulated emission, proceed to stimulate emission in other molecules. Provided enough molecules exist with populated upper states, a kind of chain reaction occurs.

Absorption, spontaneous emission, and stimulated emission occur simultaneously. Whether more photons are emitted than absorbed depends on whether more molecules exist with populated upper states relative to the lower state. The natural population distribution is for most atoms to be in the ground state, fewer in the first excited state, fewer still in the next state, and so on. Increasing temperature increases the populations in the upper states, but the ordering stays the same. Gas lasers generally require the existence of a *population inversion*, one that has the upper states of the molecule populated more than lower states.

We can understand the basic operation of a CO_2 laser by considering Fig. 3.9. The typical medium for the laser is a gas mixture of nitrogen (N_2), helium (He), and CO_2 in an approximate ratio of 6:3:1. The actual laser transition occurs as one of several possible P or R transitions (Fig. 3.9) located in the 9–11 μm region. One of the keys to laser action is for the upper level (in this case the 001 vibration state of the CO_2 molecule) to exist long enough to allow sufficient accumulation of molecules within this state, thus promoting stimulated emission. Some method is required to elevate the CO_2 molecule to the excited state (referred to as pumping of the laser). Nitrogen is used as an intermediary here as the 001 state of the CO_2 molecule very nearly coincides with the first vibrational state of the N_2 molecule. The nitrogen molecule is excited to this vibrational state through collisions with electrons which are initiated by an electric discharge in the tube. The N_2 molecule, pumped to this state, is unable to radiate to its ground state (remember this molecule is homonuclear and has no dipole moment). The vibrational energy of this molecule is thus readily transferred to the CO_2 molecule via collisions which in turn excite a vibration in the CO_2 molecule to the 001 and higher states. The CO_2 molecule then undergoes a vibration–rotation transition to a lower vibration–rotation state (such as the 100 vibration state depicted in Fig. 3.9) via stimulated emission. Collisions among the CO_2 molecules transfer vibrational energy from a molecule in this excited state to a CO_2

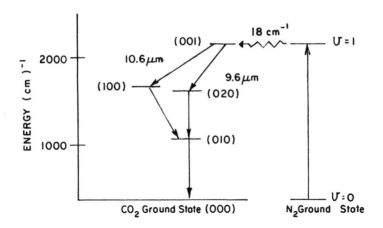

Figure 3.9 The energy levels of the CO_2 molecule relevant to the operation of an infrared CO_2 gas laser. In the example shown the CO_2 gas lasers at 10.6 μm for the $(001) \rightarrow (100)$ transition.

molecule in the ground state. This results in two molecules in the intermediate (010) state. Thus, a bottleneck occurs at (010). Lasing requires that this level be depopulated rapidly to maintain the cycle of pumping and stimulated emission. This is achieved by He atoms. The vibrational energy of the CO_2 molecules in this bottleneck state is transferred to translational energy of the lighter He atoms thus increasing the temperature of the gas mixture which is then mitigated by circulating cool water over the laser tube.

Laser light is intense because of both the multiplicative effect of stimulated emission and the fact that the gas is enclosed between two parallel mirrors at either end of the tube so that photons bounce back and forth producing even more new photons. The wavelength of the stimulated photon is precisely that of the photon that produced it so a beam of high monochromaticity results. One of the mirrors at the end of the tube is not completely reflecting; it transmits some fraction of the incident light and a beam emerges from the tube. This beam, transmitted to the atmosphere and received via appropriated collecting optics, constitutes the technique of lidar remote sensing.

3.3 Line Shapes

Discussion so far has focused only on those factors that determine the spectral positions of the absorption lines. However, the amount of radiation absorbed by molecules also depends on the strength of the absorption line. In fact, there are three main properties of an absorbing line that mathematically define molecular absorption. These properties are the central position of the line in the spectrum (ν_o), the strength of the line (S), and the shape or profile of the line.

The shape of a line characterizes its spectral "fine" structure or its "fuzziness." The *shape factor* $f(\nu - \nu_o)$ represents this property and provides an estimate of the relative absorption at a frequency which is displaced by $\nu - \nu_o$ from the line center. By convention, the shape factor is defined such that

$$\int_{-\infty}^{\infty} f(\nu - \nu_o) d\nu = 1 \qquad (3.16)$$

The broadening of absorption lines is a consequence of a number of factors. It was mentioned earlier that there is a natural line broadening because the energy levels are not sharp. Since the lifetime of a molecule in a particular energy state cannot be longer than the natural lifetime of the upper energy state of the molecule (which is determined from quantum mechanical considerations), then this limit defines the lower limit to the width of a spectral line. This is the *natural line width* and is represented as

$$\alpha_N = \frac{\Delta \nu}{2} \qquad (3.17)$$

where $\Delta \nu$ is obtained from (3.3) given an estimate of the lifetime in the upper state. Natural broadening is only a significant mechanism in the upper atmosphere (in the mesosphere and upper stratosphere) where collisions are relatively infrequent. In the lower atmosphere typical lifetimes for upper states of vibration–rotation transitions are much shorter than the natural lifetimes because of the effects of collisions with other molecules.

3.3.1   Pressure Broadening

The main forms of absorption line broadening relevant to the transmission of electromagnetic radiation through the gases of the Earth's lower atmosphere arise from both the individual motions of the molecules and their collisions with other molecules. Broadening by collisions is a complex subject and no exact theory exists. We consider this process only in a simple and somewhat heuristic way here. In this view, we consider that collisions stop the oscillator momentarily after which it starts instantaneously again with a phase that has no relationship to the phase of the oscillator just prior to the collision. In the case of a rotating molecule, this means that orientation of the molecule after collision is random (a so–called strong collision). When the collisions are frequent enough, as they are in the lower atmosphere, then a kind of equilibrium is reached between the thermal energy associated with molecular motions and the electrical energy of the dipole oscillations. This equilibrium is referred to as *local thermodynamic equilibrium* and under this equilibrium population of the various energy levels closely obey the Boltzmann energy distribution and depend on the temperature as a result.

The phase shift model just described is generally a good approximation to the collision broadened line shape. The simplest description of this collision broadening is the *Lorentz line shape*,

$$f_L(\nu - \nu_o) = \frac{\alpha_L/\pi}{(\nu - \nu_o)^2 + \alpha_L^2} \tag{3.19}$$

where $\alpha_L = (2\pi \bar{t})^{-1}$ and $\bar{t}$ is the mean time between collisions. α_L is referred to as the *Lorentz half–width* and is a measure of the frequency shift that corresponds to the half power point of the line. When typical kinetic theory is adopted to describe molecular collisions, it can be shown that

$$\alpha_L \approx \alpha_{L,s}(p/p_s)(T_s/T)^{1/2} \tag{3.20}$$

where $p_s = 1000$ mb, $T_s = 273$ K, and $\alpha_{L,s}$ is the half–width value at these "standard" temperatures and pressures. For most gases of interest, the values of $\alpha_{L,s}$ generally fall in the range 0.01–0.1 cm^{-1}. For example, a typical value of a CO_2 line is $\alpha_{L,s} \approx 0.07$ cm^{-1}. Figure 3.10 provides an illustration of the Lorentz line profile at pressures of 1, 0.5 and 0.25 bar. The regions far from the line center are referred to as *line wings* and it is relevant to note how the absorption in the wings of the lines increases with increasing pressure in contrast to the absorption in the center of the line which decreases with increasing pressure.

The relationship expressed by (3.20) bears profoundly on the transmission of radiation in the Earth's atmosphere. It states that the Lorentz line width is proportional to atmospheric pressure. Since pressure varies by about three orders of magnitude from the surface to 50 km, so then does α_L. The effect of this pressure variation has been confirmed by experiments. The dependence of α_L on temperature is both less important, since T does not vary to the same

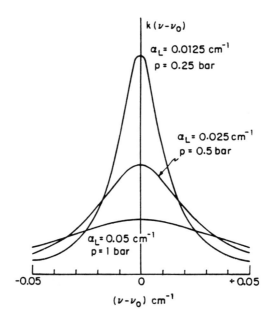

Figure 3.10 Lorentz profiles for three pressures. A line width of 0.05 cm^{-1} at a pressure of 1 bar is typical for vibration–rotation bands (from Goody and Yung, 1989). Note how the dependence of line shape on pressure changes as ν varies from line center to line wings.

extent through the atmosphere, and less well understood. Temperature dependence varies with the particular transition and is often stronger than the $T^{-1/2}$ relationship quoted. Fortunately, the temperature dependence on line shape is usually of secondary importance compared to the temperature dependence of the line strength as discussed below.

The pressure dependence of absorption has a profound effect on the remote sounding of the atmosphere, as suggested by reference to Fig. 3.11. The left part of the figure schematically presents atmospheric pressure as a function of altitude with three different levels highlighted for discussion. The right–hand portion of the diagram shows the pressure–broadened absorption line such as might occur by a transition in the CO_2 molecule at each of these levels. Consider a downward–looking radiometer that receives radiation through a filter which has a spectral width narrower than the line width and is capable of detecting radiation emitted at the three neighboring spectral regions indicated. For a line of suitable strength, when the filter position is close to the line center, the radiation detected by

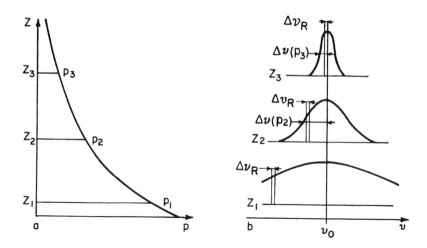

Figure 3.11 The atmospheric pressure as a function of height (left) and the variation of line shape (right) at three different heights. The position of the filter response of a hypothetical radiometer is indicated by $\Delta\nu_R$ relative to this line.

the instrument originates mainly from the higher levels of the troposphere. Further out from the line center, the radiation detected by the instrument arrives from lower down, perhaps in the middle levels of the atmosphere. Still further out into the wings of the line, little of the radiation emitted from the broadened line is absorbed by the atmosphere above owing to the narrowing of the absorption line at higher altitudes. Thus, the radiation detected at these wavelengths originates largely from emission in the lower atmosphere. In this example, we see how the variation of the absorption line width with pressure provides an intuitive explanation of the weighting functions discussed in relation to temperature sounding in Chapter 7.

The Lorentz model of an absorption line is generally accurate for pressure–broadened lines at levels typical of the middle stratosphere (around 10 mb) and below. There are, however, some limitations to this model. For instance, there are significant variations in α_L from gas to gas, from band to band, and even between lines in the same band. The line width is also a function of the type of colliding molecule, or the *broadening gas*. The principal broadening gases in the Earth's atmosphere are naturally N_2 and O_2 since these are most abundant. When these gases act to broaden the absorption lines of other gases, like H_2O and CO_2, then the line is said to undergo *foreign broadening*. Broadening by like molecules, such as a water molecule by a water molecule, is referred to as *self broadening* and α_L is usually larger than that in foreign broadening.

Departures in the Lorentz line shape occur in the line wings (i.e., where $|\nu - \nu_o| \gg \alpha_L$). This is unfortunate because these portions of the lines tend to dominate radiative transfer in the relatively transparent regions of the absorption spectrum (the so called "windows" of the spectrum discussed later). Departures in Lorentz lines in the far wings are very difficult to measure and are a major concern both to studies of atmospheric radiative transfer and to various problems of remote sensing which seek to exploit these transparent portions of the absorption spectrum.

Variations of the actual line shape from the simple Lorentz model are also important for those spectral regions where the width of rotational lines is comparable to the central frequency of the line. This situation occurs in the microwave region where the simple phase shift model is no longer strictly appropriate. At these wavelengths, the van Vleck–Weisskopf function

$$ f_{VW}(\nu - \nu_o) = \frac{1}{\pi} \left(\frac{\nu}{\nu_o} \right)^2 \left[\frac{\alpha_L}{(\nu - \nu_o)^2 + \alpha_L^2} + \frac{\alpha_L}{(\nu + \nu_o)^2 + \alpha_L^2} \right] $$

is a better approximation to laboratory measurements. Figure 3.12 is a comparison of the measured and modeled water vapor absorption line centered at 22 GHz which is asymmetrical about the central frequency of the line. The different models do depart from observations in ways that are significant especially since the absorption by this particular line is important to atmospheric remote sensing of water vapor as discussed in Chapter 7.

Excursus: Vertical Distributions of Chlorine Monoxide

The shape of an absorption line relevant to paths of varying pressure are not represented by a pure Lorentz line profile. The profile may be thought of as a superposition of several Lorentz lines at different pressures. For example, the line shape for the vertical path that varies from p_1 to p_3 in Fig. 3.11 may be thought of as a kind of superposition of the three Lorenz line profiles drawn. The net effect is a line that is narrow in the center and broad in the wings relative to a pure Lorentz line defined for a single pressure.

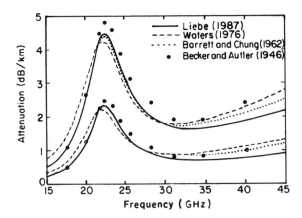

Figure 3.12 The absorption coefficient of the 22 GHz water vapor line in air (10 g H_2O m^{-3}). Comparison between measurement and the van Vleck–Weisskopf line shape is shown (after Walter, 1992a).

An important advantage of microwave spectroscopy over the spectroscopic measurements using optical systems (like those discussed at the end of this chapter) is that individual absorption lines in the microwave emission spectrum can actually be resolved. This provides the possibility of exploiting the shape information of the line to arrive at a qualitative vertical profile of the absorbing gas. An example of this approach is given in De Zafra et al. (1989) who report spectroscopic measurements of the rotational line emission of chlorine monoxide (ClO). In this study, the authors invert the emission line profile, which is just a superposition of line emissions at different pressures, to gain quantitative information on the vertical profile of ClO. The existence of elevated levels of ClO in the stratosphere over the south pole is directly implicated in the formation of the Antarctic ozone hole. Figure 3.13a presents two emission spectra measured across the 512–MHz bandwidth of the instrument. The upper spectrum is for a period four hours after sunrise and the lower spectrum represents emission at night. Ozone lines appear at 110 and 179 MHz and the ClO line is located at 278 MHz (1.1 m wavelength). Figure 3.13b shows spectra obtained over a five–day period, averaged to produce spectra at two–hour intervals after having the ozone line absorptions removed. It is important to note that the ClO line spectra shows a broadening of the daytime profiles (spectra labeled as d,e, and f) compared to the night profiles (traces i, j). The interpretation of these results is that a low level peak concentration

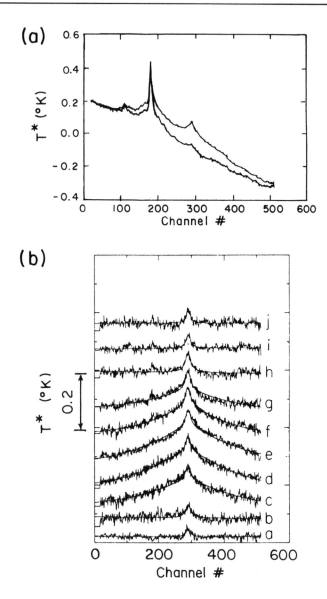

Figure 3.13 (a) Raw spectral data in 512–MHz spectral band pass showing the ClO line at 278 MHz for day (upper trace) and night (lower trace). The ordinate is in terms of brightness temperature difference between day and night and corresponds to observations in the zenith. (b) ClO spectra corresponding to two hourly intervals during the course of 24 hours showing diurnal behavior of the emission. Data represent an average over five days. Trace a, 2–0 hours before sunrise; trace b, 0–2 hours after sunrise, and so on. The midday interval is represented by trace e (from De Zafra et al., 1989).

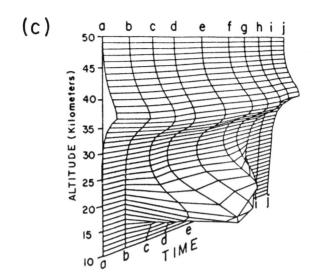

Figure 3.13 (Cont.) (c) Time evolution of the vertical profile of ClO mixing ratio obtained from a deconvolution algorithm applied to the spectra shown in (b) (from De Zafra et al., 1989).

of ClO exists during the day as well as the stratospheric peak that persists throughout both the day and night (Fig. 3.13c).

3.3.2 Doppler Broadening

Even without pressure and natural broadening effects, finite line widths arise simply due to the motions of the molecules themselves. The notion of the Doppler frequency shift was previously encountered in the discussion of wave propagation where it was discovered that the frequency shift depends on whether the dipole is moving toward or away from the observer. It is this shift in frequency that gives rise to *Doppler broadening*. However, molecules of a gas move rapidly with a range of velocities. Some are sluggish in their motion, while others move rapidly, and many move with velocities between these extremes. Experiments show that the velocity distribution of molecules obeys the Maxwellian distribution. As a result of the distributed velocities, the Doppler shifts are correspondingly represented by the same distribution function from which the *Doppler line profile* is derived

$$f_D(\nu - \nu_o) = (\pi \alpha_D)^{-1/2} \exp[-(\nu - \nu_o)^2/\alpha_D^2] \qquad (3.21)$$

where $\alpha_D = u_m \nu_o/c$, $u_m = (2k_B T/m_a)^{1/2}$, and m_a is the molecular mass; u_m is essentially the root–mean–square molecular velocity and for CO_2 at a temperature of 250 K, $\alpha_D \approx 7 \times 10^{-4}$ cm^{-1}. We note from this broadening formula that absorption lines of heavy molecules broaden less than do light molecules. This fact is actually exploited in high spectral resolution lidar measurements to separate aerosol and molecular backscattering, an aspect discussed further in Chapter 8.

The Doppler broadened line has a Gaussian shape centered at ν_o and this shape is compared to the Lorentz profile in Fig. 3.14a. According to definition, the Doppler half width depends directly on the resonant frequency and on $T^{1/2}$. Thus accurate measurements of line width provide a way of deducing the temperature of the absorbing molecule provided it can be established that Doppler broadening is the predominant effect. This temperature dependence is an important tool for mesospheric and upper stratospheric sounding.

The relative importance of Doppler broadening compared to pressure broadening can be appreciated in terms of the ratio

$$\frac{\alpha_D}{\alpha_L} \approx 10^{-12} \frac{\nu_o}{p} \tag{3.22}$$

where ν_o is in Hertz and p is the pressure expressed in millibars. Derivation of (3.22) is left as an exercise for the interested reader (Problem 3.6). This ratio is an approximation where the numerical factor is deduced for an averaged weight molecule at a temperature of $T \approx 300$K. Figure 3.14b shows the variation of line width of an oxygen and carbon dioxide molecule, expressed in frequency units, as a function of height in the atmosphere. Due to the dependence of the Doppler width on line center frequency, Doppler broadening of the O_2 line at 2.5 mm wavelength (or about 118 GHz) is two orders of magnitude smaller than that of a CO_2 line at 15 μm. As a consequence, the transition between regions dominated by pressure broadening and regions dominated by Doppler broadening occurs approximately 40 km higher up in the atmosphere for O_2 at 2.5 mm.

In the transition region where both Doppler and Lorentz broadening are important, that is where $\alpha_D \approx \alpha_L$, the shape factor may be represented as a combination of both line shapes according to a function referred to as *Voigt line shape*. This particular line shape will not be discussed further.

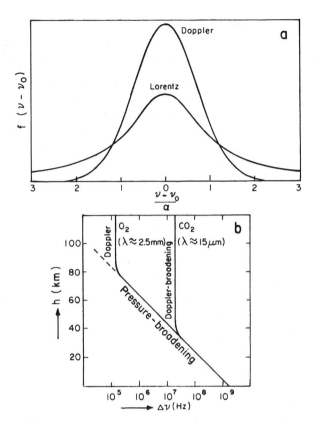

Figure 3.14 (a) A comparison of the Doppler and Lorentz line shapes. (b) Approximate relationship between atmospheric height and the linewidth for a microwave line of O_2 and an infrared line of CO_2 (from Elachi, 1987).

3.4 Absorption Coefficients and Transmission Functions

3.4.1 The Absorption Coefficient

The three factors that characterize line absorption — position, strength, and shape — combine to define the single parameter

$$k_\nu = Sf(\nu - \nu_o) \tag{3.23}$$

which is known as the *absorption coefficient*. This coefficient is an important parameter in characterizing the attenuation of electro-magnetic radiation as it propagates through an absorbing gas. The factor S is the strength of the absorption line and is a measure of how readily a given transition takes place. Line strength depends both on the properties of the single molecule and the populations of molecules in upper and lower states which in turn depend on the temperature of their environment. The line strength S is a product of two distinct factors. The first factor represents the probability that a single molecule in its original lower state absorbs a photon to jump to an upper state. This probability is measured in terms of the *line cross section* σ. The second factor represents the relative populations of the lower and upper states such that

$$S = \sigma(n_\ell - n_u) \tag{3.24}$$

where $n_{\ell,u}$ are the number of molecules in lower and upper states relative to the total number of molecules in all states. It is usually assumed that σ is the same for both the upper and the lower states and that the populations are governed by the Boltzmann distribution as discussed earlier so that the line strength is temperature–dependent. As far as rotational states are concerned, the relative populations are proportional to the factor $g(J) = (2J+1)\exp[-BJ(J+1)/k_BT]$ where B is the rotational constant as defined in relation to (3.9) and k_B is Boltzmann's constant. The shape of this factor is illustrated in Fig. 3.15a and the distribution of line intensities is shown in Fig. 3.15b for the P and R branches of a diatomic molecule. Note that the distribution is asymmetrical because of the different populations of the two branches. This asymmetry is in fact a direct measure of the temperature of the gas. Figure 3.15c presents the actual spectral transmission near 7.78 μm due to absorption by the linear N_2O molecule and demonstrates the features of the hypothetical molecule illustrated in Fig. 3.15b. Mixed amongst the regularly positioned lines of the P and R branches is a second weaker band with a slightly different band center.

3.4.2   Transmission Functions

Absorption of radiation by gases in the Earth's atmosphere is described in terms of *transmission functions*. These functions arise from elementary considerations of radiative transfer which is a topic discussed more extensively in later chapters. For present purposes, we introduce transmission functions with the aid of Fig. 3.16. This diagram provides an illustration of *Lambert's law* which states that the change in intensity along a path (of gas) ds is proportional to the amount of matter (gas) along the path according to

$$dI_\nu = -k_{\nu,v}I_\nu ds \tag{3.25}$$

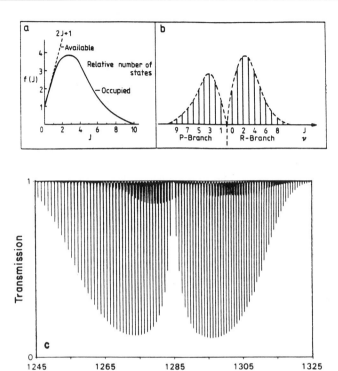

Figure 3.15 (a) The relative occupation of various rotational states. (b) Vibrational–rotational transitions of a diatomic molecule with P and R branches. For a given molecule, the asymmetry of these branches is a measure of its temperature. (c) A synthetic spectrum of N_2O near $7.7\mu m$ (after Goody and Yung, 1989).

where $k_{\nu,v}$ is the *volume absorption coefficient*. The dependence of the absorption coefficient on the density of the gas gives rise to a number of possible ways of specifying the absorption coefficient: the *molecular absorption coefficient*, $k_{\nu,n} = k_{\nu,v}/n$, where n is the number density of the absorbing molecules; the *mass absorption coefficient*, $k_{\nu,m} = k_{\nu,v}/\rho_a$, where ρ_a is the density of the absorbing gas; and *the absorption coefficient at s.t.p*, $k_{\nu,s} = k_{\nu,v}n/n_s$, where n_s is Loschmidt's number. The interrelations between these coefficients are discussed in Appendix 2 of Goody and Yung (1989). What is important is the product $k_\nu ds$ which is unitless so that for each type of coefficient there is a corresponding different measure of path length.

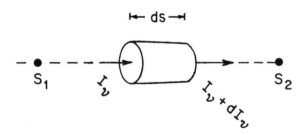

Figure 3.16 A schematic illustration of Lambert's law.

Integrating (3.25) between s_1 and s_2 provides a general solution to Lambert's law in the form

$$I_\nu(s_2) = I_\nu(s_1)\mathcal{T}_\nu(s_1, s_2) \qquad (3.26)$$

where $\mathcal{T}_\nu(s_1, s_2)$ is referred to as the *monochromatic transmission function* and is defined as

$$\mathcal{T}_\nu(s_1, s_2) = \exp\left[-\int_{s_1}^{s_2} k_{\nu,v} ds\right] \qquad (3.27)$$

Therefore measurement of the radiation flowing from the atmosphere at some level (i.e., at s_2), together with knowledge of the radiation that is incident at some other point along the same direction (i.e., at s_1), provides enough information to obtain the transmission and, with further assumptions, a relatively simple way of inferring the integrated concentration of a particular gas along the path between s_1 and s_2. This is the basic procedure used to determine the path–integrated ozone and water vapor concentrations described in later chapters.

At this point, it is convenient to introduce the quantity

$$\tau(s_1, s_2) = \int_{s_1}^{s_2} k_{\nu,v} ds \qquad (3.28)$$

and refer to this as the *optical path*. This quantity is basic to our mathematical description of how radiation interacts with matter. We

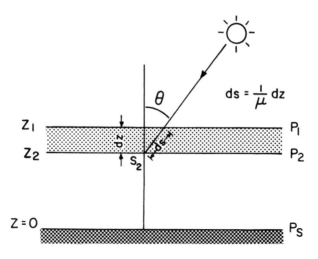

Figure 3.17 A plane parallel atmosphere and the relationship between slant and vertical paths.

shall see a variety of different forms of this quantity throughout this book. It is customary to idealize the atmosphere as a horizontally stratified medium in the manner shown in Fig. 3.17 and to define the path relative to the vertical. Thus transmission along a path tilted from the vertical by an angle θ, the zenith angle, is simply related to the transmission along the vertical path according to

$$T_\nu(s_1, s_2) = T_\nu(z_1, z_2, \mu = \cos\theta) = \exp\left[-\tau_\nu(z_1, z_2)/\mu\right] \qquad (3.29)$$

where $\tau_\nu(z_1, z_2)$ is now measured along the vertical and is referred to as the *optical depth*.

It is common to use the mass absorption coefficient in the definition of optical depth in describing the transmission along a path through an absorbing gas. Therefore combining (3.28) and (3.29) yields

$$T_\nu(s_1, s_2) = \exp\left[-\mu^{-1}\int_{z_1}^{z_2} k_{\nu,m}\rho_a dz\right] \qquad (3.30)$$

for the slant path transmission function. We define the *optical mass* as

$$u(z_1, z_2) = \int_{z_1}^{z_2} \rho_a dz \qquad (3.31)$$

which is often quoted in units of grams per square centimeter. Introduction of the hydrostatic assumption in (3.31), together with the mixing ratio $r = \rho_a/\rho$, where ρ is the density of air, produces

$$u(p_1, p_2) = \frac{1}{g} \int_{p_2}^{p_1} r\, dp \qquad (3.32)$$

where g is the acceleration of gravity, and where p_1 and p_2 are the pressures associated with the altitudes z_1 and z_2, respectively. For uniformly mixed gases, like CO_2 and O_2, r is a constant. If the path through such a gas stretches from a satellite altitude (say $p_2 = 0$) to the surface ($p_1 = p_s$), then

$$u(0, p_s) = \frac{r p_s}{g} \qquad (3.33)$$

Thus the absorption path for a uniformly mixed gas is directly proportional to the atmospheric surface pressure p_s; a relationship that has been proposed as a basis for the remote sensing of surface pressure.

For remote sensing, it is important to distinguish between *monochromatic* transmission functions and *band* transmission functions. The former represents the transmission of radiation at one selected wavelength, whereas the latter is the transmission averaged over a range of wavelengths as specified, for example, by the spectral response of a particular instrument. To illustrate this point further, let us suppose that the radiation received at a detector is of the form

$$I_{\Delta\nu} = \int_{\Delta\nu} g(\nu) I_\nu\, d\nu \qquad (3.34)$$

where $g(\nu)$ is the spectral response function of the instrument over its spectral band pass $\Delta\nu$. In terms of the transmission function, the intensity measured at s_2 is

$$I_{\Delta\nu}(s_2) = \int_{\Delta\nu} g(\nu) I_\nu(s_1) \mathcal{T}_\nu(s_1, s_2)\, d\nu \qquad (3.35)$$

where (3.26) has been substituted into (3.34). If the spectral band $\Delta\nu$ is sufficiently narrow that the incident intensity is constant across the band, then the band transmission function becomes

$$\mathcal{T}_{\Delta\nu}(s_1, s_2) = \int_{\Delta\nu} g(\nu) \mathcal{T}_\nu(s_1, s_2)\, d\nu. \qquad (3.36)$$

Therefore instrument properties [in this case the response function $g(\nu)$] directly influence the transmission derived from measurements and must be accounted for in retrieval schemes. Throughout the remainder of this book, the response function will be taken to be understood and is omitted, not because it is unimportant, but merely to simplify matters.

It is also useful to distinguish between transmission functions applicable to the atmosphere where absorption coefficients vary because of varying temperature and pressure along the path (in this case *inhomogeneous* transmission) and transmission functions which are more readily measured in the laboratory under conditions of constant pressure and temperature. A common way of relating these two types of transmission is to assume

$$\exp\left[-\int_{u_1}^{u_2} k_{\nu,m} du\right] \approx \exp\left[-k_{\nu,m}(p_o, T_o)\tilde{u}\right] \qquad (3.37)$$

where the transmission along an inhomogeneous path is approximated by a transmission function defined along a homogeneous path at a given standard temperature and pressure defined by a scaled optical mass $\tilde{u}$. The scaling typically has the form

$$\tilde{u} = u\left(\frac{p}{p_o}\right)^m \left(\frac{T_o}{T}\right)^{1/2} \qquad (3.38)$$

where u is the unscaled path, m is a constant with a value that varies according to the absorbing gas in question and p and T are taken as some representative pressure and temperature of the path, respectively. Other methods which vary in sophistication and accuracy have also been proposed to approximate nonhomogeneous transmission but their details need not concern us here.

3.5 Atmospheric Absorption Spectra

A schematic overview of the atmospheric absorption spectra is given in Fig. 3.18 a,b, and c for the ultraviolet region, the near infrared–far infrared spectral regions, and the microwave region, respectively. The absorption in the ultraviolet beyond about 0.1 μm and in the visible is dominated by transitions in the electronic bands of molecular oxygen and ozone. The absorption spectrum shown in Fig. 3.18a is expressed in terms of the penetration depth of solar radiation. This depth is the altitude at which slightly more than half of the incident solar radiation is absorbed.

The vibrational–rotational absorption bands of six absorbing gases are presented in Fig. 3.18b. The absorptions in both diagrams apply to vertical paths through the whole atmosphere. The combined infrared absorptions by individual gases are shown in the lower panel of Fig 3.18b. Of particular interest are the transparent regions (atmospheric windows) between about 8 μm and 13 μm together with several narrow regions below 4 μm. These windows are typically utilized in various remote sensing systems as a way for viewing the surface through the atmosphere.

The microwave absorption spectra of O_2 and H_2O are shown in Fig. 3.18c in terms of the optical depth. Two spectra are presented, one with and one without the effects of water vapor absorption. The water vapor lines centered at 22.235 GHz and 183 GHz are important features of the microwave absorption spectrum as are the absorption bands of O_2 centered at approximately 60 GHz and 118 GHz. Various narrow, weak absorption lines of O_3 are also evident.

It is clearly impractical to describe the general spectral distribution of molecular line absorption for all bands. Consequently, only those absorption bands considered important to various remote sensing applications are now mentioned.

3.5.1   Molecular Oxygen

Because of its symmetry, the oxygen molecule has no permanent electric dipole, but it does possess a permanent magnetic dipole moment as a result of unpaired orbital electrons. This ensures that the molecule is radiatively active through magnetic dipole transitions. These transitions are typically orders of magnitude less intense than electric dipole transitions. The great abundance of the oxygen molecule, relative to the other absorbing molecules, compensates for these weak transitions, ultimately producing a large atmospheric absorption.

Molecular oxygen is referred to as a triplet with a ground electronic state and two excited electronic states. The electronic transitions from the ground state to either of these other states, accompanied by vibrational–rotational transitions, produces bands referred to as the infrared and the red bands. The absorption by the band centered at 13120 cm^{-1} (0.76 μm), which is referred to as the oxygen A band, is illustrated in Fig. 3.19. The diagram shows the broad structure of the P and R branches (Fig. 3.19a) and the finer line structure (Fig. 3.19b).

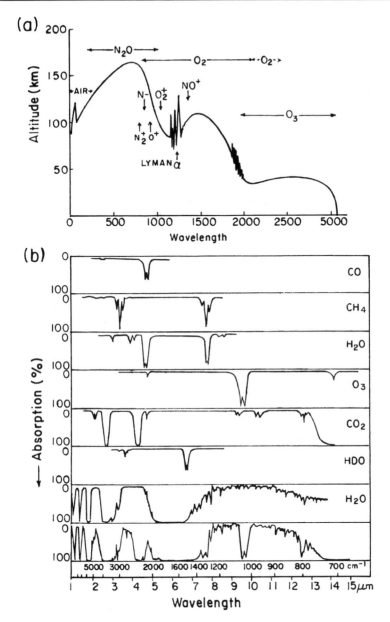

Figure 3.18 (a) The depth of penetration of solar radiation in the ultraviolet spectrum as a function of wavelength. The altitude indicated by the spectrum is the altitude of unit optical depth (after Herzberg, 1945). (b) Low resolution infrared absorption spectra of the atmosphere. The top six panels of (b) are the absorption spectra of important species. The bottom panel of (b) is a simulated absorption spectrum of the atmosphere (Valley, 1965).

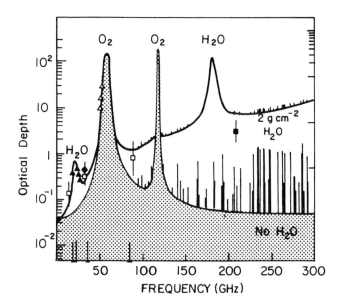

Figure 3.18 (Cont.) (c) Atmospheric optical depth in the microwave spectral region due to O_2, O_3 with and without water vapor as shown (Waters, 1976).

Molecular oxygen also possesses rotational bands in the microwave region. Important for atmospheric remote sensing are the oxygen bands located at 60 GHz and at 118 GHz; absorption by the former is measured by the MSU radiometer.

3.5.2   Ozone

Absorption by atmospheric ozone is dominated by electronic bands in the visible and ultraviolet spectral regions and by vibrational–rotational transitions in the infrared region with a strong band centered at 9.6 μm (Fig. 3.18b). The electronic spectrum of ozone has bands centered at 0.255 μm (Fig. 3.18a) referred to as Hartley bands. These consist of weak spectral structures superimposed on a very strong continuum. On the longer wavelength wing of the Hartley bands lies a series of weak bands (the Huggins bands) that appear in the spectrum of a low sun. These weak bands were responsible for the first positive identification of ozone in the atmosphere. Between about 0.45 μm and 0.75 μm are the Chappuis bands which are much weaker than the absorption bands in the ultraviolet portion of the electromagnetic spectrum.

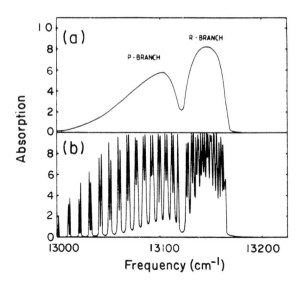

Figure 3.19 Oxygen A–band absorption at the surface for an over-
head sun. (a) spectral resolution of 5 cm^{-1} with a residual line structure
smoothed out; (b) spectral resolution of 0.5 cm^{-1} (Barton and Scott,
1986).

3.5.3 Carbon Dioxide

Vibrational–rotational absorption by carbon dioxide occurs in a
number of spectral regions (Fig. 3.18b). The infrared absorption
spectrum of carbon dioxide is dominated by the very strong 15 μm
(ν_2) and 4.3 μm (ν_3) bands. The 15 μm band is particularly important to temperature sounding techniques.

3.5.4 Water Vapor

The complex vibrational–rotational absorption spectrum of water
vapor, together with relatively large concentrations of water vapor
in the lower atmosphere, account for the dominance of this gas in the
spectrum extending from the near infrared spectral region beyond
the far infrared into the microwave region. Figure 3.20 shows the water vapor spectrum based on line absorption information compiled
from a spectral data base such as those described in the comments
included at the end of this chapter. Line absorption is represented by
the heavy black lines and the relatively smooth background absorp-

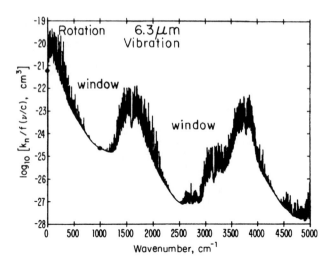

Figure 3.20 Theoretical absorption coefficients for pure water vapor
at 1 bar and at 296 K. The vertical axis is proportional to the absorption
coefficient divided by a factor that is approximately $\tilde{\nu}/c$ (modified from
Clough et al., 1980).

tion (highlighted by the shading) is the continuum. This continuum
occurs at all wavenumbers and is predominant in the window regions
indicated on the diagram. Some of the major absorption bands are
also highlighted on the diagram.

Wavelengths where continuum absorption is stronger than line
absorption are especially important to many remote sensing applications. There remains considerable debate over the mechanism for
the absorption and this general topic has been reviewed by a number
of investigators (refer to bibliography at the end of this chapter).

The analytic form and definition of the water vapor continuum,
especially in the important 8–13 μm region, is not fully agreed upon.
What has been established from laboratory measurements is the dependence of the absorption on water vapor partial pressure (and
hence the terminology *e–type* absorption from the meteorological
symbol for water vapor pressure). The effect of this dependence
is to produce strongly enhanced absorption in the lower and moist
tropical atmosphere. The presence of continuum absorption, both
in the main atmospheric window between 8–13 μm as well as in the
other windows including at microwave frequencies, looms as a significant problem in remote sensing. For example, the remote sensing
of sea surface temperature based on emission measurements in the

Table 3.1 Atmospheric Corrections to SST

T_s K	r_s^+ g/kg	k_2^* $(cm^2/g/atm)$	β	ΔT_s (K)
283	5.4	12.8	0.088	0.7
288	7.6	11.5	0.155	1.2
293	10.2	10.4	0.258	1.7
298	13.6	9.6	0.422	3.1
303	18.4	8.9	0.720	5.0

$^+$Assumes 70% relative humidity at the surface. * The absorption coefficient in the continuum may be approximated by $k = k_2 e$ where e is the water vapor partial pressure. For the definition of the absorption coefficient, see problem 3.12 and refer to Problem 7.10 for a definition of β–after Houghton and Lee (1972).

windows is significantly affected by continuum absorption. Table 3.1 provides an estimate of the sea surface temperature correction ΔT due to continuum absorption in the main atmospheric window. This correction is as large as 5 K for a sea surface temperature of 303 K. The table illustrates how this temperature correction rapidly increases as temperature and the surface mixing ratio r_s increase. The parameters in the table and their derivation are the topics of Problems 3.11 and 7.5.

Excursus: The Remote Sensing of Surface pressure

Suppose that a satellite instrument measures the solar radiation reflected in the oxygen A band. Radiation at these wavelengths is reflected by the Earth's atmosphere and the underlying ocean surface. Let us write the intensity of the radiation received by such a sensor as

$$I_\nu = I_\nu^{sea} + I_\nu^{atm} \tag{3.39}$$

where the first term is the reflection from the sea surface and the second is the intensity received by scattering from the atmosphere. Suppose that we can neglect I_ν^{atm} for the time being. We might then express the measured intensity as

$$I_\nu = I_{o,\nu} \mathcal{R}_\nu \exp\left[-\tau_\nu(0,p_s)\left(\frac{1}{\mu_o}+\frac{1}{\mu_s}\right)\right] \qquad (3.40)$$

where μ_o and μ_s are cosines of the sun's zenith angle, θ_o, and the satellite's view angle, θ_s, respectively. $\mathcal{R}_\nu$ is the ocean reflection function (this is a bidirectional reflection function as described in Chapter 4), $I_{o,\nu}$ is the inflowing solar radiation at the top of the atmosphere, and $\tau_\nu(0,p_s)$ is the optical depth associated with oxygen A band absorption along the vertical path from the satellite to the ocean surface. This optical depth may be expressed as a function of surface pressure p_s according to

$$\tau_\nu(0,p_s) = \frac{r}{g}\int_0^{p_s} k_{\nu,m}(p)\,dp = t(p_s) \qquad (3.41)$$

where $t(p_s)$ is left as an unspecified function of surface pressure (an example of this function is a subject of Problem 3.7). We recognize that the effects of pressure broadening complicate this integration and neglect the effects of temperature on absorption. Thus a ratio of two measurements at frequencies ν_1 and ν_2 in the A band provides

$$\frac{I_{\nu_1}}{I_{\nu_2}} = \frac{I_{o,\nu_1}}{I_{o,\nu_2}}\frac{\mathcal{R}_{\nu_1}}{\mathcal{R}_{\nu_2}}\exp\left[-\Delta\tau_\nu(0,p_s)\left(\frac{1}{\mu_o}+\frac{1}{\mu_s}\right)\right] \qquad (3.42)$$

where $\Delta\tau(0,p_s) = \tau_{\nu_1} - \tau_{\nu_2}$. If the frequencies ν_1 and ν_2 are close enough, then $\mathcal{R}_{\nu_1} \approx \mathcal{R}_{\nu_2}$ and assuming that I_{o,ν_1} and I_{o,ν_2} are also known, as is the viewing angle of the satellite ($\mu_s = \cos\theta_s$) and the solar elevation, then $\Delta\tau_\nu(0,p_s)$ can be retrieved by a simple inversion of (3.42). The surface pressure then follows from (3.41) provided the absorption coefficients at ν_1 and ν_2 are sufficiently well known and the function $t(p_s)$ is invertible.

This approach is essentially that proposed by Barton and Scott (1986) who recommended choosing ν_1 and ν_2 to coincide with the center frequencies of the P and R branches of the A band. However, a more detailed analysis of Mitchell and O'Brien (1987) demonstrated that this approach is only able to produce surface pressure estimates to 2 mb which is considered marginally acceptable for weather prediction requirements. One of the basic problems that arises is the temperature dependence of the absorption on the P and R branches (as discussed previously in relation to Fig. 3.14), which introduces further complications in (3.41). A second problem is the lack of discrimination between atmospheric reflections from aerosol and molecules and the reflection from the surface. Since photons reflected back to the satellite from the atmosphere do not flow through the full depth of the atmosphere, they represent a source of error in retrieving p_s. The magnitude of this error is variable, depending on the aerosol loading. Mitchell and O'Brien have proposed that observations at multiple wavelengths throughout the A band at a 1 cm^{-1} resolution might overcome these problems.

3.6 Passive Spectrometer Systems

Measurement of the Earth's radiation at a spectral resolution high enough to study details of molecular absorption bands is achieved using spectrometer sensor systems. Three major classifications of optical spectrometer systems will now be discussed. There are also various categories of microwave systems and some reference to these is given in the notes at the end of this chapter.

Optical spectrometers can be considered in terms of prism dispersion, grating diffraction, and radiation interference. Examples of each of these will now be described.

3.6.1 Prism Spectrometers

The first step in analyzing light was made by Newton in 1666 when he developed an early form of the prism spectrometer. In a prism spectrometer, the property of *dispersion* is employed to separate the various wavelengths of the light as it enters the system. Figure 3.21 shows an idealized and highly simplified laboratory arrangement of a prism spectrometer. Light from a source is made parallel by the lens L. When this parallel beam passes through a prism, rays of different wavelengths are dispersed (bent) to varying degrees depending on the refractive index of the prism (much more is said about refractive index in Chapter 4). Thus, radiation at different wavelengths is projected at different positions on the screen after being focused by a second lens L'. To each wavelength a line that is the image of the slit appears on the screen. Radiation of a particular wavelength can be then detected either by turning the prism slightly so that light of a different wavelength falls on a detector fixed at some location on the screen or by holding a prism fixed and moving the detector along the screen.

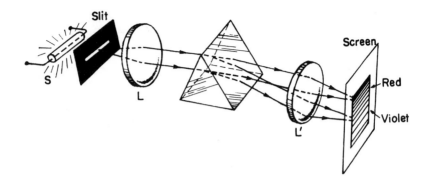

Figure 3.21 The principle of the prism spectrometer. This particular design was introduced by Joseph Fraunhofer in the nineteenth century and provided a way of sharply separating the colors of light.

Prism spectrometers are typically used in imaging devices. The spectral resolution achieved with a prism spectrometer depends on the optical layout of the instrument and the size of the prism. Typical resolutions, however, are an order of magnitude less than that of a grating spectrometer.

Prism spectrometers, despite their coarse resolution, provide important information about the atmosphere. The Dobson spectrometer is perhaps the best known example of this type of spectrometer. Dobson (1957) provides a detailed description of the spectrometer and how to operate it. A simplified layout of the spectrometer optics is shown in Fig. 3.22. Light enters a window W at the top of the instrument and, after reflection by a right–angled prism, falls on a slit S_1 of a spectroscope. The spectroscope consists of a quartz collimating lens from which the light enters a 60 % prism. A mirror then reflects the light back through the same prism and lens to form a spectrum in the focal plane of the instrument. This double–pass configuration improves the spectral resolution of the spectrometer. The required wavelengths are isolated by means of slits (S_2, S_3, and S_4) in the focal plane. In the actual instrument, the light is passed through a second spectroscope (not shown) so that the shorter wavelength ultraviolet can be separated further from longer UV wavelengths before being detected. The thick flat quartz plate (Q_1) located in front of slit S_1 is used as a fine scale wavelength adjustment. When the plates are inclined to the direction of the ray of light, the ray is

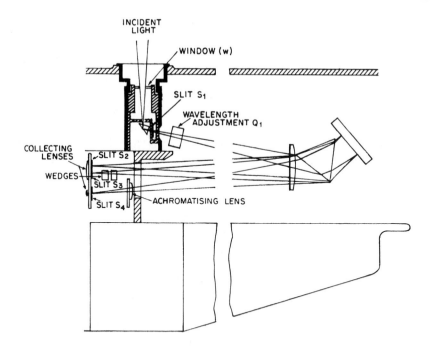

Figure 3.22 The optical layout of a Dobson spectrometer simplified.

refracted and displaced upward or downward. By this means, the wavelength of the radiation falling on the detector (a photomultipler) can be changed by accurately known amounts. The wavelength of the ray passing slit S_2 can be made of any value between about 0.305μm and 0.352μm. Slight adjustments to the instrument are needed to take account of air temperature changes, which slightly alter the refractive index of the quartz plate. The actual theory of ozone retrieval based on these measurements is discussed in more detail in Chapter 6.

3.6.2   Grating Spectrometers

The development of the diffraction grating by Fraunhofer in the nineteenth century provided a new tool for analyzing radiation that far exceeded prisms in its ability to disperse light. The working of a diffraction grating can be understood in the context of a wave flowing through a single long rectangular slit. According to Huygens' principle, when the incident wave falls on the slit all points along the wave front become secondary sources of waves producing new waves called *diffracted* waves. Observing the diffracted wave at different angles θ with respect to the direction of incidence, we find a distinct pattern of light appearing (the *diffraction pattern*) where the intensity is zero in certain directions. These zero–intensity points are referred to as null points, and their position is defined in the following way. Suppose the distance CD in Fig. 3.23a is one–half a wavelength, then the wave from A is exactly out of phase with the wave from C. In fact every ray from the first half of the slit (AC) is exactly canceled by a ray from CB, originating at a point $b/2$ away where b is the width of the slit. This condition may be stated as

$$\frac{b}{2} \sin \theta = \frac{n}{2} \lambda \tag{3.43}$$

where n is an integer, b is the width of the slit, and λ is the wavelength of the incident wave. According to this formula, the null points for radiation of different wavelengths occur at different values of θ.

Consider the diffraction pattern produced by a large number of parallel narrow slits of equal width and equal spacing (Fig. 3.23b). The pattern consists of a series of maxima and minima associated with the interference of the light from one slit to another. The location of the maxima of the interference pattern is given by

$$a \sin \theta = m \lambda \tag{3.44}$$

where $m = 0, \pm 1, \pm 2, \ldots$ Superimposed on these interference maxima is the diffraction pattern of the single slit. According to the value of m in (3.44), the principal maxima are referred to as first, second, third and so on, order of diffraction.

A system such as the one just described is called a *transmission diffraction grating*. For the purposes of analyzing infrared, visible, or ultraviolet light, transmission diffraction gratings consist of several thousands of slits per centimeter, obtained by etching a series of parallel lines on transparent film. A diffraction grating can also work by reflection in which case the grating consists of a series of parallel lines that are etched on a metallic surface. Transmission gratings generally perform poorly in comparison with reflection gratings, which are used in high performance space spectrometers.

A simple arrangement for a grating spectrometer is shown in Fig. 3.24a. When light of several wavelengths falls on the grating, diffraction orders (maxima) at different angles, prescribed by (3.44),

(a)

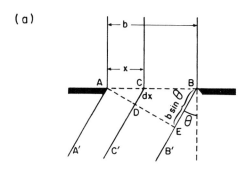

(b)

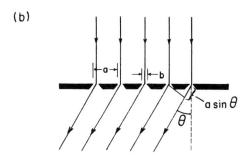

(c)

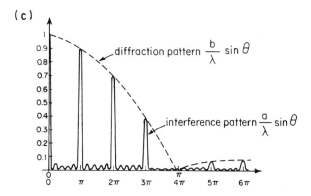

Figure 3.23 (a) The simple geometry of diffraction by a single slit and the diffraction geometry for multiple slits. (b) The cross section of a (transmission) diffraction grating. (c) The diffraction pattern from an eight slit grating. The lower abscissa is the value of θ in (3.44) corresponding to the various diffractions orders.

are produced. Each diffraction order displays the spectrum of the source. Note that the longer the wavelength, the more the radiation is deviated for a given order of the spectrum. An important point here is that the diffraction curve associated with a particular higher order n_2 of a shorter wavelength λ_2 overlaps the patterns of another longer wavelength λ_1 of order n_1 when $n_1\lambda_1 = n_2\lambda_2$. That is, first–order red light at 0.7 μm is overlapped by second–order blue light of wavelength 0.35 μm. If we are interested in sensing red light, then the low–order shortwave light must be removed which is usually achieved by a blocking filter.

The grating spectrometer has been widely used in remote sensing. The satellite infrared spectrometer (SIRS), launched in April 1989 on NIMBUS 3, was the first space–based grating spectrometer used for vertical temperature sounding experiments. Several other grating systems are now flown. One specific example is the solar backscatter ultraviolet (SBUV) instrument used for the total ozone mapping which monitors UV radiation in a number of selected wavelength bands. Another grating system is the coastal zone color scanner (CZCS) flown on the NIMBUS 7 satellite which is devoted to the study of ocean color in an effort to discriminate between organic and inorganic materials in water. A third example is the stratospheric aerosol and gas experiment (SAGE) sensor. The optical layout of the SAGE I sensor is shown in detail in Fig. 3.24b. The sensor is designed to monitor solar radiation in four bands centered at 0.385μm, 0.45μm, 0.6μm, and 1.0μm in an effort to obtain profiles of stratospheric aerosol, ozone, and nitrogen dioxide. The optical module consists of a flat scanning mirror that directs the solar radiation into the system, a telescope which directs the radiation to the grating, and four silicon detectors positioned to measure the radiation reflected from the grating at the wavelength selected. Each detector is filtered to remove contributions from unwanted grating orders.

3.6.3 Interferometer Spectrometers

The third type of optical system works quite differently from a prism or a grating and is based on the optical instrument invented by A. A. Michelson in 1880 named the *interferometer*. The role of the prism and grating in the previous systems is to separate the different spectral elements into different directions so they can be individually measured. The interferometer makes no use of dispersion of light; rather, it makes use of interference effects. In Michelson's interferom-

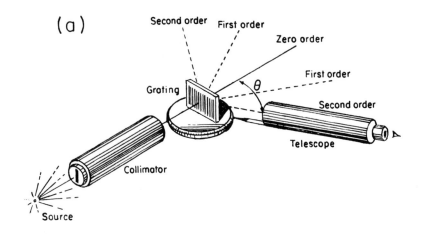

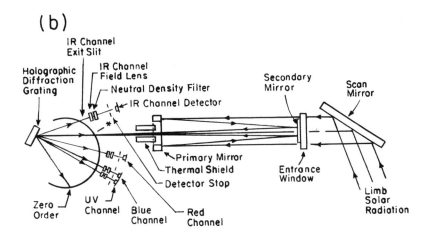

Figure 3.24 (a) A grating spectrometer. In the setup shown, light is collimated and flows onto a transmission grating. The spectra of different orders are measured by moving the detector. (b) A schematic of the SAGE I optical system including fixed, multiple detectors (channels).

eter light split into two beams by a half–reflecting glass plate follows two paths of unequal length and is recombined on the beam–splitting plate (Fig. 3.25). The path difference creates an interference of the light waves. If the path difference is varied uniformly by moving a mirror at a constant speed, then the intensity varies from bright to dark as the two component beams move in and out of phase. A recording of this intensity output is referred to as the *interferogram* and the spectrum can be reconstructed from the interferogram using

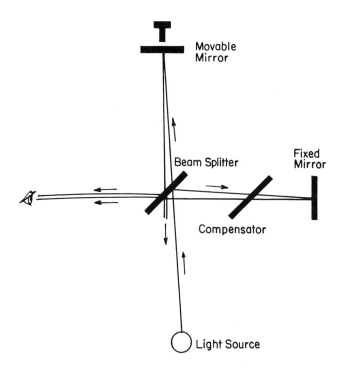

Figure 3.25 The Michelson interferometer splits the incoming light into two beams that travel different distances to the detector. A compensator plate equalizes the optical paths of the two light beams in the glass.

the Fourier transform. In a sense, the interferogram contains all the needed information about the spectrum in a coded form.

Excursus: The Resolving Power of an Interferometer

The elegance of the interferometer arises from the fact that its resolving power is no longer related to the length of the optical path or width of ruled lines on a grating. It is purely a function of the path difference of the two beams. We can appreciate this by considering a Michelson interferometer illuminated by a source of light at two adjacent wavelengths, λ and $\lambda + \Delta\lambda$. As the mirror moves, the bright interference fringes periodically appear, disappear, and then reappear again. If a small displacement of the mirror in its travel is Δx, then we can define the resolution of the instrument $\Delta\lambda$ in terms of this displacement. Here we consider Δx as the displacement of the mirror that causes a one–cycle variation in the visibility of the fringes.

We establish the relationship between Δx and $\Delta \lambda$ in the following way. The fringe visibility is high when bands of λ overlie those of $\lambda + \Delta \lambda$ and poor when the bright fringes of the former coincide with the dark fringes of the latter. This situation occurs when λ is an odd number of half–wavelengths of $\lambda + \Delta \lambda$. For a path of travel $2x$, then the condition of minimum visibility is

$$2x = n_1 \lambda = (n_2 + \frac{1}{2})(\lambda + \Delta \lambda)$$

where n_2 are odd integers. Noting that

$$n_1 = \frac{2x}{\lambda}, \qquad n_2 + \frac{1}{2} \approx \frac{2x}{\lambda}(1 - \frac{\Delta \lambda}{\lambda})$$

for $\Delta \lambda \ll \lambda$, then subtraction yields

$$n_2 - n_1 + \frac{1}{2} \approx -\frac{2x \Delta \lambda}{\lambda^2}$$

Suppose now that the integer $n_2 - n_1$ increases by 1 as x goes from x to $x + \Delta x$, and

$$n_2 - n_1 + \frac{3}{2} \approx \frac{2(x + \Delta x)\Delta \lambda}{\lambda^2}$$

If we subtract these two equations and rearrange, it follows that

$$\Delta x \approx \frac{\lambda^2}{2 \Delta \lambda}$$

In terms of wavenumber $\tilde{\nu}$, $d\tilde{\nu}/d\lambda = -1/\lambda^2$, and

$$\Delta \tilde{\nu} \approx -\frac{1}{2x}$$

In principle, the path difference Δx can be increased without limit and the resolving power of the interferometer made arbitrarily high. For example, a resolution of 0.1 cm^{-1} is achieved by moving the mirror only 5 cm. We can employ similar arguments to illustrate that the spectral range of the instrument is also broad. In this case, the spectral range of the instrument is determined by how finely we resolve the mirror position (however this range is usually practically defined by the spectral range of the detector used in the instrument).

The functioning of the interferometer is illustrated by considering the case when a monochromatic beam such as a laser source enters the instrument. If the moving mirror is adjusted so there is no path difference between the two beams, then the light will appear bright. When the mirror is moved one–fourth of the laser wavelength, the interference beam (i.e., the beam that bounces off the moving mirror) is exactly 180 degrees out of phase with the second beam so that no light is detected. Continuous movement of the mirror along one direction produces an oscillatory signal as the light changes from bright to dark. The form of this signal can be understood by considering two beams

$$\mathcal{E}_1 = \mathcal{E}_o \cos(kx - \omega t)$$

$$\mathcal{E}_2 = \mathcal{E}_o \cos(-\omega t)$$

where the first expression defines the electric field of the laser light for the path defined by the mirror position at x (the interference field) and the second expression applies to the field with the mirror determined to be in the $x = 0$ position (which is equivalent to the field of the second beam). The field combined at the detector has the form

$$\mathcal{E} = \mathcal{E}_1 + \mathcal{E}_2$$

The measured intensity is proportional to the time averaged $\mathcal{E}$ field (Chapter 2)

$$I = < \mathcal{E}^2 >$$

where we omit the proportionality constant and the angle brackets represent the time integral

$$I = \frac{1}{T} \int_t^{t+T} [\mathcal{E}_1^2 + \mathcal{E}_2^2 + 2\mathcal{E}_1\mathcal{E}_2] dt$$

Using arguments presented in chapter 2 in discussion of light intensity, and assuming the product $\omega T \gg 1$, it follows that

$$I(x) = I_{o,\tilde{\nu}}[1 + \cos(2\pi x \tilde{\nu})] \tag{3.45}$$

where $I(x)$ is the output signal expressed as a function of the distance of the mirror movement x, and $I_{o,\nu}$ is the source of radiation that enters the instrument.

An interferogram is the output signal that results from the summation of all the oscillations associated with all wavelengths. The mathematical expression of the interferogram in this case is

$$I(x) = \frac{1}{2\pi} \int_{-\infty}^{\infty} I_{o,\tilde{\nu}} \cos(2\pi x \tilde{\nu}) d\tilde{\nu} \qquad (3.46)$$

where it is assumed that the interferences are symmetric about the $x = 0$ position. An example of the interferogram obtained from measurements of the atmospheric emission by NASA's infrared interferometer spectrometer (IRIS) is shown in Fig. 3.26a. The Earth radiation spectrum is then found by taking the Fourier transform of the interferogram

$$I_{o,\tilde{\nu}} = \int_{-\infty}^{\infty} I(x) \cos(2\pi x \tilde{\nu}) dx \qquad (3.47)$$

Figure 3.26b provides an example of the optical layout of the IRIS. This instrument was designed to provide information about the vertical temperature profile, water vapor, and ozone profiles on the global scale. Radiation enters the instrument through a mirror and is divided nearly equally by the beamsplitter. The instrument is actually two coupled interferometers, with one associated with light collected from the atmosphere by scanning mirror at the top of the instrument and the other for the monochromatic source which is used to count fringes (refer to Problem 3.13 as an example of the use of this second system).

Figure 3.27 shows the emission spectra obtained from interferometers flown on different spacecraft. The infrared spectrum of Earth, under cloudless conditions, is dominated by the absorptions of water vapor and carbon dioxide. The spectra of both Mars and Venus indicate the importance of CO_2 as a source of gaseous opacity; the only major difference is the abundance in CO_2 of nearly 100%, instead of the Earth's 330 ppm.

Another example of an interferometer spectrometer is the high resolution infrared sounder (HIS) developed by the group at University of Wisconsin. This instrument has been designed for meteorological applications and measures the infrared emission of the atmosphere with a spectral resolution around 0.1 cm^{-1}. The instrument uses three detectors to sense the radiation over three broad spectral regions. An example of the output from the instrument is provided in Fig 3.28 which compares coincident spectra obtained

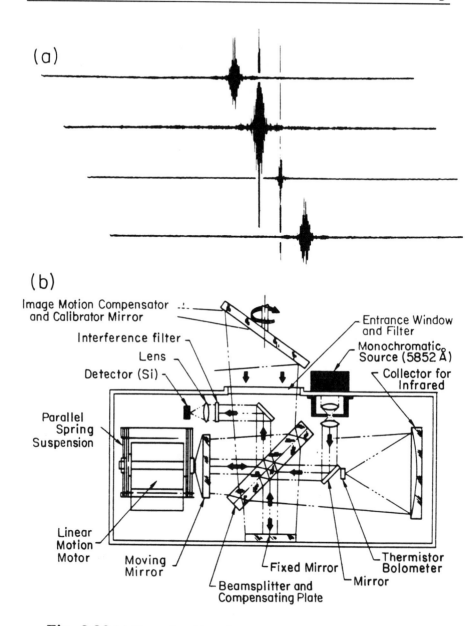

Fig. 3.26 (a) Examples of interferograms measured by the IRIS from space. (b) A schematic of the optical arrangement of the Nimbus IRIS instrument. The spectral resolution of the instrument is 5 cm^{-1}. There are actually two coupled interferometers, with one associated with light collected from the scanning mirror at the top of the instrument, the other for the monochromatic source which is used to count fringes.

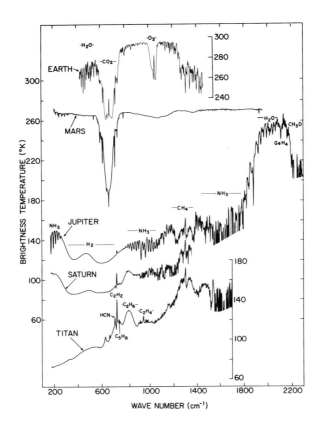

Figure 3.27 Spectrum of the infrared, expressed as brightness temperatures for four planets and Titan (Hanel, 1983).

from interferometers on an aircraft and at the ground. The project at the back of this book elaborates further on the operation of an interferometer and the emission spectra of the atmosphere measured at the ground. Interpretation of the spectra shown in Fig. 3.28 is left for Problem 7.1.

3.7 Notes and Comments

3.1 and 3.2. A wonderful elementary text on molecular spectroscopy is that of Banwell (1983) and the material described in these two sections follows his treatment of the topic.

Molecular absorption has also played a role in understanding the nature of cosmic radiation. Cosmic radiation is a topic relevant to our understanding of the origin of the universe. Radio astronomers have established that the cosmic radiation at centimeter wavelengths

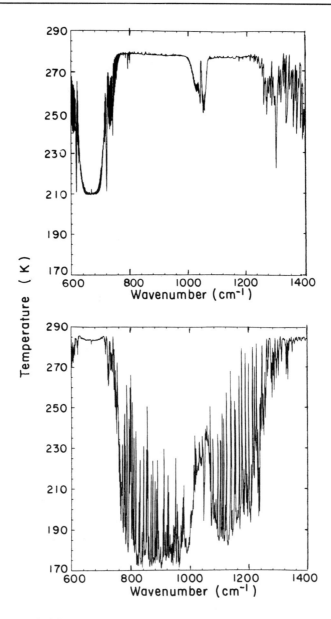

Figure 3.28 Spectra of the infrared emission obtained on March 1991 from the HIS instrument on an aircraft looking down (upper spectrum) and at the ground looking up (lower spectrum) (Smith, personal communication).

matches that of a blackbody with a temperature between 2.7 and 3 K. To establish whether or not this radiation is truly blackbody radiation we need to measure the cosmic radiation at wavelengths shorter than the microwave radiation measured by the radio astronomers to see if the energy density falls off with decreasing wavelength as the Planck function predicts. Unfortunately, the atmosphere of our planet becomes increasingly opaque at wavelengths below 0.3 cm (e.g., Fig. 3.18) and its not possible to estimate cosmic background radiation at wavelengths shorter than 0.3 cm from surface based measurements.

Interestingly enough, the amount of radiation background from space was deduced from measurements at these shorter wavelengths in 1941, long before the measurements by radio astronomers were reported. The background radiation was deduced from optical measurements of the light from the star ζ Oph which belongs to the constellation Ophiuchus ("the serpent bearer"). The light from this star observed at earth shows the existence of dark lines, indicating that a cloud of intervening gas absorbs the radiation. One of the absorption lines in the spectrum of ζ Oph occurs at a wavelength of 0.3875 μm which indicates the presence of the molecule cyanogen (CN). This absorption line is actually split into three lines with wavelengths 0.3874608, 0.3875763, and 0.3773998 μm. The first line corresponds to a transition from the ground state of the CN molecule to a vibrating state and is expected even in a zero–temperature environment. The other two lines could only be produced by transitions in which the molecule is lifted from an existing state of rotation to higher vibrating states. Thus a fraction of the CN molecules must be in this rotating state. From the known energy difference between the ground state and the rotating state, and from the observed relative intensities of the lines, it was deduced that the CN molecule was exposed to some kind of perturbation with an effective temperature of about 2.3 K. Further information about the cosmic background radiation can be found in Wienberg (1977).

3.3 and 3.4. The shape of an absorption line for paths along which pressure and temperature vary is not given by the Lorentz profile for absorption at fixed temperature and pressure. A simple way of thinking of this is to imagine the resulting line profile for a path through two cells in which the pressure and temperature are fixed but differ from one to the other. The profile is a superposition of two different profiles, one narrower than the other. This profile is broad in the wings due to contributions from the higher pressure cell and narrow and spiked at the center due to the absorption in the cell of lower pressure. In treating the effects of pressure and temperature variations along the path, it is usually assumed that the profile remains closely Lorentzian with parameters adjusted to fit the absorption in the wings and/or in the line center. These approximations are discussed in Goody and Yung (1989).

The shape of the 22 GHz water vapor line, and the matching of simple shape models to laboratory data, is a topic discussed by Walter (1992a). Yasim and Armstrong (1990) provide a theoretical study of the water vapor lines at 183 GHz and for microwave frequencies beyond.

3.5. Over the past decade or so molecular absorption data have been systematically archived. Periodic revisions are reported in the open literature and these data sources are rapidly becoming an 'industry standard'. Two of these compilations are those of the Air Force Geophysics Laboratory (AFGL) and described by Rothman et al. (1987) and another, in many respects, similar compilation has been developed by Chedin and collaborators under the acronym GEISA (Husson et al., 1992).

An abbreviated bibliography of pressure sensing via measurements in the oxygen A band is given at the end of Chapter 6.

An approximate theory of line continuum introduced in the recent work of Ma and Tipping (1992a and b) suggests that the continuum is predominantly the result of overlapping wings of distant self–broadened water vapor lines. Other schools of thought suggest that the absorption mechanism involves two water vapor molecules loosely bonded as a water vapor dimer. However, it appears that this type of absorption requires greater concentrations of dimers than are typically found at ordinary atmospheric water vapor densities. Continuum absorption in the 8–12 μm atmospheric window is important for remote sensing as well as for a number of other meteorological reasons (Burroughs, 1979).

3.6. A classic treatise on microwave spectroscopy is that of Townes and Schalow (1955). There are several ways to measure the emission spectra at microwave frequencies with enough spectral resolution to distinguish the absorbing species of interest. For example, one approach is the total power radiometer (Decker et al. 1978) which detects radiation using a parallel bank of channels, one for each frequency required. An alternative approach is the autocorrelation radiometer which provides highly resolved spectral absorption information at microwave frequencies using techniques that are directly analogous to interferometry (Ruf and Swift, 1988).

A useful text on the subject of spectrometry as it is used in satellite instrumentation is that of Chen (1985). A general reference text on interferometry is that of Hariharan (1990).

3.8 Problems

3.1. Briefly explain or interpret the following:

a. Two sealed chambers contain the same amount of water vapor and are at the same temperature. One contains only water vapor while the other holds a mixture of water vapor and air. Which has the smaller transmissivity averaged over a narrow spectral region containing a single water vapor absorption line?

b. The two sealed cells of (a) now both contain some amount of water vapor mixed in air. The concentration of water vapor in one cell is adjusted so that the transmission of $10\mu m$ radiation through one cell matches the transmission of $6.3\mu m$ radiation through the other cell. Which cell contains the most water vapor?

c. The temperature of both cells is now increased thus raising the pressure within the cell but assume no other changes occur. At which wavelength is the transmission a maximum (ignore any temperature effects on absorption)?

d. Molecules possessing a permanent electric dipole (known as polar molecules) readily absorb at infrared wavelengths. Molecular oxygen is not an electrically polar molecule but also has infrared and microwave absorption lines.

e. The rotational–vibrational spectra of CO_2 exhibits absorption lines that are regularly spaced whereas absorption lines of H_2O are more randomly distributed in the spectrum.

3.2. The wavelength of radiation absorbed during a particular spectroscopic transition is observed to be 10 μm. Express this in frequency (Hz) and in wavenumber (cm^{-1}) and calculate the energy change during the transition in both joules per molecule and joules per mole. If the energy were twice as large, what would be the wavelength of the corresponding radiation? *Hint:* Planck's constant has the value $h = 6.63 \times 10^{-24}$ joules s molecule^{-1}. Avagadro's number $N = 6.02 \times 10^{23}$ mol^{-1}.

3.3. The rotational spectrum of $^{79}Br^{19}F$ shows a series of equidistant lines spaced 0.71433 cm^{-1} apart. Calculate the rotational constant B and hence the moment of inertia and the bond length of the molecule. Determine the wavenumber of the $J = 9 \rightarrow J = 10$ transition.

3.4. Using your answers to Problem 3.3, calculate the number of revolutions per second which a BrF molecule undergoes when in (a) the $J = 0$ state, (b) the $J = 1$ state. [*Hint:* Use (3.9) but remember that ω is in radians per second.]

3.5. The masses of the H,CL,C, and O atoms are 1.6×10^{-27} kg, 58.8×10^{-27} kg, 20×10^{-27} kg, and 26.5×10^{-27} kg, respectively.
   a. Calculate the reduced masses of the HCl and CO molecule.
   b. If the spring constants of the HCl and CO molecules are 4.78 and 1907 kgs^{-2}, respectively, determine the wavelength of the vibrational transition $0 \rightarrow 1$.

3.6. Derive a relationship between the central frequency ν_o of a line and the pressure (in atmospheres) at which the half–widths of a Lorentz line and a Doppler line are the same. Estimate this pressure for a CO_2 and O_2 molecule for the frequencies and temperature used to produce the curves shown in Fig. 3.14b. Assume the reference value of the Lorentz half–width at the ground is that given in Fig. 3.14b.

3.7. Develop a relationship between the vertically integrated water vapor path through the entire vertical extent of the atmosphere (precipitable water) and the sea surface temperature. Assume
   a. the vertical profile of specific humidity has the following form $q_s(p/p_s)^\lambda$ where q_s is the surface specific humidity.
   b. $e_s \approx b \exp[a(T_s - T_o)]$. Derive your answer in terms of the surface relative humidity, λ, and the SST T_S.

3.8. Compute the optical path for:
   a. Water vapor of a 100 mb thick homogeneous layer of mixing ratio r.
   b. Total atmospheric CO_2 if the mixing ratio is 330 ppm by volume.

3.9. The following function

$$r(\psi) = r_p \frac{4a\psi^2}{(1 + a\psi^2)^2}$$

reasonably resembles the vertical profile of ozone mixing ratio such that with $a = 1600$, the maximum occurs at $\psi = p/p_s = 0.025$. Assuming a value $r_p = 1 \times 10^{-5}$ kg/kg, derive the total column ozone and express your answer in Dobson units (the density of ozone at S.T.P. is 2.14 kgm^{-3}).

3.10. The rationale for the surface pressure measurement using two frequencies in the O_2 A band is discussed in Section 3.5. Given the definition of optical thickness, obtain an explicit form of the function $t(p_s)$ given in (3.41) assuming (1) a Lorentz line and frequencies at the line center ($\nu = \nu_o$), and (2) frequencies in the line wing $| (\nu - \nu_o) | >> \alpha_L$. Neglect the effects of atmospheric temperature on line intensity and half–width. Express your answers in terms of S, the line strength; α_o the line half width defined at some reference pressure p_o, the mixing ratio r of the gas, p_{sat} the satellite pressure, and the acceleration by gravity g.

3.11. Absorption in the atmospheric window between 8 and 13 μm is represented by an absorption coefficient of the form $k_2 e$ where e is the water vapor pressure (in kPa), $k_2 \cong 10^{-1}$ (g cm^{-2})$^{-1}$ kPa^{-1}. If the water vapor pressure near the surface is 1 kPa, calculate (1) the transmission of a horizontal path 1km long near the surface, and (2) the transmission of a vertical path of atmosphere assuming that the distribution of water vapor pressure is proportional to pressure (in units of atmospheres) raised to the fourth power.

3.12. The absorption coefficient in the continuum has the form

$$k_\nu \approx k_{2,\nu} e$$

where e is the water vapor partial pressure in units of atmosphere. Assuming a hydrostatic atmosphere

$$p = p_s e^{-z/H}$$

where $p_s =$1013.13 mb, and assuming that the mixing ratio profile of water vapor is similarly exponential with

$$H_r = H/3$$

where H_r is the scale height of vapor

   a. Derive an expression for the optical mass u for the vertical path from $\tilde{p} = 0$ to $\tilde{p}$ where $\tilde{p} = p/p_s$ is the pressure in atmospheres. Express your answer in terms of r_s, the surface mixing ratio of water vapor, and $\tilde{p}$.

  b. Assume that the temperature dependence of the absorption
     parameter $k_{2,\nu}$ has the form

$$k_{2,\nu} = k_{2,\nu,s}/\tilde{p}$$

  show that

$$T_\nu = \exp[-\beta \tilde{p}^7]$$

  where

$$\beta = \frac{p_s r_s^2 k_{2,\nu,s}}{4.354g}$$

  where $e = r\tilde{p}/0.622$.

3.13. A Michelson interferometer is illuminated by a monochromatic
      source of light with a wavelength $\lambda = 0.5825$ μm (as in the ex-
      ample of the IRIS instrument) and this is used to measure the
      position of the moving mirror with respect to some fixed posi-
      tion of the same mirror. As the mirror moves continuously, the
      fringe pattern fades in and out in a periodic fashion. Compute
      the mirror travel corresponding to a shift in visibility from a
      maximum to minimum.

3.14. A grating with 20,000 lines has a length of 4 cm.
      a. Find the angular separation of the whole visible spectrum
         for first and second order diffraction. Assume the wave-
         length range is from 0.39 to 0.77 μm.
      b. Can the grating resolve the two yellow lines of sodium
         whose wavelengths are 0.5890 μm and 0.5896 μm? (*Hint:*
         The resolving power of a grating is defined as $r = \lambda/\Delta\lambda$
         where $\Delta\lambda$ is the minimum wavelength difference between
         two sources for which for a given order the principal max-
         imum of one falls on the first zero of the other. The in-
         terested student can show that this leads to the condition
         $r = Nn$ where N refers to the number of slits.)

4
Macroscopic Interactions — Optical Properties

The previous chapter dealt with the topic of molecular absorption. This absorption is a type of "resonance" between the radiation and the molecule in the sense that radiation is absorbed over a very narrow and specific range of frequencies. Actually, atoms and molecules react to radiation of all frequencies, and not just to the resonant frequency. This nonresonant reaction is subtle and cannot be described in terms of quantum jumps from one level to another. These subtle reactions are nonetheless important because the very appearance of solid objects, like atmospheric particulates, are based on the responses of matter to nonresonant radiation. In this chapter, nonresonant interactions are described in terms of bulk macroscopic properties of the matter known as *optical properties.*

Optical properties are relevant to many problems of remote sensing. The difference between the optical properties of water, and ice at microwave frequencies, for example, is used in radar studies of clouds and precipitation, and in the study of sea–ice and its distribution. Optical properties of vegetation, water and dry soil are also important to the remote sensing of the surface, and the optical properties of atmospheric particles are crucial in developing remote sensing tools to study these particles.

4.1 Polarization of Matter

The *polarization* of matter, in contrast to the polarization of radiation, is a property that relates to the ability of the material to form dipoles. For instance, in certain material the charge distribution of an atom placed in an electric field aligns with the field such that the atoms acquire an induced electric dipole (Fig. 4.1). Alternatively, molecules like the water molecule that possesses a permanent electric dipole moment aligns itself with the dipole moment parallel to the applied electric field. As a consequence of either induced or permanent dipoles, a piece of matter placed in an electric field becomes *electrically polarized,* and the material polarized in this way is called a *dielectric.* While this process occurs on the microscopic

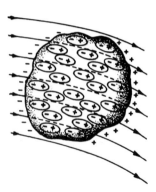

Figure 4.1 Polarization of matter under the influence of an electric field.

level, either by charge displacement or molecular orientation, it is conveniently represented by macroscopic properties of matter in the following way. The polarization per unit volume of matter is defined as

$$\vec{P} = (\epsilon_r - 1)\,\epsilon_o \vec{\mathcal{E}} \qquad (4.1)$$

where ϵ_o is the electric permittivity in a vacuum. This macroscopic expression states that the electric field and polarization are directly related and the proportionality constant, ϵ_r, is referred to as the *relative permittivity* or alternatively as the optical or dielectric constant.

Various mechanisms cause displacement of charge in matter and therefore contribute to its polarizability. Under the influence of oscillatory fields of different frequency, the constituents of matter vibrate on different time scales and thus contribute to the observed properties in different portions of the electromagnetic spectrum. Figure 4.2 provides a schematic depiction of the three principal polarization mechanisms that are relevant to the topics of this book. The mechanism that takes place on the shortest time scale displaces the charges associated with the lighter part of matter, namely electrons in the atoms. Since the oscillators are light, the oscillations occur rapidly, and the frequencies associated with electronic transitions occur in the ultraviolet. This mechanism contributes to the electronic polarization of the material. The next mechanism occurs on the atomic level where the atoms of molecules perform vibrations. Because the mass of an atomic nucleus is larger than that of an electron, the oscillations are more sluggish and the frequencies associated with the atomic polarization mechanism are lower than those are associated with electronic polarization. For the H_2O molecule, these frequencies occur in the infrared portion of the electromagnetic spectrum. Even more sluggish are the oscillations that occur as a result of the tendency of the permanent dipole of a molecule to become oriented relative to the applied field. Orientational or dipole polarization occurs more slowly than do the two previous mechanisms because of the larger vibrating mass. The characteristic frequencies associated with this mechanism tend to predominate at the longer infrared and microwave spectral regions. At the same time, thermal agitation of the molecules in condensed matter attempts to return the polarized region of the material towards a more random orientation of the dipoles and a kind of thermal buffeting of the orientation mechanism results. The optical properties established by this orientational mechanism are thus temperature dependent.

Let us now consider what happens to an individual dipole when an electric field is applied to it. The dipole moment of an individual atom or molecule $\vec{p}$ can be related to the locally active electric field $\mathcal{E}'$ by

$$\vec{p} = \alpha \vec{\mathcal{E}}' \qquad (4.2)$$

where α is the *polarizability*[1] of the material. If there are N of these molecules per unit volume of matter, then the polarization per unit volume of the material is $\vec{P} = N\vec{p}$.

We cannot yet combine (4.2) and (4.1) to establish the link between the macroscopic parameter ϵ_r to the microscopic parameter α. The problem is that in condensed matter where molecules are tightly packed, the field $\mathcal{E}'$ acting locally on the dipole is not the same as the external field $\mathcal{E}$ applied to the material. We will not discuss the way that we can express the local field in terms of the applied field here; references that elaborate on this topic are given at the end of this chapter. Suffice it to say that the field at the dipole may be derived by imagining that it sits in a spherical hole in a surrounding dielectric material. The field in such a hole is increased over a uniform static field $\mathcal{E}$ by an amount $P/3\epsilon_0$. The same argument applies for an electric field in the form of a wave

---

[1] There are different forms of polarizability that can be defined. The polarizability introduced in (4.2) is referred to as the atomic polarizability, the ratio of P to $\mathcal{E}$ defines the volume polarizability (i.e., $N\alpha$), and the quantity $N_o\alpha$ is the molar polarizability where N_o is Avogadro's number.

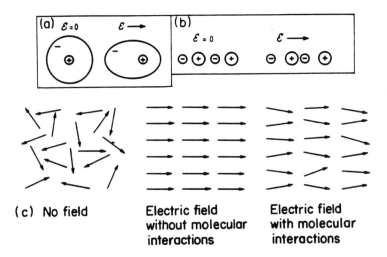

Figure 4.2 The three main mechanisms of polarization under consideration: (a) electronic, (b) atomic, and (c) orientation.

provided that the wavelength of the wave is much longer than the spacing between atoms and molecules. In this case the field locally is increased by the fields associated with the neighboring dipoles according to

$$\mathcal{E}' = \mathcal{E} + \frac{P}{3\epsilon_0} = \frac{\mathcal{E}}{3}(\epsilon_r + 2) \tag{4.3}$$

Combining of (4.1),(4.2), and (4.3)produces

$$N\alpha = 3\epsilon_0 \frac{\epsilon_r - 1}{\epsilon_r + 2} \tag{4.4}$$

which is known as the *Clausius–Mosotti* equation.

4.2   Classic Theories

The relative permittivity ϵ_r, a property relating the response of dense matter to the action of an electric field, is obviously related to the properties of atoms and molecules of the material as suggested by our discussion of Fig. 4.2. In this section we will provide a more quantitative, albeit phenomenological, account of how this quantity relates to these properties.

Actual calculation of ϵ_r reduces to the calculation of the polarizability of atoms or molecules. This amounts to determining the effects of an external field on the motion of charge in the material following the laws of quantum mechanics. For our purposes, simplified mechanical models suffice to approximate the permittivity.

4.2.1 The Lorentz model

We often picture in our minds a model of an atom represented by electrons whirling around a nucleus in a kind of fuzzy orbit. So far as problems involving nonresonant interaction with radiation, these electrons behave as though they are attached to springs producing a distortion of charge in response to an oscillating electric field. These electrons react to electromagnetic radiation in such a way that they vibrate just like a classical harmonic oscillator (Fig. 4.3) vibrates. H. A. Lorentz introduced his model of electronic and atomic polarization around the beginning of this century based on the principle of a classical harmonic oscillator.

The equation of motion of such an oscillator is

$$m\frac{d^2x}{dt^2} + \gamma\frac{dx}{dt} + kx = q\mathcal{E}' \tag{4.5}$$

where m is the mass of the oscillator, $\gamma dx/dt$ is the damping force exerted by neighboring dipoles, and k is the "spring" constant. In this expression, $q\mathcal{E}'$ is the driving force produced by the local electric field $\mathcal{E}'$, and x is the displacement of the mass from its equilibrium position. This is not really a legitimate model of an atom, but simple cases of correct quantum mechanical theory give results equivalent to this model. In a crude sense, the effects of quantum theory are accounted for by the appropriate choice of the properties of oscillators.

If the electric field acting on the dipole vibrates with a frequency ω, then the displacement x of the charge oscillates at the same frequency. Assuming that $x = x_o e^{i\omega t}$, then x can be solved for in terms of $\mathcal{E}'$ producing

$$x = \frac{(q/m)\mathcal{E}'}{\omega_o^2 - \omega^2 - i\gamma\omega} \tag{4.6}$$

where $\omega_o = \sqrt{k/m}$ is referred to as the resonant frequency of the oscillator. This displacement is complex and a convenient form for it is $Ae^{i\Phi}(q/m)\mathcal{E}'$ where $A(q/m)\mathcal{E}'$ is the amplitude of the oscillation and Φ is its phase relative to the driving force of the electric field.

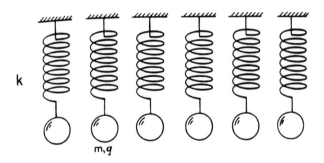

Figure 4.3 The Lorentz model of matter.

Simple algebra provides us with

$$A = \frac{1}{[(\omega_o^2 - \omega^2)^2 + \gamma^2\omega^2]^{1/2}} \quad (4.7a)$$

$$\Phi = \tan^{-1}\frac{\gamma\omega}{\omega_o^2 - \omega^2} \quad (4.7b)$$

which follow from (4.6). An interpretation of these results is provided in Fig. 4.4a and b where A and Φ are shown as a function of frequency ω. How these properties of the oscillator vary with frequency depends on the value of ω relative to the resonant frequency ω_o of the oscillator. For $\omega \gg \omega_o$, the nonresonant oscillations are weak and out of phase with the driving force of the light. The amplitudes of the oscillation for this range of frequencies, according to (4.7a), decrease at a rate proportional to $1/\omega^2$ (Fig. 4.4b). In the spectral range of low frequencies $\omega \ll \omega_o$, the nonresonant oscillations are again weak, but, in this case, in phase with the driving force (Fig. 4.4a). In this spectral range, the amplitude approaches a constant value as ω is decreased from resonance. Only the resonance case ($\omega = \omega_o$ and $\Phi = 0$) corresponds to a transition from one quantum state to another.

Given the response of the single oscillator to a time–harmonic electric field, the relative permittivity can be derived using the definition of the dipole moment for a single oscillator as $p = qx$, and since $p = \alpha\mathcal{E}'$, then

$$\alpha = \frac{q^2/m}{\omega_o^2 - \omega^2 - i\gamma\omega}$$

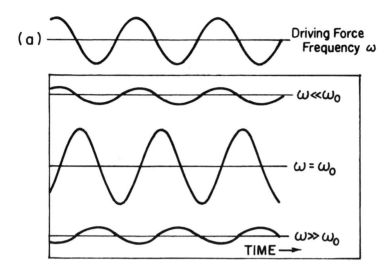

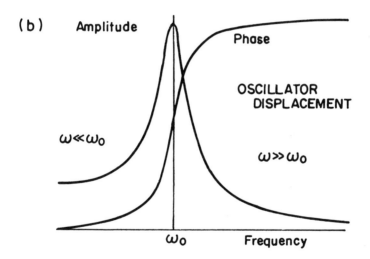

Figure 4.4 (a) The response of an oscillator to a periodic driving force serves as a model of how charges in matter react to an electromagnetic driving force. The response of the oscillator depends on the frequency of the forcing ω relative to the oscillator's resonant frequency ω_o. (b) The oscillator amplitude and phase as a function of the frequency ω of the applied electromagnetic field. The amplitude approaches a constant value when the frequency of the driving force is much below resonance such as in the case of N_2 and O_2 molecules exposed to visible light.

and the polarization per unit volume, P, for N oscillators in a unit volume follows as

$$P = \frac{\omega_p^2}{\omega_o^2 - \omega^2 - i\gamma\omega}\epsilon_o\mathcal{E}' \tag{4.8}$$

where $\omega_p^2 = Nq^2/\epsilon_o m$ is the *plasma frequency*. The difference between the local field and the external field is ignored since a proper treatment of local field effects here only complicates matters without adding further insight. With this assumption, it follows by matching (4.1) to (4.8) that

$$\epsilon_r = 1 + \frac{\omega_p^2}{\omega_o^2 - \omega^2 - i\gamma\omega} \tag{4.9}$$

which has the following real and imaginary parts

$$\epsilon_r' = 1 + \frac{\omega_p^2(\omega_o^2 - \omega^2)}{(\omega_o^2 - \omega^2)^2 + \gamma^2\omega^2} \tag{4.10a}$$

$$\epsilon_r'' = \frac{\omega_p^2\gamma\omega}{(\omega_o^2 - \omega^2)^2 + \gamma^2\omega^2} \tag{4.10b}$$

respectively. The frequency dependence of each of these components is schematically shown in Fig. 4.5a. The complex component provides the dampening of the oscillations and is a maximum at resonance which coincides with the most rapid change of the real part of the relative permittivity with frequency.

Quantum mechanical solutions provide similar results, but with the following modifications. Atoms and molecules have several natural frequencies and each has its own dissipation constant. The effective strength of each mode is also different and we represent this by the strength factor f. Summing over all modes leads to a modification of (4.9) of the form

$$\epsilon_r - 1 = \frac{Nq^2}{\epsilon_o m}\sum_i \frac{f_i}{\omega_i^2 - \omega^2 - i\gamma_i\omega} \tag{4.11}$$

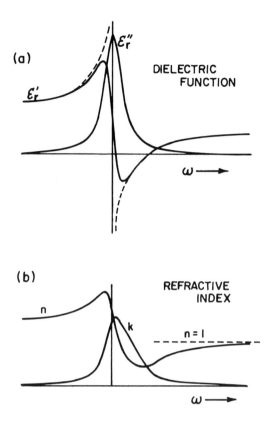

Figure 4.5 (a) The frequency dependence of the real and complex parts of the relative permittivity. Note that when the damping term is neglected, $\gamma = 0$ and $\epsilon_r'' = 0$ and the unphysical result occurs at the resonant frequency (dashed curve). Damping is not a result of the viscous movement of the oscillators; rather, it represents transitions from one state to another and therefore represents absorption processes. (b) The frequency dependence of the real and complex parts of the refractive index.

4.2.2 Orientational Polarization—Debye Relaxation

Lorentz's classical model describes polarization arising from the distortion of charge in nonpolar molecules. In solids and liquids composed of polar molecules, the orientation of the dipoles with respect to an electric field produces an additional low frequency contribution to the polarization. The ability of a molecule to orient itself depends on its shape and its interactions with the environment. The nearer to sphericity and the lower the dipole moment, the more easily and faster the molecule reorients itself in a changing electric field. An asymmetrical molecule like H_2O has several stable orientations and changes direction relatively slowly from one stable orientation to another. The average time between these changes is the *relaxation time*.

The polarization that results via orientation of dipoles can be computed from methods of statistical mechanics. We consider only very simple aspects of these methods here. Consider a molecule with a permanent dipole moment p_o aligned at some angle θ to the electric field. The potential energy of the dipole (e.g., Kittel, 1971) is

$$U = -p_o \mathcal{E}' \cos \theta_o$$

Statistical mechanics tells us that in a state of equilibrium, the relative number of molecules with a potential energy U is

$$e^{-U/k_B T}$$

and the number of molecules oriented at an angle θ

$$n(\theta) = n_o e^{p_o \mathcal{E}' \cos \theta_o / k_B T}$$

where k_B is Boltzmann's constant and T is the temperature. For normal temperatures and $\mathcal{E}$ fields, this approximates to

$$n(\theta) = n_o \left(1 + \frac{p_o \mathcal{E} \cos \theta_o}{k_B T} \right)$$

where n_o is $N/4\pi$ [we find this by integrating $n(\theta)$ over θ which should just be N the total number of molecules]. The net dipole moment per unit volume follows from the integration of the moment $p_o \cos \theta_o$ over solid angle $d\Omega = 2\pi \sin \theta d\theta$,

$$\bar{P} = 2\pi \int_0^\pi n(\theta) p_o \cos \theta \sin \theta d\theta$$

resulting in an average dipole moment

$$\bar{P} = \frac{N p_o^2}{3 k_B T} \mathcal{E}' \qquad (4.12)$$

and by combining (4.4) and (4.12) leads to

$$\alpha_o = \frac{p_o^2}{3 k_B T}$$

Debye (1929) elegantly discusses the dielectric relaxation of polar molecules in liquids. He supposes that dipoles initially align themselves in the direction of a field only to relax their orientations back to an equilibrium state as defined by the average dipole moment $\bar{P}$ relevant to a static field. The characteristic time scale for this relaxation is τ. The central result of Debye's theory is that the orientational part of the polarizability depends on the applied frequency ω such that

$$\alpha = \frac{p_o^2}{3k_BT}\frac{1}{1+i\omega\tau} \tag{4.13}$$

Using the Mosotti field for $\mathcal{E}'$, then

$$\frac{N\alpha}{3\epsilon_o} = \frac{N}{3\epsilon_o}\left(\frac{p_o^2}{3k_BT}\frac{1}{1+i\omega\tau}\right) = \frac{\epsilon_r - 1}{\epsilon_r + 2} \tag{4.14}$$

From this expression, the complex permittivity may be given in terms of the permittivity defined at the limits $\omega \to 0$ (ϵ_{rs}, the static permittivity) and $\omega \to \infty$ (ϵ_{rh}, the high frequency permittivity) and the effective relaxation time constant which is,

$$\tau_e = \tau\frac{\epsilon_{rs} + 2}{\epsilon_{rh} + 2}$$

It follows that

$$\epsilon_r = \epsilon_{rh} + \frac{\epsilon_{rs} - \epsilon_{rh}}{1 + i\omega\tau_e} \tag{4.15}$$

This expression is the Debye relaxation formula for the permittivity of a friction–dominated medium in which the internal field is assumed to be the Clausius–Mosotti field. The relaxation time is lengthened from τ to τ_e due to the difference between the internal field and this applied field.

The real and imaginary parts of ϵ_r follow from (4.15) as

$$\epsilon' = \epsilon_{rh} + \frac{\Delta}{1 + \omega^2\tau_e^2}$$
$$\epsilon'' = \frac{\Delta\omega\tau_e}{1 + \omega^2\tau_e^2} \tag{4.16}$$

where $\Delta = \epsilon_{rs} - \epsilon_{rh}$. The imaginary part of the dielectric function, according to (4.16), is a maximum at $\omega = 1/\tau_e$, and its behavior with frequency is broadly similar to ϵ_r'' predicted for the Lorentz oscillator. The real part behaves quite differently: it has no maxima or minima, but it decreases monotonically with increasing frequency from a value of ϵ_{rs} at low frequencies to ϵ_{rh} at high frequencies. At low frequencies, permanent dipoles react to the more slowly oscillating electric field in enough time that they become aligned producing a significant polarization and large values of ϵ_r'. At higher frequencies, this part of the matter is unable to respond quickly enough to produce any polarization.

The Debye relaxation model has been successfully used to describe measured values of the dielectric function at microwave frequencies as demonstrated in Fig. 4.6a. Both the real and complex parts of ϵ_r for water at microwave frequencies are compared to the Debye theory on this diagram. The parameters $\epsilon_{rs}, \epsilon_{rh}$ and τ are chosen to provide the best fit to the data. An especially relevant consequence of the relaxation spectrum of H_2O to remote sensing lies in the change of the spectrum of ϵ_r with the phase transition from liquid to solid water. To understand the differences in ϵ_r as this transition occurs it is helpful to consider the simple classical expression Debye derived for τ:

$$\tau = \frac{4\pi\eta a^3}{KT} \tag{4.17}$$

for a sphere of radius a in a fluid of viscosity η. This time constant is a ratio of the viscous–restoring torque applied to the sphere which maintains alignment to the thermal forces that act to disrupt this alignment. When numerical values are substituted into (4.17), the derived relaxation time corresponds approximately to that estimated from measurement. A naïve interpretation of the phase transition from liquid water to ice is to consider a large discontinuous increase in viscosity that occurs when water freezes. Thus, the permanent electric dipoles that were free to rotate in the liquid are now immobilized. The relaxation time for ice is significantly larger than it is for water leading to smaller values of ϵ_r'' and a dramatic shift in the maximum of ϵ_r' to smaller frequencies. The consequences of such large changes in ϵ_r as ice melts are observed when microwave radiation transmitted by a radar system is backscattered by melting ice particles producing the "bright band" in vertical profiles of radar reflectivity.

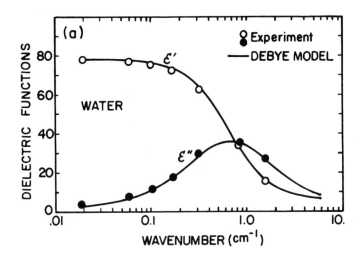

Figure 4.6 The dielectric function of water at room temperature calculated from the Debye relaxation model with $\tau = 0.8 \times 10^{-11}$ sec, $\epsilon_{rs} = 77.5, \epsilon_{rh} = 5.27$. Data were obtained from three sources (after Bohren and Huffman, 1983).

4.2.3   Summary

We learn from both models that when a sinusoidal electric field acts on a dieletric material, there is an induced dipole moment that is proportional to the electric field. The proportionality constant $\epsilon_r - 1$ depends on the frequency of the oscillating field and is a complex number which means that the polarization does not follow the electric field but is shifted in phase. A schematic diagram summarizing the frequency dependence of ϵ'_r and ϵ''_r for an ideal nonconducting substance is shown in Fig. 4.7. At the low frequency end, ϵ'_r is composed of contributions by all three mechanisms with the largest contributions resulting from dipole orientation processes. As the frequency increases, the dipoles are unable to respond fast enough, and this mechanism ceases to contribute to ϵ'_r; instead, the atomic polarization processes that produce vibrational motions contribute. For the water molecule, the resonances associated with these processes are found at infrared wavelengths. At even higher frequencies, interatomic vibrations cannot respond fast enough to the applied field. At these frequencies, the electronic oscillations that are induced by the electric field now contribute to ϵ'_r, and the resonant frequencies associated with these oscillators are typically found at UV wave-

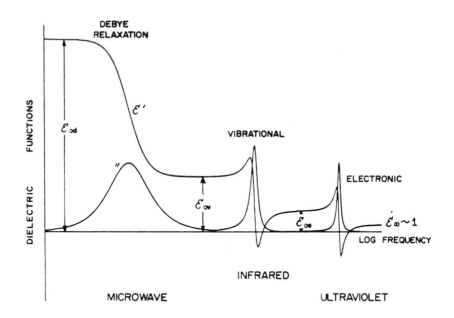

Figure 4.7 Schematic diagram of the frequency variation of the dielectric function of an ideal nonconductor (Bohren and Huffman, 1983).

lengths. Finally, as the frequency increases beyond the point where all electronic modes are exhausted, ϵ'_r approaches unity.

Where ϵ'_r changes most dramatically with frequency there is an associated peak in ϵ''_r which characterizes the absorption of radiation by the substance. This absorption arises from the resonances associated with the vibrations of atoms and molecules of matter. In dense matter, the molecules are so tightly packed together that significant interactions exist between them. The internal modes of the oscillations are therefore modified and the natural frequencies of the atomic oscillations are spread out by the interactions producing a broadening of the absorption lines in much the same way as pressure broadening occurs in gases. For example, in place of the precisely defined characteristic energy levels associated with the vibration and rotation states of the individual molecules are *energy bands* composed of a continuum of levels. Thus the energy levels of the vibration and rotation states of, for instance, a water molecule form a continuous absorption band resulting in a broad absorption spectrum as indicated in Fig. 4.7. Figure 4.8 provides a schematic illustration of the electron energy bands of two different types of material.

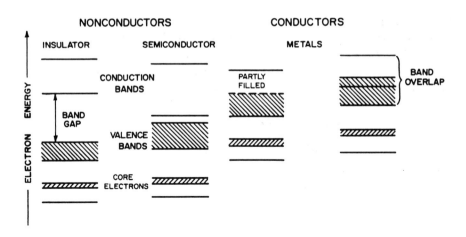

Figure 4.8 Electron energy bands in nonconductors and conductors. The filled bands are shown hatched (Bohren and Huffman, 1983).

Since the energy bands in a solid or liquid form as a superposition of the energy levels of the individual molecules, the spectral positions of the more continuous absorption bands roughly overlap the absorption spectrum of the individual molecules. Thus the infrared absorption spectra of liquid water and solid ice, for instance, occur at approximately the same wavelengths where absorption bands of water vapor lines are found.

There are features of the energy bands that have a significant bearing on the way radiation interacts with condensed matter and which are therefore important to our understanding of particle scattering. The energy bands of certain materials overlap, as depicted in Fig 4.8, and the electrons in such a material have a continuous distribution of energy within these overlapped bands. If one of the overlapping bands is partially empty, then application of an electric field readily excites electrons into adjacent unoccupied states and an electric current results. The material is said to be a good conductor of electricity and its electrical behavior is determined by both the energy band structure and how the bands are normally filled by electrons. This is the case for metals which can absorb radiation at any wavelength. When a photon is absorbed in a metal, the electron jumps to an excited state. A photon of the same energy is immediately re-emitted and the electron returns to its original state. Because of this rapid and efficient re-radiation, the surface of the metal appears reflective rather than absorbent (consider Problem 4.5a). Another type of material is the nonconductor which possesses energy bands that are separated by intervals referred to as *forbidden bands*; absorption of radiation by such material is therefore only likely for photons possessing energies greater than this energy gap.

4.3   The Refractive Index

The two sets of quantities that are often used to describe optical properties of matter are the relative permittivity ϵ_r and the *refractive index* m^2. Both are related according to

$$\epsilon_r' = n^2 - \kappa^2$$

$$\epsilon_r'' = 2n\kappa \tag{4.18}$$

where n and κ are used here to denote the real and imaginary parts of the refractive index, respectively. The spectral variations of both n and κ from the near infrared to the microwave regions are depicted in Fig. 4.9. Certain features of the hypothetical spectra of ϵ_r' and ϵ_r'' shown in Fig. 4.7 can be identified in the refractive index spectra. Readily apparent are the relaxation spectra extending from about the millimeter wavelength range into the centimeter range. For water and ice, the values of κ lead to significant absorptions in clouds when wavelengths are greater than about 1 μm. For ice, κ decreases again beyond wavelengths of about 100 μm. At microwave frequencies, ice particles in the atmosphere are more effective scatterers of radiation than absorbers, whereas the reverse is true of water drops. There are also significant differences between values of κ for water and ice in the near infrared especially around 1.6 and 3.7 μm, which also happen to be channels associated with radiometers flown (or to be flown) on meteorological satellites. The consequence of the different values of κ to the transfer of solar radiation through clouds at these wavelengths has been proposed as a way of discriminating ice clouds from water clouds.

Determining the refractive indices of atmospheric aerosol is quite a complex problem and a topic of apparent controversy. In Fig. 4.10, the spectra of the imaginary parts of the refractive index

---

[2]   The refractive index is sometimes written as $m = n + i\kappa$ and other times as $m = n - i\kappa$. The latter applies when the time dependence factor of the wave is $\exp(i\omega t)$ rather than $\exp(-i\omega t)$. Both will be used in this book.

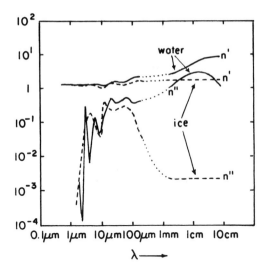

Figure 4.9 Typical values of the refractive indices for water and ice.

of several materials that exist in atmospheric particles are shown. Results are given for water, ammonium sulfate, crystalline quartz, sulfuric acid, carbon, sodium chloride, and hematite over selected spectral regions. As we have come to expect from our previous discussions, κ is large (around unity) in the infrared and ultraviolet spectral regions and small at visible wavelengths for all materials except for carbon and hematite both of which significantly absorb visible light. To emphasize the transparency of the material in the visible region, the dashed line is the value of κ corresponding to a 1% transmission through a 1cm thick homogeneous slab of material. Only carbon, which has metalic overlapping electronic energy bands (e.g., Fig. 4.8), has high values of κ throughout most of the spectrum. The mineral hematite, although a very minor constituent of the atmospheric aerosol, is one of the few known materials that is also highly absorbing at visible wavelengths.

The hatched region in Fig. 4.10 shows the values of κ obtained from remote measurements using a retrieval scheme based on the particle scattering theories discussed in Chapter 5. These derived values of κ clearly do not seem to match those of any of the pure materials that make up the particle. They are presumably some kind of average of a mixture containing a small amount of a highly

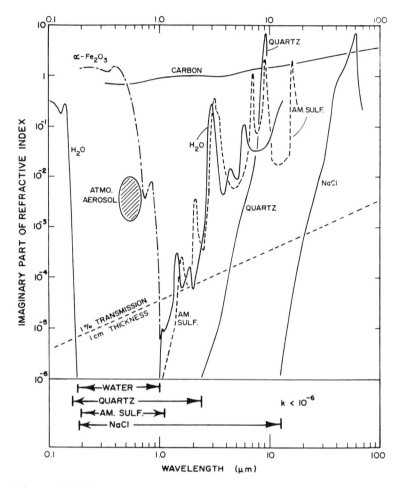

Figure 4.10 The imaginary part of the refractive index of several solids and liquids that are found as atmospheric particles (Bohren and Huffman, 1983).

absorbing material. The meaning of such an average value and its direct application to theories of particle scattering must be treated with caution. Measurement of the refractive index of a substance in a pure homogeneous slabform is difficult enough. These results highlight the complexity of estimating the refractive index when such material is broken up into small particles of heterogeneous material.

Both sets of optical constants (n, κ) and $(\epsilon'_r, \epsilon''_r)$ will be used in this book. Generally ϵ'_r and ϵ''_r are used in discussions of the microscopic interactions that take place in matter under the influence of radiation. The refractive index $m = n + i\kappa$ is related to the phase velocity and attenuation of plane waves in matter and tends to be used in discussions of wave propagation. We will now consider the relevance of the latter in the context of wave propagation through a homogeneous slab of material.

Excursus: The Refractive Index of Air and the Concentration of Atmospheric Oxygen

We would be correct in thinking that the mechanisms that produce polarizability in dense material also polarize the atoms and molecules of gases. In fact the problem of describing the refractive index of a gas is simpler than that for dense materials on at least two counts. First, we often do not need to distinguish between the local and applied fields for gases since the interactions between atoms and molecules in the gaseous state are generally negligible. Second, weak interactions imply less damping of oscillations. If we suppose that $\gamma = 0$ in (4.11), then it follows that

$$m^2 = 1 + N\alpha \tag{4.19}$$

where

$$\alpha = \frac{q^2}{\epsilon_o m} \frac{1}{\omega_o^2 - \omega^2}$$

If N is small enough as it is for a gas, then m is close to unity and we can write

$$m = 1 + \frac{1}{2}N\alpha$$

which states that the refractive index of the gas depends on the volume concentration of the gas. For a gas mixture, it follows that

$$m - 1 = \frac{1}{2}N\sum_i y_i(m_i - 1) \tag{4.20}$$

where y_i is the mole fraction of species i (the ratio of the number of moles of the gas in question to the total number of moles of the mixture), and m_i is the refractive index of species i at the density N. The quantity $m-1$ is known as the *refractivity*.

For air, (4.20) may be written in the form (Owens, 1967)

$$m - 1 = k_1\frac{p}{T}z_a^{-1} + k_2\frac{e}{T}z_w^{-1} + k_3\frac{e}{T^2}z_w^{-1}$$

where e is the partial pressure of water vapor, p is the partial pressure of dry air, T is the absolute temperature, the parameters k_i at a given wavelength are constants, and $z_{a,w}$ represent deviations from the ideal gas law (Owens, 1967). This refractive index depends on the humidity of the air and so the propagation of the pulse also depends on the humidity. This dependence is actually important for some remote sensing applications and is explored further in Problem 4.6.

An interesting application of (4.19) is described by Keeling (1988) who proposed that the relationship between refractive index and concentration could be used to measure small changes in the mole fraction of oxygen in air. Measurement of this fraction is extremely difficult but it is an important measurement in the study of the CO_2 budget of the atmosphere. Keeling's approach requires measurement of small changes in the refractivity of dry air at two wavelengths. The parameter relevant to his analysis is the refractivity ratio

$$r = \frac{m(\lambda_1) - 1}{m(\lambda_2) - 1}$$

and it follows from (4.20) that this ratio varies with species abundance according to

$$\delta r = s_i \delta y_i \qquad (4.21)$$

where

$$s_i = \frac{m_i(\lambda_2) - 1}{m_{air}(\lambda_2) - 1} \frac{r_i - r_{air}}{1 - y_i}$$

In deriving (4.21), Keeling assumes that the change in mole fraction δy_i of species i occurs in such a way that the relative abundances of all other species (i.e., the ratios y_j/y_k where $j, k \neq i$) remain constant. Measurement of the changes in r in dry air corrected for changes in CO_2 and other gases then yield the change in mole fraction of O_2. To measure the change in refractivity ratio δr, Keeling designed a novel interferometer illuminated by an Argon lamp filled with a trace of ^{198}Hg for which the positions of two lines at 4360 and 2537 Å are known to within 0.0001 Å. The refractivity is determined from the optical path difference between beams that pass through two cells, one filled with the flowing air sample and one filled with a reference gas. By combining these beams at a recombining plate, the path difference in the interferometer is determined by counting fringes formed by the interference of the beams. Keeling proposes that this approach is capable of yielding changes in the mole fraction

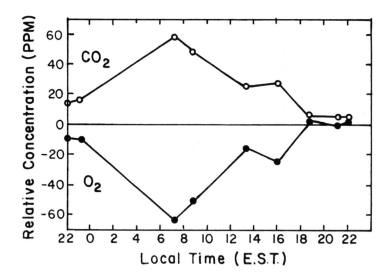

Figure 4.11 Diurnal trend in oxygen and carbon dioxide. The data are expressed as a difference between ambient air and a reference gas. The approach to near background levels of both gases after 1800 local time is due to the mixing by strong winds associated with the passage of a storm system (Keeling, 1988).

of oxygen with a precision of ± 2 ppm and used this measurement approach to show how changes in O_2 are anticorrelated with CO_2 which is expected for combustion processes (Fig. 4.11).

4.4 Reflection and Transmission at a Plane Boundary

Packing oscillators in dense materials like solids and liquids, in contrast to gases, dramatically impacts on the way radiation interacts with this type of matter. To appreciate how the arrangement of dipoles in bulk matter affects these interactions, it is useful to consider the reaction of electromagnetic radiation with a regularly arranged, large number of oscillators. We consider these reactions here in the context of reflection and transmission at a plane interface. This is more than a tutorial exercise as reflection and transmission properties of homogeneous slablike materials are widely used in remote sensing of land and ocean surface properties.

4.4.1 Propagation in a Homogeneous Slab

Every oscillator influenced by incident radiation emits an electromagnetic wave. Because the oscillators are packed together in a homogeneous slab, these emitted waves (which we refer to as "secondary waves") interfere with each other and with the incident wave in a very orderly way. For example, the wave incident on a slab of water is completely cancelled by interference with these secondary waves and is replaced by a wave which propagates along a direction that is different from the original incident direction. The individual emissions from the dipoles build up a wave referred to as the *refracted* wave propagating in a forward direction (i.e., into the material) and a wave referred to as the *reflected* wave propagating in a direction away from the material.

The refracted wave also travels in matter with a speed v which is different than the ordinary velocity of light c. One conceptual interpretation of refraction is offered by Huygens, who considered the wave front to slow down as it enters the slab, ultimately altering its direction of propagation (Fig. 4.12). The distance between wave fronts in water is shorter than in air, and this distance is proportional to the respective refractive indices. As we shall see in Chapter 5, the actual change in phase of the wave in a particle also gives rise to interference effects which dominate the way radiation reacts to the presence of particles.

Refraction at a plane surface is described in terms of an approximate relation referred to as *Snell's law of refraction* that can be simply written as

$$\frac{\sin \theta_i}{\sin \theta_r} = \frac{v_1}{v_2} = m_{21} \tag{4.22}$$

where θ_i and θ_r are the angles of incidence and refraction, respectively, relative to the surface normal, and v_1 and v_2 are the speeds of the wave in medium 1 and medium 2. The ratio of the speeds of the wave m_{21} is the relative refractive index, which indicates the relative difference in the speeds of the wave in medium 2 to medium 1. In a vacuum, v_1 is simply the speed of light, $m_1 = 1$ and $m_{21} = m_2$.

In describing the properties of the reflected wave, we need to consider a thin layer of oscillators near the surface (about as deep as half a wavelength). In this thin layer, the radiation scattered in the back direction (i.e., in a direction opposite to that of the incoming wave) is not completely cancelled by interference. The scattered

(a) (b)

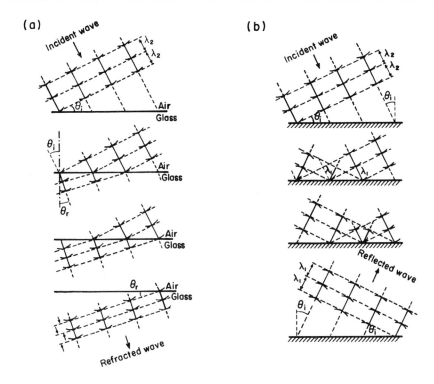

Figure 4.12 (a) Refraction and (b) reflection of plane waves at an interface with ($m_2 > m_1$).

waves superimpose to produce the reflected wave, which is also related to the refractive index of the material. We arrive at such a relationship in the following way. Consider an electromagnetic wave incident on a plane boundary as shown in Fig. 4.13. In describing the reflection and refraction of this wave it is convenient to think of each field as having a component parallel and perpendicular to the plane of reference. Maxwell's equations provide certain relations among the parallel and perpendicular components of the electric and magnetic fields on both sides of the interface. From these relationships, and by matching the components at the interface, reflection (and transmission) coefficients can be derived (e.g., Born and Wolf, 1964). These coefficients are

$$r_\ell = \frac{m_1 \cos \theta_r - m_2 \cos \theta_i}{m_1 \cos \theta_r + m_2 \cos \theta_i}$$

$$r_r = \frac{m_1 \cos\theta_i - m_2 \cos\theta_r}{m_1 \cos\theta_i + m_2 \cos\theta_r}$$

$$t_\ell = \frac{2m_1 \cos\theta_i}{m_1 \cos\theta_r + m_2 \cos\theta_i}$$

$$t_r = \frac{2m_1 \cos\theta_i}{m_1 \cos\theta_i + m_2 \cos\theta_r} \tag{4.23}$$

We now define the *reflectivity* and *transmissivity* as coefficients for the energy reflected and transmitted at the interface rather than as coefficients that relate to the reflection and transmission of wave amplitudes. The reflectivities for intensities are

$$\mathcal{R}_r = |\, r_r \,|^2$$

$$\mathcal{R}_\ell = |\, r_\ell \,|^2 \tag{4.24}$$

For an opaque slab, the emissivities are

$$\varepsilon_r = 1 - \mathcal{R}_r$$

$$\varepsilon_\ell = 1 - \mathcal{R}_\ell. \tag{4.25}$$

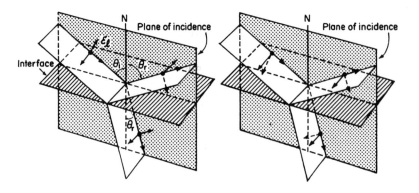

Figure 4.13 The electric fields of incident, reflected, and refracted waves relative to the plane of reference for the two polarization components.

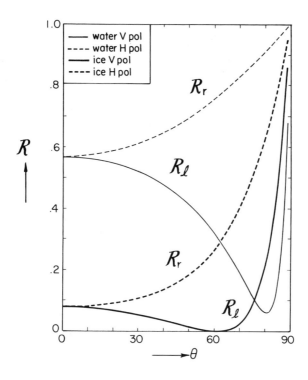

Figure 4.14 Reflection coefficient of water and ice at 19GHz. The angle corresponding to a zero parallel component is the Brewster angle.

There is one very important consequence of (4.23). Unpolarized waves incident on a homogeneous layer of material can become partially or totally polarized on reflection. In the example of microwave radiation, a portion of the unpolarized radiation emitted from the atmosphere is reflected by the ocean surface by an amount determined by (4.24), and another portion of this atmospheric radiation is absorbed and subsequently re–emitted by the ocean to the atmosphere. The components of this oceanic microwave emission are given by (4.25). Therefore, the amount of radiation emitted with a perpendicular polarization state is different from the component emitted with a parallel component for a plane water surface when viewed at an oblique angle. This property provides a way of discriminating surface water from rainfall in the microwave.

Figure 4.14 shows the amplitudes of the reflection coefficients for water as a function of incident angle. These were calculated using (4.23) and (4.24) with values of m at 19 GHz for both water and ice surfaces. According to these formulas, there is one special case corresponding to the situation in which $r_\ell = 0$, that is the reflected wave has no component parallel to the incident plane. This occurs when the direction of this would–be parallel component parallels the direction of the incoming beam. Since there is no oscillation along the latter direction, the reflected component can have no oscillation in this direction. According to (4.23), this happens when $m_2 \cos\theta_i = m_1 \cos\theta_r$. Using Snell's law, we obtain the condition that $\theta_i + \theta_r = \pi/2$ for the case of total polarization, and θ_i is referred to as *Brewster's angle* or the *polarizing angle*.

Another consequence of polarization by reflection, and one exploited in the design of certain types of systems (such as in some lidar systems), is that when circularly polarized light is incident on the slab, the reflected light is also partially circularly polarized but in the opposite sense. Right–handed polarized light transmitted from an instrument arrives back left–handed after reflection from particles in the atmosphere. In a lidar system as shown in Fig. 4.15, parallel–polarized light from the laser is transmitted directly through a transmit/receive (T/R) switch which is a plate oriented relative to the beam at the Brewster angle (assuming the plate is lossless). The beam transmitted to the atmosphere is first made circularly polarized by passing it through a quarter wave plate. The scattered radiation returns circularly polarized but with opposite handedness. This light, when passed through the same quarter wave plate, is transformed to perpendicularly polarized light, which is then reflected by the T/R switch into another part of the instrument for detection and further analysis.

4.4.2   Attenuation of Radiation in a Homogeneous Slab

In our discussion of oscillating dipoles we view absorption as a kind of dampening process. We will now consider the issue of absorption in the context of wave propagation. The general expression for a propagating wave is written in the form

$$\mathcal{E} = \mathcal{E}_o e^{i(\vec{k}\cdot\vec{r}-\omega t)} \tag{4.26}$$

where in a slab of condensed matter

$$\vec{k}\cdot\vec{r} = m\vec{k}_o\cdot\vec{r} \tag{4.27}$$

which is defined relative to the wavenumber $\vec{k}_o$ in a vacuum. Since m is complex in an absorbing slab, $\vec{k}$ too is complex. For the simplest

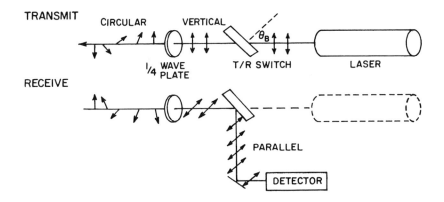

Figure 4.15 A simplified optical arrangement of a transmit/receive switch which uses the properties of light in a slab at Brewster's angle to discriminate between transmitted and returned light. Light is transmitted totally in one direction and reflected when entering from the other direction because of change in the handedness of polarization by particle scattering.

case of plane wave propagation along the z direction and with $m = n + i\kappa$ in (4.27), equation (4.26) may be written as

$$\mathcal{E} = \mathcal{E}_o[e^{-2\pi\kappa z/\lambda_o}]e^{i(n\vec{k}_o z - \omega t)} \tag{4.28}$$

The first of the exponential factors describes the rate at which the radiation is attenuated in the slab. The second exponential factor represents the oscillatory part of the wave and we observe that the real part of the refractive index determines the phase speed of the wave. The attenuation factor can be written in terms of a bulk absorption coefficient $\varrho = 4\pi\kappa/\lambda_o$ such that the attenuation of intensity of the radiation is

$$I = I_o e^{-\varrho z} \tag{4.29}$$

A useful and convenient way to interpret this attenuation is in terms of the penetration depth $d_I = 1/\varrho$, is the depth to which the incident intensity is reduced by $1/e$.

4.5   Selected Applications of Surface Remote Sensing

For practical reasons, the focus of this book is directed toward the remote sensing of the atmosphere. Appreciation of the importance of the ocean and its coupling to the atmosphere has loomed as one of the major developments in atmospheric research over the past two decades.[3] The advent of the Seasat–A and Nimbus 7 satellites, both launched in 1978, opened up new opportunities for evaluating various radiometric techniques for monitoring the ocean surface from space. Two applications that exploit the differences in the microwave optical constants of both water and ice will now be described.

4.5.1 Monitoring Sea Ice Extent

One of the more useful consequences of the large emissivity difference between ice and water in the microwave (the real parts of the dielectric constants are approximately 3 and 80, respectively) lies in its application to mapping sea ice. Understanding the variability and extent of sea ice is considered important to understanding climate variability.

To illustrate how sea ice is mapped using passive microwave measurements, consider the case of microwave radiation upwelling from the atmosphere at an angle θ. The equivalent microwave temperature for the given polarization state ℓ or r of such a (natural) surface can be expressed for a general observation angle θ as[4]

$$T_{\ell,r}(\theta) = \mathcal{T}r(\theta)[(1 - \mathcal{R}_{\ell,r}(\theta))T_{ocean} + \mathcal{R}_{\ell,r}(\theta)T_{sky}]$$

$$+T_{sky} \tag{4.30}$$

where $\mathcal{T}r$ is the transmission through the atmosphere along θ, and T_{ocean} and T_{sky} are the temperatures associated with emission from the ocean surface and atmosphere, respectively. The first term is

---

[3] The Tropical Ocean Global Atmosphere (TOGA) Program is a program under the auspices of the World Climate Research Program (WCRP) which focuses on improving the medium to long–range predictions of the tropical ocean–atmosphere system, including the El Nino–Southern Oscillation. One of the problems that emerged from TOGA was an appreciation of the complexity of the coupled system over the warm pool region of the equatorial west Pacific Ocean.

[4] A more complete derivation of this equation is presented in Chapter 7. Equation (4.30) is merely a simplified version of (7.21).

the emission from the surface, the second is the reflection of the sky emission at the surface, and the last term is the direct emission from the atmosphere to a sensor on a satellite.

Consider two adjacent surfaces, one referred to by the subscript i (ice) and the other by the subscript w (water). The temperature contrast between these surfaces (neglecting reference to polarization) is

$$\Delta T = T_i - T_w = (\mathcal{R}_i - \mathcal{R}_w)(T_{sky} - T_{water}) \qquad (4.31)$$

where for simplicity we take $Tr \approx 1$. It follows from (4.23) and (4.24) that for vertical incidence

$$\mathcal{R}_{i,w} = \left| \frac{m_{i,w} - 1}{m_{i,w} + 1} \right|^2 \qquad (4.32)$$

and assuming typical values of m for ice ($m = \sqrt{3}$) and water ($m = \sqrt{80}$), it follows that

$$\Delta \mathcal{R} = \mathcal{R}_w - \mathcal{R}_i = 0.57.$$

Given this estimate, together with values $T_{sky}(\approx 50K)$ and $T_i(\approx 273K)$, we then obtain $\Delta T \approx -127K$. Thus, the water surfaces appear substantially darker than the more emissive ice surface. It is this basic difference in the emission of water and ice that is used in satellite monitoring of sea ice.

Microwave brightness temperature distributions over the southern polar regions for four months of the year are shown in Fig. 4.16. In the ice covered area, the changes in brightness temperature are mainly due to changes in the nature and composition of the ice. In fact, sea ice is quite a complex dielectric material. Its composition includes all three phases of matter: a solid phase consisting of ice crystals and salt precipitates; a liquid phase consisting of brine solution; a gaseous phase in the form of air pockets in the ice.

Changes in the composition of sea ice occur continuously, producing changes in the optical properties of ice. Certain ice types have been identified as having distinguishable radiometric signatures: new ice, first–year ice, multiyear ice, and summer ice. A basic difference between first–year ice and multiyear ice is the presence of brine in the former and the replacement of brine by air pockets in the latter. The change in dielectric properties of ice associated with this change in composition produces radiometric effects such as that presented in Fig. 4.17a and b. Brine in first–year ice is more absorptive, so

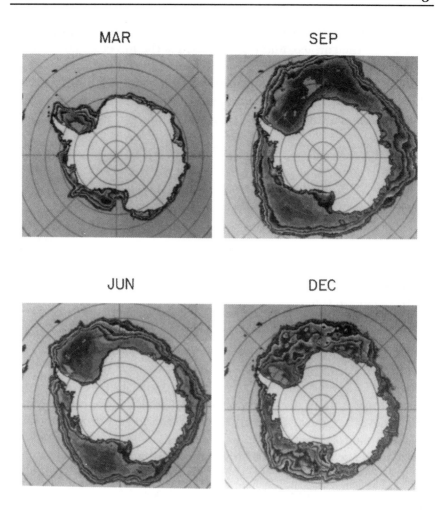

Figure 4.16 The extent of Antarctic sea ice varies from season to season. The satellite microwave images for the four months clearly show the minimum sea ice extent in March and the maximum sea ice extent in September. The grayscale variations represent different brightness temperatures due to different ice composition and thus age of the ice (from Gordon and Comiso, 1988).

this ice appears warmer than multiyear ice, especially for frequencies exceeding 30 GHz. This difference is further exaggerated by the air bubbles in multiyear ice. These increase the forward scattering of microwave radiation in multiyear ice, especially at 31 GHz as shown in Fig. 4.17a, leading to decreased values of brightness temperature. Tooma et al. (1975) used the multispectral behavior of brightness temperature as a means of discriminating different types of ice. They proposed an identification procedure based on a cluster approach in which temperature differences measured at two frequencies are correlated in the manner shown in Fig. 4.17b.

4.5.2   Measurement of Near—Surface Wind Speed

We know from observational experience that the microwave brightness temperature measured above a wind roughened water surface increases in a somewhat systematic way with increasing wind speed. This observation forms the basis for deriving wind speed from microwave brightness temperature data. However, the brightness temperatures are only an indirect indication of wind speed and the physical basis for the retrieval of surface wind is not firmly established.

A water surface becomes sloped when it is roughened by wind, and the direction of the reflected and refracted radiation at the surface is therefore tilted according to the amount of surface slope. This is what alters the brightness temperature of the surface and it has been empirically deduced that the amount of sloping of a wind–roughened ocean surface varies according to the wind speed. Cox and Munk (1955) demonstrate that the full range of slopes of waves is empirically specified by a Gaussian distribution with a variance

$$\sigma_{CM}^2 = 0.003 + 0.0048v \qquad (4.33)$$

where v is the wind speed in meters per second defined at a 20 m height above the sea surface. Cox and Munk use visual data to determine wave slopes so we might expect that only the more sloped surfaces influence microwave radiation. The slope variance corresponding to these longer wavelengths is therefore expected to be less than that determined by Cox and Munk from visible radiation. Wilheit (1979) empirically deduced that

$$\sigma_\nu^2 = (0.3 + 0.02\nu)\sigma_{CM}^2 \qquad \nu < 35\text{GHz}$$

$$\sigma_\nu^2 = \sigma_{CM}^2 \qquad \nu \geq 35\text{GHz} \qquad (4.34)$$

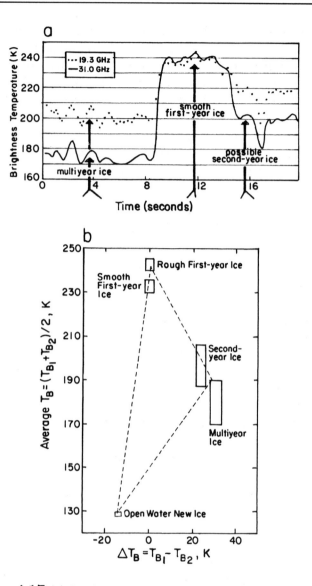

Figure 4.17 (a) Passive microwave profiles for possible second–year ice, smooth first–year ice, and multiyear ice obtained from radiometer measurements as the instrument was flown over an ice surface (from Tooma et al., 1975). (b) A plot of the brightness temperature averaged for two frequencies versus the temperature difference. The nadir brightness temperatures are measured at 19.3 GHz and 31 GHz, respectively (from Tooma et al., 1975).

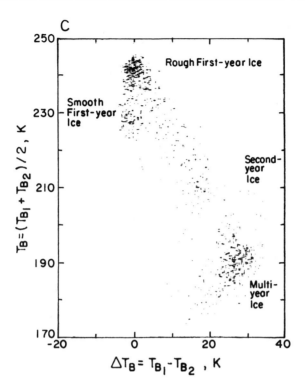

Figure 4.17 (Cont.) (c) An example of the data of (a) clustered on the bispectral diagram drawn in (b) (from Tooma et al., 1975).

With this information about wave slopes, the reflectivity of a rough surface is obtained simply by averaging the Fresnel relations over the distribution of wave slopes. We will not discuss this derivation in detail here, but we will examine a relatively simple method developed by Wilheit (1979). Consider our simple model of the interaction of microwave radiation at the surface as expressed by (4.30). On rearrangement, we obtain

$$\mathcal{R}_{\ell,r}(\theta)[T_{sky} - T_{ocean}] = T_{\ell,r}(\theta) - T_{ocean} \qquad (4.35)$$

when $T_{sky} = [1 - Tr(\theta)]T_{ocean}$ and the ratio of two polarization measurements is

$$\mathcal{F} = \frac{T_r - T_{ocean}}{T_\ell - T_{ocean}} = \frac{\mathcal{R}_r}{\mathcal{R}_\ell} \qquad (4.36)$$

A priori knowledge of the sea surface temperature and measurement of the parallel and perpendicular components of the polarized radiation upwelling from the surface provides us with an estimate of the surface reflectivity ratio. For a perfectly smooth slab observed slightly off Brewster's angle (say at 50 degrees), $\mathcal{R}_\ell < \mathcal{R}_r$ and $\mathcal{F} > 1$. As the surface roughens, $\mathcal{R}_\ell$ increases because this reflection now includes reflections associated with a range of angles defined over a spectrum of sloped wave surfaces. The quantity $\mathcal{F}$ is a measure of surface roughening and thus wind speed. Since it is a ratio, $\mathcal{F}$ is also independent of foam effects which are described shortly. Data from a microwave radiometer have been analyzed in this empirical way and compared to wind speeds measured by operational buoys. The results of this comparison are presented in Fig. 4.18, together with the relationship predicted using the Cox and Monk sea surface slope model (dashed curve). The approach is then to use the measured brightness temperatures and some value of the sea surface temperature to estimate $\mathcal{F}$ and then deduce the wind speed from the relation depicted in Fig. 4.18

Very strong winds also produce foam on the ocean surface and the emissivity and reflectivity of the surface is thus altered depending on the fraction of the surface covered by foam. The measured brightness temperature is found to increase with wind speed in a linear fashion as v exceeds about 7 ms^{-1}. This increase is thought to be largely due to the effects of foam on surface reflection. A number of models varying in complexity have been introduced to deal with the effects of foam on emissivity. Wilheit (1979) provides a simple, albeit empirical, representation of surface reflectivity according to

$$f = 0.006(1 - e^{\nu/7.5})(v - 7), \quad \text{for} \quad v \geq 7 \text{ ms}^{-1}$$

$$f = 0 \quad \text{for} \quad v < 7 \text{ ms}^{-1} \quad (4.37)$$

where ν is in GHz. This function represents the amount by which the surface reflectivity is reduced by foam. In the context of this model the emissivity of the partially covered surface is

$$e_{\ell,r} = 1 - \mathcal{R}_{\ell,r}[1 - f] \quad (4.38)$$

where the effects of roughness are included through $\mathcal{R}_{\ell,r}$ and the effects of foam through f. Here the effects of foam are assumed to be independent of polarization so that $\mathcal{F}$ remains independent of the amount of foam cover.

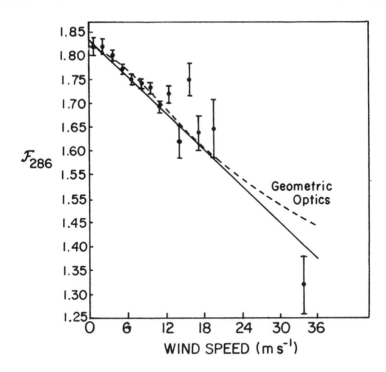

Figure 4.18 Observed values of the reflectivity ratio $\mathcal{F}$ as a function of wind speed. The relationship predicted by the Cox and Munk wave slope model is also shown by the dashed curve and a simple linear regression is given by the solid curve (Wilheit, 1979).

Excursus: Reflection from surfaces— Bidirectional Reflection Functions

Atmospheric remote sensing methods based on measurements of reflected sunlight require information about the reflection properties of the underlying surface. To set the stage for later discussion, these surface properties are introduced using the geometrical framework illustrated in Fig. 4.19. A surface is illuminated with a collimated beam of radiation of flux density F_o which is measured on the surface perpendicular to the direction of flow. For a flux F_o incident on a horizontal surface at an angle θ_o, the incident irradiance at the surface is $F_o \cos\theta_o$. An amount of this radiation is reflected by the surface along the direction $\vec{\xi}$ and confined to the solid angle $d\Omega$. We

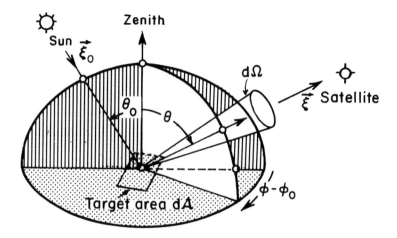

Figure 4.19 Radiation of flux density F_o incident at an angle θ_o on an area element dA and reflected into a solid angle $d\Omega$ along the direction $\vec{\xi}$ which is defined in terms of the zenith angle θ and the azimuth angle ϕ_r defined relative to the azimuth of the sun.

also note here [5] how $\vec{\xi}$ may be represented in terms of the polar angle θ and the azimuth angle ϕ.

If $I(\vec{\xi})$ is the intensity of the radiation reflected from the surface along the direction $\vec{\xi}$ and F_o is the incident flux density, then

$$R(\vec{\xi}, \vec{\xi_o}) = \pi I(\vec{\xi})/F_o \cos\theta_o$$

defines a surface reflectivity factor that is a function of two directions; the directions associated with incidence and reflection. This reflection function is sometimes referred to as the *bidirectional reflection function* (BDRF) or alternatively as an anisotropic reflection function. This function is a particularly important parameter in the analysis of satellite radiance data. For example, we need to know $R(\vec{\xi}, \vec{\xi_o})$ for all satellite observation angles $\vec{\xi}$ and all sets of incident solar angles $\vec{\xi_o}$ in order to convert the satellite measurement into a hemispheric flux measurement. Models of $R(\vec{\xi}, \vec{\xi_o})$ for a variety

---

[5] A more detailed discussion of the directional vector $\vec{\xi}$ and its relation to the angular pair (θ, ϕ) is given in Appendix 1.

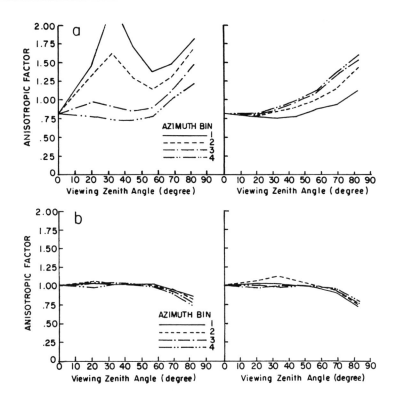

Figure 4.20 (a) The ERBE BDRF of a clear ocean scene as a function of viewing zenith angle for solar zenith angles between 25.84 and 36.87 degrees for eight azimuth angle bins (in degrees) defined as bin 1: $0 < \phi_r < 9$, bin 2: $9 < \phi_r < 30$, bin 3: $30 < \phi_r < 60$, bin 4: $60 < \phi_r < 90$, bin 5: $90 < \phi_r < 120$, bin 6: $120 < \phi_r < 150$, bin 7: $150 < \phi_r < 171$, bin 8: $171 < \phi_r < 180$. (b) Same as (a) but for a clear snow surface.

of surfaces (that is for a variety of different scenes) have been constructed for this purpose as part of the Earth Radiation Budget Experiment (ERBE) analysis method.

Figure 4.20 presents the BDRF of two surface types constructed from a large number of satellite radiance observations obtained for a given range of solar zenith angles. The reflection function is presented as a function of the viewing angle θ, for eight sets (bins) of azimuth angles ϕ_r. The large values of the BDRF over clear oceans

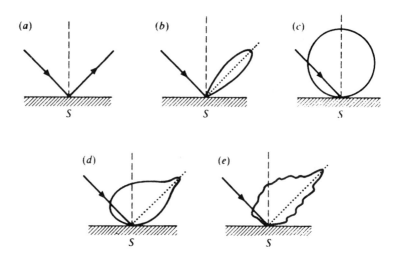

Figure 4.21 A schematic of various types of surface reflection expressed in the form of polar diagrams. The length of an imaginary line joining the point S on the surface to the lobe is proportional to the reflected intensity: (a) specular, (b) quasi–specular, (c) Lambertian, (d) quasi–Lambertian, and (e) complex.

(Fig. 4.20a) for bins 1 and 2 around the view angles of 30 degrees are a result of the sunglint on the ocean surface. Another feature of the ocean BDRFs is the limb brightening (i.e., the increase in reflectivity) toward the horizon when viewed away from the sun (bins 5 to 8). The BDRFs corresponding to snow surfaces are shown in Fig. 4.20b. The angular reflection properties of snow are seen to be more uniform than are the angular reflection properties of water.

We will now consider the BDRFs for two special surface types, the *specular reflector* which is perfectly smooth, and the *Lambertian reflector* which is perfectly rough. These are two important limiting cases of surface reflection. The specular reflector has the property that if the radiation is incident along the direction (θ_o, ϕ_o), then it will be reflected only into the direction $\theta = \theta_o$ and $\phi = \phi_o - \pi$. In this case, the BDRF is expressed by a delta–function as shown schematically in Fig. 4.21a. The ERBE BDRF for an ocean surface is more like Fig. 4.21b which we might consider to be a quasi–specular reflector. The Lambertian reflector represents the case of isotropic reflection from a surface (Fig. 4.21c). In this case the BDRF is defined by a constant value of unity and the ERBE BDRF for a snow surface closely approximates Lambertian reflection.

Excursus: Spectral Reflectance Properties of Land Surfaces

In addition to the angular characteristics of reflected sunlight, the spectral properties of surface reflection also contains useful information about the surface. Unfortunately, the land surface is a highly complex boundary, and so it follows that the spectral properties of that boundary are much harder to define than is the relatively simple case of the ocean surface. The Earth's surface can be heterogeneous in character covered by almost any combination of rocks and sand, grass and vegetation, swamps, mud, and water. The coverage of various types of surface material influences the spectral properties of the surface. Discussions of the spectral properties of the surface can be found in numerous references, some of which are cited at the end of this chapter. Because the land surface is so highly variable in character, analysis of its spectral properties tends to involve a large amount of empiricism; it is beyond the scope of this book to provide a detailed overview of this research.

Instead we focus on the visible and near–infrared spectral properties of certain types of vegetation. An important characteristic of the reflection by vegetated surfaces is the sharp transition in the reflection spectra at a wavelength of about 0.7 μm. Figure 4.22a presents an example of the reflection from corn, soybean, and bare soil. The presence of chlorophyll in vegetation leads to strong absorption at wavelengths shorter than 0.7 μm; the green color of vegetation arises from the rapid rise in the reflectivity in the green portion of the spectrum which continues into the near infrared. The amount of green biomass also affects the reflectance signature at wavelengths greater than 0.7 μm. Figure 4.22b illustrates this feature and presents the spectral reflectance of alfalfa at different stages of its life cycle. The bare field signature corresponds to the zero biomass curve.

A characteristic property of the reflection spectra of vegetation, namely the rapid change with wavelength near 0.7 μm, can be usefully represented in terms of an index

$$\mathrm{NDVI} = \frac{I(NIR) - I(VIS)}{I(NIR) + I(VIS)} \qquad (4.39)$$

which is referred to as the normalized difference vegetation index (NDVI) and is an index of biospheric activity more so than as a measure of vegetation amount. In the definition of this index, $I(NIR)$

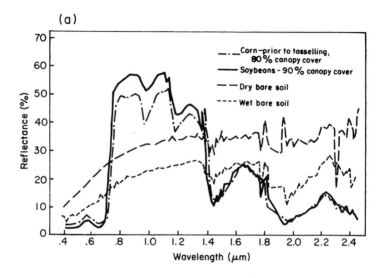

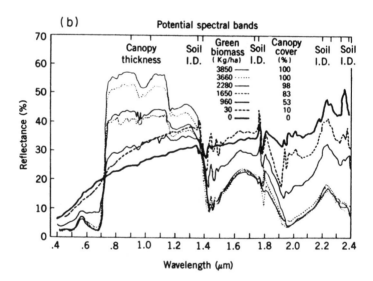

Figure 4.22 (a) The green edge in the reflectance spectrum of vegetation compared to reflection of dry and wet soils. (b) Progressive changes in the spectral response of a surface as the amounts of green biomass and percentage of canopy cover are changed (Short, 1982).

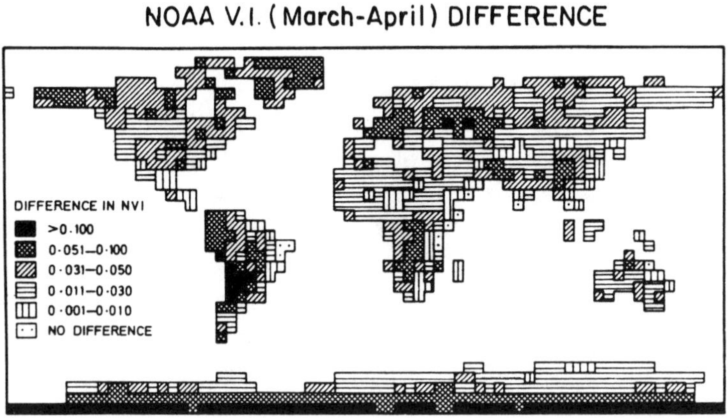

NOAA V.I. (March-April) DIFFERENCE

Figure 4.23 Global distribution of the absolute difference between the NDVI on a 5°×5° grid derived from the NOAA AVHRR satellite data for two consecutive months (from McGuffie and Henderson–Sellers, 1986).

and $I(VIS)$ are the intensities of sunlight reflected at a near–infrared and a visible wavelength, respectively. In this way, the NDVI can be used to relate the intensities measured by a satellite sensor to the amount of live vegetation in the field of view of the space sensor. Figure 4.23 is an example of a satellite archive of the NDVI. The map shows the absolute difference between the values of NDVI taken from a seven–day composite for two example weeks from two consecutive months, March and April 1983. The data are displayed as averages over a 5×5 degree area. The map indicates significant changes in the NDVI in the high latitudes over North America, over much of central Europe, and central and northern South America, and small changes in the NDVI for regions of sparse vegetation such as over deserts. The map also hints at the problems associated with interpretation of satellite imagary over polar regions covered by ice and snow among others. While large vegetation changes have occurred in some regions, some of the mapped differences may be due to noise introduced by clouds and aerosols, as well as by changing values of the bidirectional reflectances associated with different viewing and solar geometries. Aspects of these complications are discussed later in Chapter 6.

4.6   Notes and Comments

4.1. The topic of optical properties of condensed matter is a field of study that is too extensive to be covered adequately by a single volume, let alone a single chapter of a book. The aim of this chapter is to provide the reader with a conceptual interpretation of the optical properties that appear in various remote sensing applications. Some of the material given in Sections 4.1 and 4.2 are taken from Feynman et al. (1977, Volumes I and II). More detailed discussions on the topic are found in Bohren and Huffman (1983) and references located therein. A good elementary discussion of local electric field effects is provided in the book of Kittel (1971).

4.2 Basic presentation of the Lorentz and Debye models are given by Born (1965) and von Hippel (1954). One might hesitate to use the simple single oscillator model described in this chapter to represent the optical properties of solids and liquids. Bohren and Huffman (pp. 241–43) provide an impressive "textbook" example of how this one–oscillator model fits the reflection spectrum of α–SiC extremely well. They also show how a multiple oscillator model accurately represents the reflection spectrum of MgO (their Fig. 9.7, p. 246).

4.3. The optical properties of water and ice have been studied for centuries. Warren (1984) provides perhaps the most extensive review of the refractive index data for water extending from the ultraviolet to the microwave. Liebe et al. (1991) describe a model of the optical properties of water at microwave frequencies.

The delay of radio waves propagating in the moist atmosphere is a topic of some importance to microwave remote sensing (e.g. Elgered, 1992). For example, a microwave pulse sent from a satellite and returned to the satellite after being reflected from the ocean surface suffers a delay due to the underburden of water vapor. It is important to deduce this delay in order to map the sea surface with a precision of approximately 1 cm. In this space application, the correction for the effects of water vapor on the refractive index amounts to a delay of several centimeters. Walter (1992b) describes an experimental set up to measure these delays.

Bricaud and Morel (1986) introduce a method for deriving the refractive index based on the inversion of particle scattering measurements. The procedure utilizes the anomalous diffraction formulas that will be introduced in Chapter 5.

4.5. Microwave sensing of surface properties is discussed extensively in the reference text of Ulaby et al. (1986). The reader interested in a more detailed theoretical and observational account of ocean and land surface remote sensing, including wave slope effects on microwave emission, sea–ice remote sensing among many other topics, should consult that book. Another text relating to microwave sensing, primarily of surface properties, is that of Tang et al. (1985). A more recent example of a microwave land classification scheme using a space–borne microwave imager is described in the study of Neale et al., (1990).

There are even more empirical approaches to the estimation of surface wind speed (e.g., Goodberlet et al., 1989) and more sophisticated models of microwave surface emissivity (e.g., Petty, 1990).

The remote sensing of land surface properties is a topic extensively covered in several texts on remote sensing. For example, Elachi (1987) heavily emphasizes surface–property remote sensing as do the texts of Rees (1990) and Sabins (1982).

The application of AVHRR data to land–cover classification and the monitoring of vegetation dynamics was introduced by Tucker (1978, 1979). Spectral measurements of reflected sunlight are also extensively used in geological exploration using spectrometers flown on aircraft. The reflection spectra of geological materials is strongly influenced by crystal lattice structure and composition and the wide variety of geological material complicates the interpretation of these spectra.

4.7 Problems

4.1. Explain or interpret the following in just one or two sentences:

    a. Would you expect sound waves to obey the laws of reflection or refraction obeyed by light waves?

    b. Why does a diamond sparkle nore than a glass imitation cut to the same shape?

    c. Can (i) reflection phenomona or (ii) refraction phenomena be used to determine the wavelngth of light?

    d. Observed during daylight hours at 3.7 μm by a sensor on satellite, ice clouds appear darker than an equivalent thickness water cloud.

    e. Measurements of upwelling 19 GHz polarized microwave radiation incident on a satellite detector discriminates cold rain from surface water.

f. Microwave brightness temperatures of sea–ice vary according to the age of the ice.

g. The relative permittivity for water is strongly temperature dependent in the microwave region but not so in the infrared and visible regions.

h. The spectral reflection from vegetated surfaces increases sharply as the wavelength increases from 0.6 to 0.8 μm.

i. The reflection from snow surfaces is well approximated by a Lambertian surface whereas the reflection from a smooth water surface is not.

4.2. The density and refractive index of liquid benzene (C_6H_6) at 20°C are 0.88 gcm^{-3}, and 1.5 (for $\lambda = 0.589$ μm), respectively. Use the Clausius–Mossotti equation to compute the refractive index of benzene vapor at 20°C where its vapor pressure is 0.1 atm and also at 80°C, its boiling point. (*Hint: remember N is the number of molecules per m^{-3} and the equation for an ideal gas is $N = p/k_B T$*).

4.3. Using the parameters of the Debye relaxation model listed in the caption to Fig. 4.6 (assume room temperature to be 20°C), calculate and plot the spectrum from 0.2 to 5 cm^{-1} of the dielectric function for temperatures of 0° C and 30° C.

4.4. Show that

$$\epsilon'_r + i\epsilon''_r = 1 + \frac{4\pi\alpha_o N}{1+\omega^2\tau^2} + i\frac{4\pi\alpha_o\omega\tau N}{1+\omega^2\tau^2}$$

for orientational polarization where $\alpha_o = p_o^2/3k_B T$ and where the factor 4π is used instead of $1/\epsilon_o$. Plot these as a function of $\omega\tau$ and show that the maximum in ϵ''_r occurs at $\omega = 1/\tau$ which is also the frequency where ϵ'_r decreases most rapidly with increasing ω.

4.5. Why are some metals shiny? To answer this question, consider a metal with a refractive index m$=-i\kappa$. Calculate the reflection coefficient assuming normally incident light.

4.6. Rewrite the expression for the amplitude reflection coefficients given by (4.20) and express them as functions of θ_i and θ_r only (i.e., remove the explicit dependence on m).

4.7. A dielectric of thickness d and relative permittivity $\epsilon_r = 4$ is placed in air. Find the magnitude squared of its reflection coefficient for a plane wave normally incident upon it. Plot this as a function of kd where k is the wavenumber in the dielectric.

4.8. The wavelength of yellow sodium light in air is 5890 Å. What is its wavelength in glass whose index of refracrion is 1.52? What is its frequency in glass and find its speed in glass. If the speed of this light in a certain liquid is measured to be 1.92×10^8 ms^{-1}, what is its index of refraction in this liquid with respect to air?

4.9. The refractivity of air is represented by the formula

$$(m-1) \times 10^6 = k_1 \frac{p}{T} z_a^{-1} + k_2 \frac{e}{T} z_w^{-1} + k_3 \frac{e}{T^2} z_w^{-1}.$$

Calculate the water vapor partial pressure in mb for a difference in the time of arrival of two pulses expressed in terms of a path delay of 10 cm for one pulse at 36 GHz and another at 0.83 μm for a path of 13.35 km with $p=1013$ mb and $T = 293.16$ K. For 36 GHz, the constants are: $k_1 = 77.60$ Kmb^{-1}, $k_2 = 72$ Kmb^{-1}, and $k_3 = 3.754 \times 10^5$ K^{2}mb^{-1}; for 0.83 μm: $k_1 = 79.43$ Kmb^{-1}, $k_2 = 67.4$ Kmb^{-1}, and $k_3 = 0$ K^{2}mb^{-1} and you may assume the factors z_a and z_w are unity. (*Hint: The relevant speed of the pulse is the group velocity v_g and you may assume the refractive index relevant to the group velocity is that given*).

4.10. Water and ice possess refractive indices that are strongly frequency dependent and thus have a penetration depth that varies significantly from wavelength to wavelength. Calculate the depth of penetration d_I in a water and ice slab for the following wavelengths and refractive indices. What inferences would you make about scattering versus absorption processes by water and ice particles at each wavelength? [*Hint: reflect on your answer to 4.1(a)*].

Wavelength	Instrument	Refractive Index (n, κ)	
		water	ice
0.7 μm	AVHRR	(1.33,0)	(1.31,0)
1.6 μm	AVHRR	(1.317, 8×10^{-5})	(1.31,0.0003)
3.7 μm	AVHRR	(1.374,0.0036)	(1.40,0.0092)
10.8 μm	AVHRR	(1.17,0.086)	(1.087,0.182)
0.8 cm	k–Band radar	(8.18,1.96)	(1.789,0.0094)
10 cm	S-Band radar	(5.55,2.85)	(1.788, 0.00038)

4.11.   The dielectric properties of dry and moist soils in the microwave
        have also been exploited in methods of remote sensing of soil
        moisture. Using the relation between ϵ_r and the volumetric
        moisture content m_v of soil, calculate the change in nadir bright-
        ness temperature of dry soil (i.e., $m_v = 0$), moist soil (say,
        $m_v = 0.2$), and wet soil (say, $m_v = 0.35$). Assume a surface
        temperature of 300 K and the following for the dielectric con-
        stants:
        dry soil $\epsilon_r' = 3.2$, $\epsilon_r'' = 0.33$
        moist soil $\epsilon_r' = 9.5$, $\epsilon_r'' = 1.8$
        wet soil $\epsilon_r' = 20.8$, $\epsilon_r'' = 3.75$

4.13.   For the values of the refractive index of water corresponding to
        the three microwave frequencies given, calculate the wavelength
        of this radaition in water. What is the value of Brewsters angle
        at each frequency and how does this compare with the 54 degree
        view angle of the SSMI?

Frequency GHz	Refractive Index 0°C
10	7.08, -2.91
19	5.37, -2.96
37	3.93, -2.39
85	2.88, -1.47

4.13.   Consider a dielectric medium with $\epsilon_r = 2$ at a temperature of
        300 K. Plot the brightness temperatures T_ℓ and T_r as a func-
        tion of emergent zenith angle θ between 0 and 90 degrees. For
        which polarization does the brightness temperature possess a
        maximum, why is there a maximum, and at what value of θ
        does this maximum occur?

4.14.   A two–channel microwave radiometer is placed at the ground
        looking vertically upward. The instrument measures the radia-
        tion emitted by the atmosphere at 19 and 22 GHz. The objective
        is to retrieve the precipitable water from these measurements.
        First consider the radiation received at the surface as

$$T_\nu = T_\nu \; T_{\text{cosmic}} + (1 - T_\nu) \; T_{\text{sky}}$$

        where T_{cosmic} is the cosmic radiation (which you can ignore) and
        T_{sky} is the sky temperature and where T_ν is the transmission at
        frequency ν which has the form $T_\nu = \exp(-k_\nu u)$. For simplicity,

assume an isothermal atmosphere (perhaps reasonable for a water vapor atmosphere confined to the lowest layers). Derive the brightness temperature difference as a function of water vapor path length u stating any assumption you make. Calculate these brightness temperature differences for values of u varying from 5, 10, 15, 20, 25, 30, 35, and 40 kgm^{-2}. In your calculations use $k_{19GHz} = 3.75 \times 10^{-3}$ kg^{-1}m^2 and $k_{22GHz} = 9.121 \times 10^{-3}$ kg^{-1}m^2.

5

Macroscopic Interactions— Particle Absorption and Scattering

The tiny vibrations of the electrons of atoms and molecules, when exposed to electromagnetic radiation, are fundamental to the way radiation interacts with matter. The ability of condensed matter to produce these vibrations is characterized by the optical properties discussed in the previous chapter. Optical properties alone, however, do not provide us with a complete description of these interactions. Further complications arise by the way these tiny vibrators are arranged to make up particles. For instance, changing the macroscopic structure of a solid sheet of glass by smashing it into a fine powder, results in a dramatic change in the visual appearance of the material and, therefore, in the way radiation interacts with the glass, despite the fact that the composition of both configurations is identical. We can view a small piece of this powdered glass as being composed of a large collection of tiny oscillating dipoles as schematically portrayed in Fig. 5.1. These dipoles, oscillating at the frequency of the applied electromagnetic field, produce a secondary field that radiates out in all directions (we will call this the *scattered* field). Particle scattering is a complex problem because the secondary waves generated by each dipole also act to stimulate oscillations in neighboring dipoles. Thus radiation scattered out from the particle in a particular direction, say to point P, is a superposition of all the scattered wavelets and the precise details of this superposition depend on the way dipoles are arranged to form a particle.

Particle scattering is germane to many atmospheric remote sensing problems, from radar estimates of rainfall, to the estimation of cloud droplet size from the measurement of solar radiation reflected by clouds. This chapter discusses the scattering and absorption properties of particles and attempts to build up an understanding of these properties by first considering scattering by a single radiating dipole. The scattering properties of larger particles, formed as an arrangement of many individual dipoles, is considered later in the chapter.

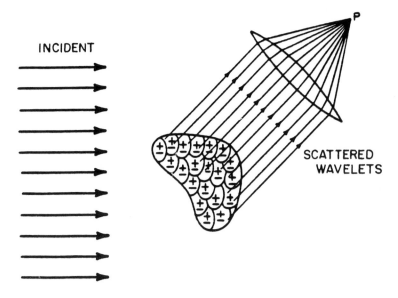

INCIDENT

P

SCATTERED
WAVELETS

Figure 5.1 The radiation scattered by a particle and observed at P results from the superposition of all wavelets scattered by the subparticle regions (dipoles) (from Bohren and Huffman, 1983.)

5.1 A Single Oscillating Dipole: Rayleigh Scatter

A logical place to begin a discussion of particle scattering is to consider the radiation scattered by a single oscillating dipole. We might consider an individual dipole as a small spherical particle, much smaller in comparison to the wavelength of the incident radiation, and refer to the scattering by such a particle as *Rayleigh scattering*, after Lord Rayleigh, who first described the properties of scattered sunlight by air molecules. In fact, an understanding of Rayleigh scattering has more than just heuristic appeal as it applies to many scattering phenomena of relevance to the atmosphere. Rayleigh scattering describes how visible radiation is scattered by atmospheric gases, how the longer wavelength infrared radiation is scattered by aerosol particles a few tenths of a micron in size, and how the even longer microwave radiation is scattered by cloud droplets and small rain drops. The properties of Rayleigh scattering have been exploited in both active and passive remote sensing applications. Rayleigh scattering is used directly in the analysis of weather radar (Chapter 8) and exploited in microwave sensing of cloud water (Chapter 7).

The backscatter of ultraviolet radiation by O_2 and N_2 is described by Rayleigh scattering and is important in the estimation of atmospheric ozone using observations from satellites (Chapter 6) as well as in the calibration of the backscatter of laser radiation in lidar systems (Chapter 8).

We start with a spherical particle small enough that the electric field within the particle is constant. This single spherical dipole oscillates under the influence of an oscillating electric field. This field has two orthogonal components: one parallel and the other perpendicular to a particular *plane of reference*. According to (4.1), each component of the electric field, in turn, induces a dipole moment parallel to the incident electric field. Figure 5.2 provides a reference to the geometry of the interaction under discussion. The components of the electric fields and the associated dipole moments are, respectively, represented by $\mathcal{E}_{\ell,r}$ and $P_{\ell,r}$ and are defined relative to the plane of reference. This plane contains the directions of propagation of both the incident and the scattered radiation.[1] The angle formed by these directions, represented as Θ in Figs 5.2 and 5.3, is referred to as the *scattering angle*.

We may visualize the dipole, defined by its moment P_r, as a line of fixed length separating two charges (as in Fig. 2.2). In this case, the line is perpendicular to the plane of reference and its length remains unchanged when viewed by an observer who moves around the particle on the plane of reference. Thus, the vertical (or perpendicular) component of the dipole moment and the vertically polarized component of the electric field is independent of Θ. Using the same arguments, the magnitude of the dipole that lies parallel to the plane of reference varies with the direction of observation and disappears completely when viewed end on at $\Theta = 90$ degrees. Based on these simple geometric considerations, the horizontal (or parallel) compo-

---

[1] The scattering plane remains arbitrary for both forward scattering ($\Theta = 0$ degrees) and backscattering ($\Theta = 180$ degrees) and thus requires an additional direction to define this plane. Radar studies, for example, define the scattering plane as the plane that contains both the transmitted and received beams and the local vertical. The polarization parallel to this plane is then referred to as vertical polarization even though this component might not actually point along the vertical direction. In some lidar applications, the scattering plane is defined as the plane containing both the transmitted and received beams and the direction of the (linear) polarization of the transmitted beam (refer to section 5.7 for further discussion).

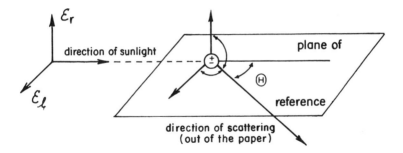

Figure 5.2 Geometry for scattering by a single dipole. The plane of reference and the orthogonol components of both the electric field and the dipole moment which lie parallel and perpendicular to the plane are shown. The scattering angle Θ is defined on this plane.

nent of the electric field varies as $\cos \Theta$. The two components of the electric field are therefore

$$
\begin{aligned}
\mathcal{E}_r &= \mathcal{E}_{or}\left[\frac{e^{-ik(R-ct)}}{R}\right]k^2\alpha \\
\mathcal{E}_\ell &= \mathcal{E}_{o\ell}\left[\frac{e^{-ik(R-ct)}}{R}\right]k^2\alpha\cos\Theta
\end{aligned}
\tag{5.1}
$$

where $\mathcal{E}_{or}$ and $\mathcal{E}_{o\ell}$ are the respective amplitudes of the fields incident on the particle, α is the polarizability introduced in Section 4.1, k is the wavenumber, and where radiation scattered by a single oscillating dipole is a spherical wave expressed by the factors in parentheses. The two intensity components of polarized radiation therefore take the form

$$
\begin{aligned}
I_r &= I_{or}k^4\,|\,\alpha\,|^2\,/R^2 \\
I_\ell &= I_{o\ell}k^4\,|\,\alpha\,|^2\cos^2\Theta/R^2
\end{aligned}
\tag{5.2}
$$

A special and important case applies to an unpolarized beam of radiation, like sunlight, scattered by small particles. Unpolarized radiation can be viewed as a mixture of two independent linearly polarized beams of the same intensity. Therefore, $I_{or} = I_{o\ell} = I_o$, and

$$
I = \frac{1}{2}(I_\ell + I_r) = I_o[1 + \cos^2\Theta]k^4\,|\,\alpha\,|^2\,/R^2
\tag{5.3}
$$

describes the scattered intensity of unpolarized radiation by small particles. The scattering pattern predicted by (5.3) is shown in Fig. 5.3a. The polarized components of the intensity, expressed by (5.2), are shown as dashed curves. The two independent beams that compose the incident radiation are scattered in unequal portions by a small sphere with the greatest difference occurring at scattering angles of 90 degrees where the parallel component of the scattered beam completely vanishes. In this way, the originally unpolarized incident light becomes polarized by scattering. At Θ =90 degrees, the scattered light is completely polarized along the direction perpendicular to the scattering plane whereas the light is unpolarized at Θ =0 degrees and has a mixture of parallel and perpendicular polarizations at other scattering angles. This amount of polarization is conveniently expressed in the form of the following ratio

$$LP(\Theta) = \frac{I_r - I_\ell}{I_r + I_\ell} \qquad (5.4)$$

which is referred to as the *degree of linear polarization* (refer to Table 2.1). Substitution of (5.2) into (5.4) provides

$$LP(\Theta) = \frac{\sin^2 \Theta}{1 + \cos^2 \Theta} \qquad (5.5)$$

which is shown in the form of a polar diagram in Fig. 5.3b. Since the scattered light is completely polarized at Θ =90 degrees, then $LP = 1$ in that direction. The points on the scattering plane for which $LP = 0$, in this case at Θ =0 degrees and Θ =180 degrees, are special and are referred to as *neutral points*.

Scattering of visible light by a pure molecular atmosphere is described by Rayleigh scattering. This scattering, according to Figs. 5.3a and b, polarizes sunlight. Based on the pattern displayed in Fig. 5.3b, the degree of polarization is expected to be greatest at 90 degrees to the sun and least when viewed directly toward and away from the sun. Actually the positions of the neutral points vary from these theoretical positions and the maximum degree of polarization never reaches 100 %. One of the reasons for these departures from theory is due to the effects of multiple scattering. Multiple scattered light is a mixture of waves that have been scattered a different number of times. Each time a scattering occurs, the radiation acquires a different degree of polarization depending on the direction of the scattering. Mixing multiply scattered waves leads to a

(a)

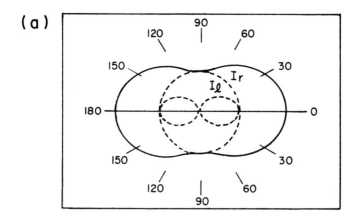

(b)

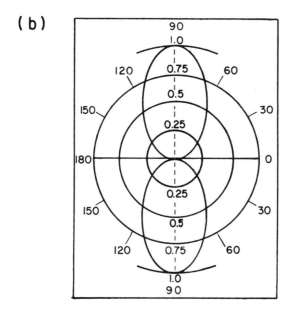

Figure. 5.3 (a) The scattering pattern of a Rayleigh particle for unpolarized incident radiation. The parallel and perpendicular components are shown as dashed curves. (b) The angular distribution of the degree of polarization of radiation scattered by a Rayleigh particle for the case of unpolarized incident radiation.

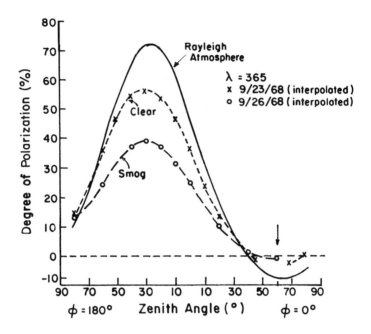

Figure 5.4 The degree of polarization of skylight as a function of zenith angle for a sun elevation of 30 degrees from measurements at $\lambda=0.365$ μm for clear and smoggy conditions. The more aerosol, the greater is the multiple scattering, and the less distinct is the maximum of the degree of polarization (from Coulson, 1988).

decrease in the degree of polarization. Figure 5.4 provides an example of how much multiple scattering reduces the maximum degree of polarization in a pure Rayleigh atmosphere. Other factors, such as the scattering by non–Rayleigh particles, also reduce the degree of polarization because of the intrinsic scattering properties of these particles and the multiple scattering between these particles.

Multiple scattering also alters the pattern of polarized skylight, producing three neutral points (Fig. 5.5) instead of the predicted two points. These three points are the *Babinet* point, the *Brewster*, point and the *Arago* point. The positions of these in the sky can be predicted using a suitable model of multiple scattering and are a function of the amount of scattering material in the atmosphere. Since multiple scattering is produced both by air molecules and by the aerosol particles in the Earth's atmosphere (thick clouds provide an extreme example of multiple scattering), then the change in

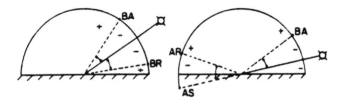

Figure 5.5 A schematic diagram showing the positions of the neutral points of Babinet (BA), Brewster (BR), and Arago (AR). The plane of the paper represents the principal plane of the sun and the antisolar point is denoted as AS.

the position of the neutral points from their normal positions determined for a pure molecular atmosphere is indicative of the amount of aerosol in the atmosphere (i.e., the *turbidity*, which is a concept discussed in more detail in Section 6.1). Measurement of the shift of the Arago point, for instance, has been proposed as a way of quantitively deducing atmospheric turbidity (Coulson 1983).

Excursus: Haidinger's Brush

It is difficult to believe that we can see with our naked eye, unaided by any instrument, both polarized skylight and the direction of this polarization. It does, however, require some practice and a certain amount of cunning. We begin by observing the twilight sky at 90 degrees from the setting sun. As we observe this region of sky almost directly above us for a minute or two, a kind of marble effect appears. This is followed by a remarkable feature known as "Haidinger's brush," which resembles the illustration in Fig. 5.6. It is a yellowish brush with a small blue cloud on either side. The yellow brush is perpendicular to the direction of light vibration and appears as a consequence of the polarization of skylight. Haidinger's brush takes some practice to see (Minneart, 1954) as the brushes fade rapidly due to the adjustments made by our eye. The reason the brushes appear is a result of the dichroism of the yellow spot of our retina.

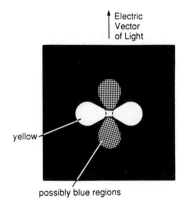

Figure 5.6 Haidinger's brush is a remarkable feature that can be seen in the blue sky and is an indication of the polarization of skylight. The light brush is yellowish and the clouds at each side are blue (Minneart, 1954). The brushes tend to get lost in a cluttered background, so viewing an illuminated white sheet of paper through a linear polarizer (like that used for sunglasses) works well. The size of the pattern is about 3/4 in. across when viewed at normal reading distance (about 18 in.).

5.2   Radiation from Multiple Dipoles

Whereas an understanding of the interaction of electromagnetic waves by small particles is relevant to many remote sensing applications, scattering by particles comparable in size to, or larger than, the wavelength of the incident plane wave is perhaps relevant to an even more diverse range of topics. In this section we will consider the radiation field scattered by two neighboring dipoles as an analogy to scattering by such particles. This provides us with a simple, but conceptually rich, view of particle scattering and offers a framework for extending the discussion to even larger particles composed of more than two dipoles.

5.2.1   Scattering by Two Isolated Dipoles

Consider the simple hypothetical case of two identical dipoles separated by some distance r as in Fig. 5.7. Suppose that these dipoles radiate independently of one another and consider radiation scattered along some direction defined by the scattering angle Θ. The phase difference $\Delta\phi$ between the two scattered waves at some point P is simply the result of the difference in the path length

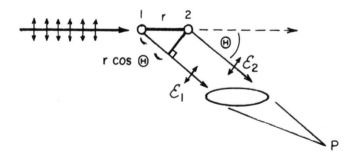

Figure 5.7 Two isolated dipoles emit waves in all directions. At some point P far from the "particle," these waves superimpose to create the scattered wave along the direction Θ. These waves either constructively or destructively interfere depending on their relative phase difference $\Delta\phi$.

traversed by one wave relative to another, that is

$$\Delta\phi = x\left(1 - \cos\Theta\right) \tag{5.6}$$

where $x = 2\pi r/\lambda$ is a quantity that is hereafter referred to as the *size parameter*. The two waves superimpose at some distance from the 'particle' to produce a field

$$\mathcal{E}_{1+2} = \mathcal{E}_1 e^{i\phi} + \mathcal{E}_2 e^{i\phi+\Delta\phi} \tag{5.7}$$

Suppose the detector that receives this scattered radiation has a time constant which is long enough that at least one full cycle of the oscillation is sampled. Then integration of (5.7) over a complete cycle produces a detected intensity of the form

$$I_{1+2} = B[\mathcal{E}_1^2 + \mathcal{E}_2^2 + 2\mathcal{E}_1\mathcal{E}_2 \cos\Delta\phi], \tag{5.8}$$

where B is a constant. If $\mathcal{E}_1 = \mathcal{E}_2$, then the two fields exactly cancel when $\Delta\phi = \pi, 3\pi, ...$; that is they are exactly out of phase. On the other hand, for $\Delta\phi = 0, 2\pi, 4\pi, ...$, the two fields reinforce each other exactly and are in phase. In both cases, the "scattering" is said to be *coherent*. But the phase difference depends on both the separation of the dipoles and the direction of scatter. For an arbitrarily shaped particle, the value of $\Delta\phi$ may vary in an arbitrary way except when $\Theta = 0°$ (i.e. except in the forward direction). The scattered radiation at $\Theta \neq 0°$ is thus a result of complex superpositions of waves of many different relative phase differences (Fig. 5.8a). The forward direction, however, is special as the fields at the detector remain in phase. For two dipoles the intensity in this direction is four times that of the incident wave (Fig.5.8b). Hence a particle of N such oscillators produces light in this forward direction that has an intensity proportional to N^2. It follows that scattering by large particles is predominantly in the forward direction and more so as the particle size increases. The situation is much more complicated in the backwards direction. Scattering does not increase nearly so rapidly with an increase in the number of dipoles since not all dipoles are in phase. Furthermore, the phase difference depends on the scattering direction and the scattering pattern (that is the scattered intensity as a function of Θ) undergoes excursions from a maximum in the forward direction through one or more minima as Θ varies and the number of these minima depends on the separation of the two dipoles.

The arguments presented here for two dipoles extend to a huge array of them and thus to a larger particle. The larger the particle, the more it scatters forward and the greater the forward to backward asymmetry. Also, the larger the particle, the more complicated is the pattern of scattered light. Returning to our simple two dipole particle, we note that when the phase difference between two scattered waves is an odd multiple of π, the two waves are out of phase producing a zero intensity at the scattering angle that produces these phase differences. We infer from this that as x increases the number of null points in the scattering pattern increases.

So far, the effects of interactions among the dipoles have been ignored. These interactions complicate matters but do not change the nature of our discussions substantially. The expectations about particle scattering arrived at here are supported by detailed calculations from a more precise theory (such as from the Lorenz–Mie theory described later). Example calculations from this theory are shown in Fig. 5.9 in the form of polar plots of the perpendicular and parallel components of the intensity of the light for scattering by different sized water droplets (represented by different values of x). For small size parameters ($x < 1$; not shown), the Rayleigh scattering patterns predicted by (5.2) are obtained. Deviations from Rayleigh scatter appear as x increases producing forward to backward asymmetries in both I_r and I_ℓ. The scattering pattern of a

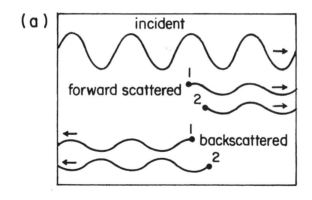

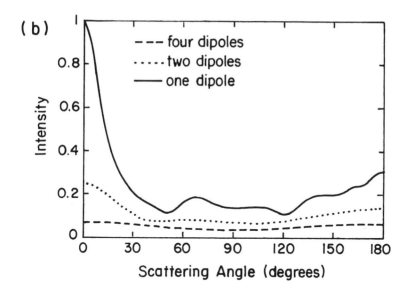

Figure 5.8 (a) Excited by an incident wave, two dipoles scatter in all directions. In the forward direction, the two waves are exactly in phase regardless of the separation of dipoles. (b) The greater the number of dipoles in the particle array, the more they collectively scatter toward the forward direction. For the example shown here all dipoles lie on the same line and are separated by one wavelength and interact with each other. The scattered intensity is obtained as an average over all orientations of the line of dipoles (from Bohren, 1987).

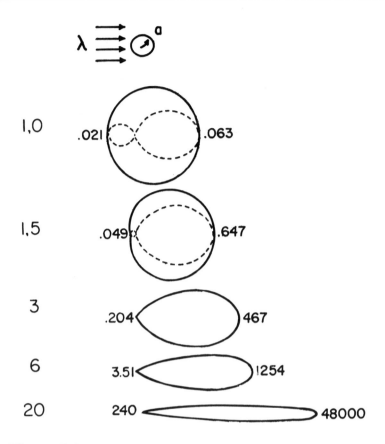

Figure 5.9 Polar plots of the scattered intensity for selected values of the size parameter. The numbers indicate relative magnitudes in the forward and backward directions. Note the scale change (Bohren and Huffman, 1983).

large particle represented in this diagram at $x = 20$ is dominated by a large forward scattering lobe and has a much more complicated shape. For 10μm cloud droplets, the size parameter x corresponding to visible light of wavelength 0.6μm is about 100 so scattering of this radiation is mainly in the forward direction.

Excursus: Phased–Array Antennas

A very practical application of dipole array scattering is provided in the example of a dipole antenna array. We mentioned in Chapter 2 how a single dipole, radiating at microwave frequencies, can be constructed. Let us suppose now that we have a linear array of N

independent dipoles as shown in Fig. 5.10a. The contribution of the nth radiator to the total far field along the direction Θ is

$$\mathcal{E}_n \sim (a_n e^{i\phi_n}) e^{-ikd_n \sin \Theta} \tag{5.9}$$

and the total field, which is the sum of each dipole field, is

$$\mathcal{E}(\Theta) \sim \sum_{n=1}^{N} a_n e^{i\phi_n - ikd_n \sin \Theta} \tag{5.10}$$

where a_n is the relative amplitude of the signal radiated by the nth element and ϕ_n is its phase. If all the radiators are identical in both amplitude and phase, and equally spaced, then the total far field emitted by the array is

$$\mathcal{E}(\Theta) \sim a e^{i\phi} \sum_{n=1}^{N} e^{-inkd \sin \Theta} \tag{5.11}$$

which we represent as a vectorial sum of N equal vectors each separated by a phase $\Delta\phi = kd \sin \Theta$. As seen in Fig. 5.10b, the sum is strongly dependent on the value of $\Delta\phi$. For example, with $\Theta = 0°$ and $\Delta\phi = 0$, all vectors add coherently just as in our example of forward scattering by particles. As $\Delta\phi$ increases, the vectors are out of phase relative to each other, leading to a decrease in the total sum. Where $\Delta\phi$ is such that $N\Delta\phi = 2\pi$, the sum is then equal to 0. This corresponds to

$$Nkd \sin \Theta = 2\pi \tag{5.12}$$

or

$$\Theta = \sin^{-1} \frac{2\pi}{Nkd} \tag{5.13}$$

At this angle, there is a null in the total radiated field and the shape of the field between the null points is referred to as the *lobe*. These occur when $N\Delta\phi = 2m\pi$ for integer values of m, leading to a series of nulls at

$$\Theta = \sin^{-1} \frac{2m\pi}{Nkd} \tag{5.14}$$

Peaks that lie between these nulls correspond to the condition

$$N\Delta\phi = 2m\pi + \pi \tag{5.15}$$

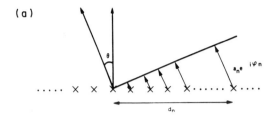

(a)

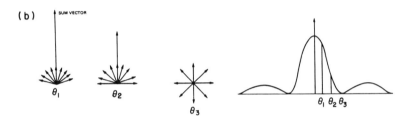

(b)

Fig. 5.10 (a) The geometry of a linear array of oscillators. (b) The radiation pattern emitted from a linear array of oscillators shown in vector form and as a function of angle Θ. One of the main challenges of antenna design is to minimize the side lobes and keep the main lobe as narrow as possible.

and to the angle

$$\Theta = \sin^{-1} \frac{(2m+1)\,\pi}{Nkd} \tag{5.16}$$

Suppose the linear array of dipoles is used to direct a beam to a target P that is not vertically above the antenna. Depending on the mechanism of combining the signal from each antenna, the array can be focused or unfocused on our target of interest. By engineering a specific phase difference between each dipole, which is done electronically in modern systems, we are able to steer the beam along a chosen direction.

We also pointed out previously that the radiation emitted from a dipole antenna will be linearly polarized. The orthogonal components of polarization are obtained when two linear arrays of radiating dipoles are placed perpendicular to one another. This configuration is typical of the wind profiling antenna systems used to measure atmospheric wind and is described further in Chapter 8.

5.2.2   The Discrete Dipole Approximation

Scattering of electromagnetic radiation by irregularly shaped particles is a topic of relevance to many atmospheric remote sensing problems. Unfortunately, our ability to solve Maxwell's equations precisely and thus compute the scattering and absorption properties of atmospheric particles is limited to very idealized geometries (i.e., spheres, circular cylinders, and spheroids). A number of approximate methods for calculating scattering by nonspherical particles exist. One approach, developed to study the scattering of starlight by interstellar dust grains, is the discrete dipole approximation (DDA). The basic concept, which was first introduced by Purcell and Pennypacker (1973), may be viewed as a generalization of the ideas presented earlier in our discussion of a two–dipole particle.

We can attempt to understand the concept of the DDA by considering the situation shown in Fig. 5.11 where an incident field $\mathcal{E}_{in}$ is scattered by a particle composed of many dipoles. A response field $\mathcal{E}_{res}$ is then created by this scattering which we imagine to be very far from the particle at P. $\mathcal{E}_{res}$ at P is the (vector) sum of the fields produced by the external source and the fields produced by each of the dipoles in the particle, that is

$$\mathcal{E}_{res} = \mathcal{E}_{in} + \sum_{dipoles} \mathcal{E}_{dipoles} \qquad (5.17)$$

The electric field at each dipole, defined by the position index j, is characterized in terms of the dipole moment

$$p_j = \alpha_j \mathcal{E}_{dipole,j} \qquad (5.18)$$

according to (4.2). The field at each dipole results from the superposition of the external field from our source at S with the fields associated with the other $N-1$ dipoles. That is

$$p_j = \alpha_j [\mathcal{E}_{in,j} - \sum_{j \neq k} A_{jk} p_{k,}] \qquad (5.19)$$

where $-A_{jk} p_k$ is the contribution to the electric field at dipole position j due to the dipole oscillating at position k. The precise form of A_{jk} does not need to concern us and the interested reader should

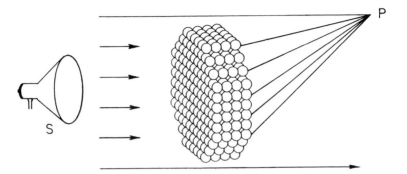

Fig. 5.11 A hexagonal plate particle represented by 480 dipoles. Incident light from a source S is scattered by the particle to produce a response field at P. This response field has a component that is represented by the superposition of the fields from each dipole. The complication arises from the mutual interference of the dipoles producing a distortion of the incident field.

attempt Problem 5.6 as well as consult the references given at the end of this chapter to learn more about the DDA.[2]

The DDA method solves (5.19) for all $p_j; j = 1, \ldots, N$ by rearranging this equation in matrix form which is explored further in Problem 5.6. The details are omitted here, but it is relevant to note that this involves solution of a $3N \times 3N$ complex matrix. As a rough guideline, the size d of the individual dipoles needed to represent the particle follows from

$$| m | \, kd < 1/3$$

---

[2] Equation (5.19) is actually a numerical simplification of the integral solution to Maxwell's equations. The summation term represents a discretization of the integral

$$\nabla \times \nabla \times \int_V [\epsilon(\vec{r'}) - 1]\mathcal{E}(\vec{r'})G_o(\vec{r}, \vec{r'})dV'$$

where $G_o(\vec{r}, \vec{r'})$ is the free space Green's function, the integral is carried out over the entire volume of the particle, and the curl operator, $\nabla \times$, is from standard vector calculus (e.g., Arfken 1985).

(a)

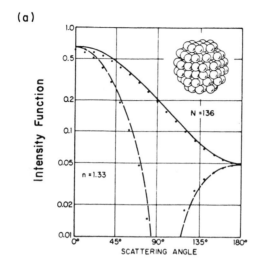

(b)

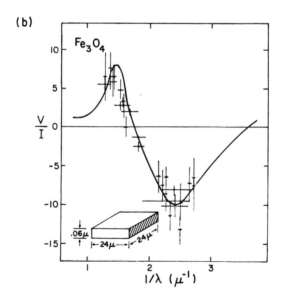

Figure 5.12 (a) Scattered light (arbitrary units) as a function of scattering angle for a dielectric sphere with $x = 1.5$, m= 1.33. The curves correspond to Lorenz–Mie solutions, the points to DDA solutions for a "sphere" consisting of 136 dipoles. Solid curve – scattering normal to electric field; dashed curve – scattering normal to the magnetic field (after Purcel and Pennypacker 1973). (b) DDA derived circular polarization V/I for magnetite platelets of the size given. Plotted are circular polarization observations (after Shapiro, 1975).

or

$$d < \frac{\lambda}{6\pi \mid m \mid}$$

where λ is also expressed in micrometers. For example, scattering of 0.5 μm radiation by a water sphere of radius 1 μm can be calculated to better than 5% accuracy when represented by a "sphere" of 65,400 dipoles.

The matrix that must be inverted to solve (5.19) rapidly becomes large as the size parameter increases. An example of a scattering calculation for a sphere composed of 136 dipoles is shown in Fig. 5.12a. The curves in this diagram provide a measure of the scattered intensity as a function of scattering angle which is defined in two planes: one normal to the electric field (solid curve); the second normal to the magnetic field (dashed curve). Other parameters are defined on the diagram. The curves represent the scattering from spheres as derived from the Lorenz–Mie scattering solution (Section 5.6) and the points are the results of calculations obtained using the DDA approach for the sphere represented by the cluster of dipoles shown in the inset. Figure 5.12b is a diagram of the degree of circular polarization (refer to the discussion of Table 2.1 for the definition of this parameter) derived from the DDA calculations for a magnetite platelet shown in the inset and plotted as a function of wavelength. These calculations are compared to actual observations of the circular polarization of starlight. One of the most important properties of interstellar scattering of starlight is the way this light is polarized. For dust grains to polarize light, they must be nonspherical in shape and collectively aligned in some organized fashion. The predicted behavior of V/I by DDA is at least consistent with the idea that these dust grains are flat, oriented platelets of magnetite.

5.3   Particle Extinction

The special properties of forward scattering have already been noted. Scattering in this direction is also a measure of the totality of the interaction between the radiation and the particle. To illustrate how forward scattering is used in such a way, suppose that particles exist between a source of radiation like the sun and a detector like that flown on a satellite (Fig. 5.13). The detector receives less radiation when particles or molecules exist along the path to attenuate the radiation than when no particles or molecules are present. The incident beam is said to have undergone *extinction* or *attenuation* when

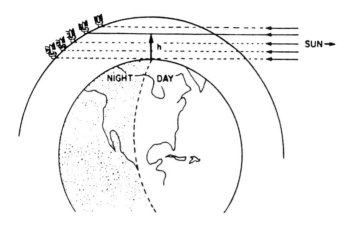

Figure 5.13 A sunset as viewed by the SAGE instrument on the Nimbus 7 satellite. The instrument starts to scan the solar disk at a tangent height h and follows it down until the sun disappears. Different layers are successively sampled during the event (McCormick et al., 1979).

such a reduction occurs. A number of both active and passive remote sensing methods make use of direct extinction by both particles and molecules as discussed further in Chapter 6.

Excursus: Extinction–It is Black and White

The experimenter who observes radiation scattered in the forward direction, as for the example considered earlier, cannot determine if the radiation is decreased because it is absorbed or decreased because it is scattered. A simple illustration of this elementary point is well described by Bohren (1987) and is highlighted in Fig. 5.14. One cannot distinguish between the images of two water–filled glass petri dishes projected on a screen, yet their darkness arises from different mechanisms. Light incident on the inky water is attenuated mainly by absorption, whereas light incident on the milky water is mostly scattered. It is only by looking at the dishes that this difference between them becomes apparent. An important scattering parameter that helps quantify these differences is the *single scattering albedo* $\tilde{\omega}_o$. This parameter is the ratio of the amount of scattering that attenuates the light to the total extinction (absorption *plus* scattering). For the milky water, $\tilde{\omega}_o \approx 1$ since light is primarily scattered

Figure 5.14 The images of two water filled petri dishes projected on a screen are identical yet their darkness arises from different mechanisms. Light incident on the inky water is attenuated mainly by absorption whereas light incident on the milky water is attenuated mostly by scattering. It is only by looking at the dishes that this difference becomes apparent.

in all directions from the dish. On the other hand, $\tilde{\omega}_o \approx 0$ for inky water as little light is scattered and most of the extinction occurs through absorption. We will see how the parameter $\tilde{\omega}_o$ is fundamental to problems of multiple scattering and, for example, to the remote sensing of clouds from measurements of reflected sunlight.

5.3.1 Efficiency Factors and Cross–Sectional Areas

Particle extinction is conveniently defined in terms of a quantity called the *extinction efficiency* Q_{ext}. One way of visualizing this quantity is to consider radiation as a stream of photons that flow into a volume containing the scattering particles. Each particle within the volume blocks a certain amount of radiation resulting in a reduction of the amount of radiation directly transmitted through the volume. The reader must be cautioned at this point as this is not an entirely correct depiction of how extinction takes place. For now, though, this simple view of extinction is a convenient visualization of particle extinction.

The extinction of light as it traverses a volume of spheres can be expressed in terms of a cross–sectional area C_{ext} which is generally different from the geometric cross–sectional area of the particle. For spherical particles of radius r, the definition of Q_{ext} then follows as

$$Q_{ext} = \frac{C_{ext}}{\pi r^2} \tag{5.20}$$

When C_{ext} exceeds the value of the geometrical cross–sectional area of the particle, $Q_{ext} > 1$ and more radiation is attenuated by the particle than is actually intercepted by its physical cross–sectional area. Since this extinction occurs by absorption or by scattering or by a combination of both, it follows that

$$Q_{ext} = Q_{abs} + Q_{sca} \tag{5.21}$$

Given the discussion so far, we expect that Q_{ext} depends on the refractive index of the material, the wavelength of radiation, and the size and shape of the particle. Figure 5.15 highlights some of these dependences in the form of a plot of Q_{ext}, calculated from Lorenz–Mie theory, as a function of the size parameter x for a water sphere illuminated by light of a wavelength of 0.5 μm. Somewhat obvious are the large maxima and minima with a superposition of finer scale variations (referred to as ripples). Another familiar phenomenon that may be deduced from Fig. 5.15 is the *reddening* of white light as it passes through a collection of small particles. For fixed refractive index, reddening is depicted by the rapid rise in extinction as x increases from near zero as the wavelength of the radiation decreases and is a general characteristic of nonabsorbing particles smaller than the incident wavelength. For this case, blue light is extinguished (scattered) more than red light, leaving the transmitted light reddened in comparison to the incident light. This reddening is a phenomenon that is not limited to sunlight in the Earth's atmosphere. Interstellar dust particles also redden starlight transmitted through a cloud of such particles. Extinction is obviously highly dependent on the size of the particle.

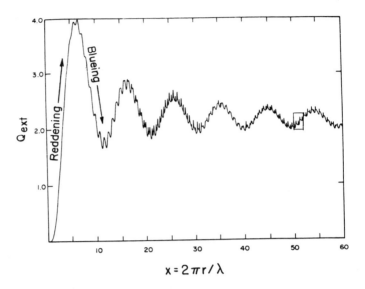

$$x = 2\pi r / \lambda$$

Figure 5.15 Extinction efficiency for water droplet in air calculated for $\lambda = 0.5$ μm as a function of size parameter x. There are two ways of presenting results of this type. For the example shown, the size parameter varies because particle size varies and the wavelength is kept fixed. The size parameter can also be varied by changing the wavelength while fixing the size of the particle. The results are not the same because as wavelength varies so does the refractive index, and extinction depends not only on the size parameter but also on the refractive index (adapted from Bohren and Huffman, 1983).

5.3.2   Extinction by a Cloud of Many Particles

The opposite spectral effect of reddening is the *blueing* of transmitted white light. This occurs when the extinction decreases with increasing x on the high x side of the extinction peaks shown in Fig. 5.15. Unlike reddening, this blueing phenomenon is highly dependent on the character of the particle size distribution and occurs rarely: "once in a blue moon." In fact, this extinction feature and others that depend on particle size are obscured, if not totally obliterated, when the extinction is determined from observations of light scattered by a small volume of air containing particles of a variety of sizes. Under atmospheric conditions, the intensity of radiation scattered by such a volume of particles may be simply obtained as the addition of the intensities of light scattered by individual particles. Particles in the atmosphere are typically separated by a distance that is much larger than the size of the particle[3] and it can be assumed that there is no correlation between the scattered waves (i.e., phase shifts are randomly distributed and change rapidly so that interference between the light scattered by different particles does not occur).

In Chapter 1, we introduced a convenient mathematical expression for a particle size distribution which we write in the form

$$n(r) = \text{constant} \quad r^{(1-3b)/b} e^{-r/ab} \tag{5.22}$$

where $n(r)$ is the number of particles per unit volume with a size that falls in the radius range r to $r + dr$ ($n(r)dr$ has units of the inverse of volume) and the parameter b defines the variance of the distribution. Figure 5.16 shows the effect of increasing b on the extinction of the polydispersion determined from a series of Lorenz–Mie calculations. This extinction efficiency is defined as the ratio $\bar{Q}_{ext} = \int_0^\infty n(r)Q_{ext}(r)dr/N_o$ where N_o is the total particle concentration per unit volume. The very fine ripple structure in extinction for the monodispersed cloud (i.e., $b = 0$) disappears as b is systematically increased from zero and the interference structure (i.e., the broad maxima and minima) eventually fades away as the distribution widens. The only remaining features for the widest distributions chosen in this illustration ($b = 0.5$) are reddening at small size parameters and the asymptotic approach to the limiting value 2. The pertinence of both limits to the remote sensing of clouds is discussed shortly.

A dramatic example of the smearing of the interference patterns caused by the superposition of the scattering by different particle sizes is given in Fig. 5.17 which is taken from the work of Roth et al. (1991). In that study, the angular scattering pattern produced by droplets in the path of a laser is recorded by a charged coupled array device. The upper panel of Fig. 5.17 is the intensity distribution produced by a single droplet where the incident beam comes from the right hand side of the diagram and hits the drop in the

---

[3] The actual volume of air that is typically occupied by cloud droplets is estimated as $V = N_o \frac{4}{3}\pi r^3$ where N_o is the number of droplets per unit volume of air and r is the radius of the spherical droplet. Using the values of $N_o = 100$ droplets cm^{-3} and $r = 10\mu$m, which is typical of water clouds, $V \sim 10^{-7}$.

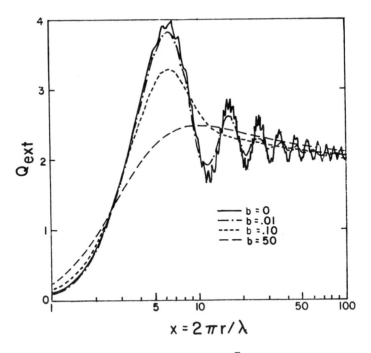

Figure 5.16 The extinction efficiency, $\bar{Q}_{ext}$, as a function of the effective size parameter $x = 2\pi a/\lambda$ for the values of effective variance b given. Mie theory was used with a refractive index $n = 1.33, \kappa = 0$ (after Hansen and Travis, 1974).

position indicated by the white spot. Light is primarily scattered in the forward direction and many fringes are observed between $\Theta = 0°$ and 180°. We deduce that the size parameter relevant to the experiment shown is large. Rainbows are also visible on the right–hand side of the photograph which represents the backward hemisphere. The lower panel shows what happens to the scattered light when scattering patterns by different sized droplets are superimposed on one another. This superposition smears the fringe pattern of single droplets leaving only the rainbows, the enhanced forward scattering and some slight brightening in the backward direction. Parameters that contain the smearing effect of the size distribution are the volume coefficients for extinction, scattering, and absorption defined as

$$\sigma_{ext,sca,abs} = \int_0^\infty \pi r^2 n(r) Q_{ext,sca,abs}\, dr \qquad (5.23)$$

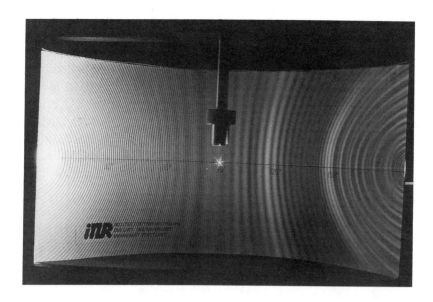

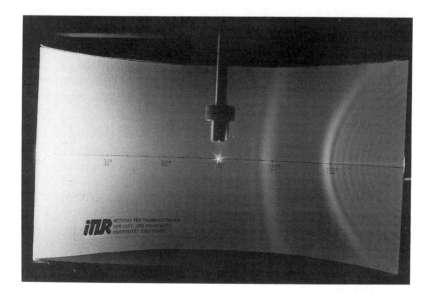

Figure 5.17 (a) The intensity field of light scattered by droplets of a single size with a size parameter ≈ 250 (upper panel). The laser enters from the right and the white spots marks the position of the drop. (b) The superposition of the scattered intensity by droplets of different sizes (from Roth et al., 1991).

5.3.3 The Rayleigh Limit and Its Application to Cloud Liquid Water Retrieval

The reddening of white light by the scattering of small particles is one of the observable consequences of Rayleigh scattering. In fact, the term Rayleigh scattering tends to be used to refer to extinction by all particles that are much smaller than the incident wavelength even though such terminology lacks historical accuracy. Nevertheless, we adopt this loose terminology in reference to extinction as $x \rightarrow 0$. In this limit, the efficiency factors for absorption and scattering are shown to be (van de Hulst, 1957)

$$Q_{abs} \approx -4x\Im m \left[\frac{m^2 - 1}{m^2 + 2} \right]$$
$$Q_{sca} \approx \frac{8}{3}x^4 \mid \frac{m^2 - 1}{m^2 + 2} \mid^2$$

(5.24)

and from these expressions we deduce that Q_{abs} is directly proportional to particle radius (through x) whereas Q_{sca} varies as the fourth power of particle radius. Since $x \ll 1$ in the Rayleigh limit, it generally follows that $Q_{abs} > Q_{sca}$ for m complex.

The retrieval of cloud water from measurements of microwave emission exploits this particular property of particle extinction (Chapter 7). The physical basis for these methods is as follows. Scattering by the small cloud droplets (remember these droplets are a few microns in size compared to wavelengths of, say, a few centimeters) is negligible compared to absorption. Substitution of (5.24) into (5.23) produces

$$\sigma_{ext} \approx \sigma_{abs} \approx \frac{-8\pi}{\lambda}\Im m \left[\frac{m^2 - 1}{m^2 + 2} \right] \int_0^\infty n(r)\pi r^3 dr \qquad (5.25)$$

Therefore the volume extinction coefficient is a function of the cloud liquid water content,

$$\ell = \frac{4\pi\rho_L}{3} \int_0^\infty n(r)r^3 dr$$

where ρ_L is the density of water and the optical depth of the cloud is

$$\tau = \int_{\Delta z} \sigma_{abs} dz = \frac{-6\pi}{\lambda\rho_L}\Im m \left[\frac{m^2 - 1}{m^2 + 2} \right] \int_{\Delta z} \ell dz \qquad (5.26)$$

where the latter integral is the liquid water path (W) of the cloud. Given that we know the composition of the cloud particles and thus their refractive index and density ρ_L, and given that we know the relationship of the radiation emitted by clouds to the optical thickness (this is discussed in more detail in chapter 7), then the vertically integrated liquid water path can be determined without prior knowledge of the cloud droplet size distribution.

5.3.4 The Large Particle Limit and the Extinction Paradox

Another important feature of the $Q_{ext} - x$ spectrum is the tendency for the extinction Q_{ext} either to oscillate around the value of 2 as $x \to \infty$ as illustrated in Fig. 5.15 or to converge to the value of 2 as in the cases of Fig. 5.16. This behavior is referred to as the *extinction paradox*. Why is it a paradox? Intuition suggests that if we consider extinction as just the radiation that is blocked by the particle, then the extinction cross–section is just the shadow projected by the very large particle. This geometrical view of extinction implies that the limiting value of Q_{ext} is 1 and not 2. However, no matter how large the particle, it still has an edge, and in the vicinity of the edge rays do not behave according to simple geometrical arguments. The energy removed from the forward direction can be thought of as being made up of a part that represents the amount blocked by the cross–sectional area of the particle and another part that is diffracted around the particle's edge.

The diffracted amount eventually fills in the shadow area when viewed far enough from the particle. The total amount removed from the incident beam by diffraction is therefore also characterized by the particle cross–sectional area. The net result is that an amount twice the cross–sectional area of the particle is scattered out of the incident beam. The important point we learn from both this, and from further discussion later, is that particle extinction is both a process of blocking of light as well as actually a result of more subtle interference effects.

Figure 5.18 offers a graphic example of the extinction paradox and emphasizes how diffraction fills in the shadow area with light. This diagram shows the diffraction pattern observed when a small spherical ball bearing is placed between the telescope and a light source that is far removed from the ball. The bright spot in the middle of the pattern is referred to as the *Poisson spot* and shows how

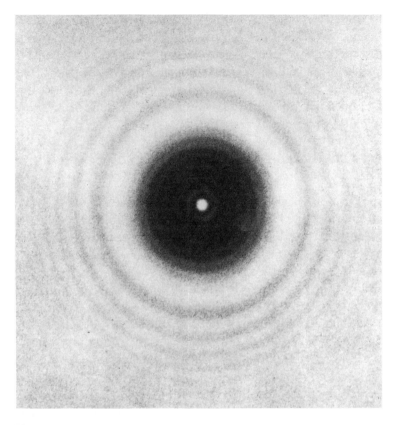

Figure 5.18 Illustration of the extinction paradox showing the
diffraction pattern created when observing a light source obscured at
some distance away by a ball bearing of 10mm diameter.

the light is diffracted into the shadow area of the particle, ultimately
filling this area.

   An important consequence of the large particle extinction limit
is that optical thicknesses of many clouds at solar wavelengths are related directly to the vertically integrated liquid water path in a manner analogous to extinction of microwave radiation by cloud droplets.
The optical thickness is

$$\tau = \int_{\Delta z} \sigma_{ext} dz \tag{5.27}$$

For wavelengths corresponding to $x \gg 1$ (this occurs when submi-
cron wavelength solar radiation is scattered by cloud particles a few
microns in size), we then invoke the asymptotic limit $Q_{ext} = 2$ so

that

$$\tau \approx 2 \int_{\Delta z} [\int_0^\infty n(r)\pi r^2 dr] dz \qquad (5.28)$$

Using the definition of the effective radius (see Section 1.6), and the cloud liquid water content ℓ, the optical thickness then becomes

$$\tau \approx \frac{3}{2\rho_L} \int_{\Delta z} \frac{\ell}{r_e} dz$$

Suppose that r_e does not vary significantly throughout the depth of the cloud, then τ simplifies to

$$\tau \approx \frac{3W}{2\rho_L r_e} \qquad (5.29)$$

This is a relationship that seems to apply for clouds illuminated by visible and near–infrared radiation. Unlike the case for microwave radiation, the optical thickness predicted by (5.29) depends on particle size. Attempts have been made to use measurements of the spectral reflection of solar radiation by clouds, together with some a priori relationship between optical thickness and reflectance, to obtain estimates of both W and r_e. These methods are examined in further detail in Chapter 6.

Excursus: Ship Tracks

Concentrations of cloud droplets are determined, to a large extent, by the concentration of cloud condensation nuclei (CCN) in air. Anthropogenic sources of pollution affect these nuclei concentrations. It has been proposed that the climatic impact of pollution might be amplified through its effect on CCNs which in turn alters the cloud droplet concentrations and subsequently alters the cloud albedo. A unique opportunity for testing this hypothesis is provided in the example of ship tracks observed in satellite visible imagery. Under certain conditions, ships influence the structure of shallow cloud layers and the reflection properties of these clouds. Radke et al. (1989) report aircraft measurements of cloud droplet concentrations and other microphysical properties in low level water clouds which are affected by ship stack effluents. An example of data collected during one aircraft flight is shown in Fig. 5.19. These are data on the droplet concentration, liquid water content, interstitial aerosol particle concentration, and radiative fluxes along the flight track. A

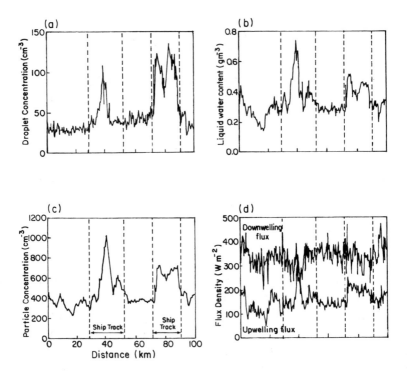

Figure 5.19 Aircraft transects through two ship tracks showing changes in (a) total droplet concentrations, (b) the liquid water content, (c) total concentrations of cloud interstitial particles, and (d) measured radiation fluxes (from Radke et al., 1989).

key feature is the increased droplet concentration and liquid water content in the vicinity of the ship tracks which combine to increase the optical depth of the cloud and thus its albedo.

We can make some estimate of the effects of the observed increases of droplet concentrations and liquid water content on the optical thickness of clouds by considering the following definitions

$$\ell = \frac{4\pi}{3} \rho_L \int_0^\infty n(r) r^3 dr$$

$$N_o = \int_0^\infty n(r) dr$$

and

$$\bar{r} = \int_0^\infty n(r) r \, dr / N_o$$

for the liquid water content, total droplet concentration, and mean radius, respectively. Suppose that

$$\ell \approx \frac{4\pi}{3} \rho_L N_o \bar{r}^3$$

then it follows that

$$\bar{r} \approx \left(\frac{3\ell}{4\pi N_o \rho_L} \right)^{1/3}$$

From the definition of optical depth,

$$\tau = \int_{\Delta z} dz \int_0^\infty n(r) \pi r^2 Q_{ext} \, dr$$

it follows that

$$\tau \approx h 2\pi N_o \bar{r}^2 \approx \text{constant} \times h N_o^{1/3} \ell^{2/3}$$

where we make use of the large particle limit to Q_{ext} and assume the cloud is vertically homogeneous with a geometric depth h. The ratio of the cloud optical depths in the ship tracks to those in the unperturbed cloud may be approximated by

$$\left(\frac{\tau_{st}}{\tau_{cl}} \right) \approx \left(\frac{N_{o,st}}{N_{o,cl}} \right)^{1/3} \left(\frac{\ell_{st}}{\ell_{cl}} \right)^{2/3}$$

assuming all other factors being equal. If we take values of the ratios of droplet concentrations and liquid water contents from Fig. 5.19 to be 2 and 1.5, respectively, then the ratio of optical depths is 1.65. This increase in the optical depth of clouds in the vicinity of ship tracks is approximately equally governed by the increase in N_o and the increase in ℓ.

5.4 Scattering Functions

The discussions so far focus on the importance of the attenuation of radiation along the forward direction of propagation. We note in reference to Fig. 5.9, however, that radiation is scattered in all directions although in unequal proportions and in a way that depends on

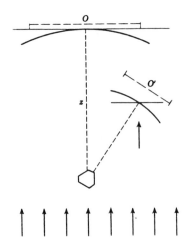

Figure 5.20 An arbitrary particle that scatters a plane wave to the point O'.

particle size, shape, and composition. Many remote sensing methods actually exploit the radiation scattered in directions other than at $\Theta = 0°$. Active sensing by radar and lidar, for instance, makes extensive use of scattering in the backward direction $\Theta = 180°$. A number of passive methods also make use of scattering of radiation by the atmosphere away from the forward direction to derive information about the atmosphere. Thus a mathematical expression for the angular scattering pattern is needed.

5.4.1   Amplitude Functions

The scattering pattern is described in terms of a quantity referred to as the *amplitude function* $S(\Theta)$. This may be introduced via the following experiment. A particle of arbitrary shape is illuminated by a plane wave traveling from the negative z direction as shown in Fig. 5.20 and has the form

$$\mathcal{E}_{inc} = \mathcal{E}_o e^{-ikz+i\omega t} \tag{5.30}$$

The scattered wave in the distant field at the point O' is a spherical wave with an amplitude inversely proportional to the distance R from the particle to the observer (Section 2.2) and directly proportional to the incident amplitude. Thus we can write this wave in the form[4]

$$\mathcal{E}_{sca} = S(\Theta)\frac{e^{-ikR+i\omega t}}{kR}\mathcal{E}_o \tag{5.31}$$

which defines the amplitude function $S(\Theta)$ which is introduced here to describe the scattering pattern as a function of scattering angle. Combining (5.30) and (5.31) yields

$$\mathcal{E}_{sca} = S(\Theta)\frac{e^{-ikR+ikz}}{kR}\mathcal{E}_{inc} \tag{5.32}$$

and in terms of intensities it follows that

$$I_{sca} = \frac{|S(\Theta)|^2 I_o}{k^2 R^2} \tag{5.33}$$

We discussed earlier how extinction is defined in terms of scattering in the forward direction ($\Theta = 0°$). An important mathematical relationship thus follows from the amplitude function, namely that

$$C_{ext} = \frac{4\pi}{k^2}\Re e[S(0)] \tag{5.34}$$

and this is referred to as the *fundamental extinction formula*.

A more complete treatment of the effects of scattering requires a proper treatment of polarization and this requires four amplitude functions which define an amplitude matrix $\mathbf{S}(\Theta)$ under the linear basis

$$\begin{pmatrix} \mathcal{E}_{sca,\ell} \\ \mathcal{E}_{sca,r} \end{pmatrix} = \begin{pmatrix} S_2 & S_3 \\ S_4 & S_1 \end{pmatrix}\frac{e^{-ikr+ikz}}{kr}\begin{pmatrix} \mathcal{E}_{o,\ell} \\ \mathcal{E}_{o,r} \end{pmatrix} \tag{5.35}$$

where the matrix $\begin{pmatrix} S_2 & S_3 \\ S_4 & S_1 \end{pmatrix}$ is known as the *amplitude scattering matrix*, and the unusual, nonsequential numbering of the matrix elements follows established convention. For homogeneous spherical particles, $S_3 = S_4 = 0$.

---

[4] In the derivation of van de Hulst (1957, p. 29), a complex factor i appears in the denominator for reasons that become evident in his derivation of cross-sectional areas. We will not outline these derivations here and choose to omit this factor from our discussion.

5.4.2   The Scattering Phase Functions

It is more usual to describe the angular patterns of scattered light in terms of a quantity known as the *scattering phase function*.[5] We can consider the relation of the phase function to the amplitude function in the following way. Consider an instrument located at the position at O' in Fig. 5.20 and suppose the area of the detector is dA such that the amount of radiation received by it is confined to the set of directions defined by a small solid angle element $d\Omega = dA/R^2$. The detector is moved to all positions around the particle and measurements are made at each position. The integral of these measurements represents the total energy per unit time at a given wavelength scattered by the particle, namely

$$R^2 \int_\Xi I_{sca} d\Omega = \frac{I_o}{k^2} \int_\Xi |S(\Theta)|^2 \, d\Omega \qquad (5.36)$$

where Ξ denotes the entire sphere of directions. This total scattered power can also be defined in terms of the scattering cross sectional area C_{sca},

$$I_o C_{sca} = R^2 \int_\Xi I_{sca} d\Omega \qquad (5.37)$$

such that

$$C_{sca} = \frac{1}{k^2} \int_\Xi |S(\Theta)|^2 \, d\Omega \qquad (5.38)$$

At this point, it is convenient to introduce the quantity

$$\frac{1}{4\pi} P(\Theta) = \frac{|S(\Theta)|^2}{k^2 C_{sca}} \qquad (5.39)$$

where $P(\Theta)$ is the scattering phase function. This is a unitless quantity that conveniently represents the variation of the scattered intensity as a function of angle. When integrated over solid angle, $P(\Theta)$ obeys the following condition of energy conservation

$$\frac{1}{4\pi} \int_\Xi P(\Theta) d\Omega = 1 \qquad (5.40)$$

---

[5] The use of the word phase to name this function has no relation to the phase of the wave but originates from the astronomical literature and refers to lunar phases.

Equation (5.40) states that, in the absence of absorption, the energy scattered in all directions around the particle equals the amount removed from the incident field. An alternative and convenient way to think about the phase function is in terms of probability of scattering such that the quantity $P(\Theta)d\Omega$ is the probability that a photon is scattered between Θ and $\Theta + d\Theta$ where the scattering angle increment $d\Theta$ determines the solid angle increment $d\Omega$.

In Chapter 2, four Stokes parameters were introduced to describe the polarization of the intensity field. Whereas we characterize the scattering of the electric field in terms of the amplitude matrix and its four elements S_1, S_2, S_3, and S_4, scattering of intensities is described in terms of a 4×4 matrix, called the *phase matrix* **P**. Each of the 16 coefficients are quadratic expressions of the coefficients S_1, S_2, S_3, and S_4 as explicitly derived by van de Hulst (1957). The phase matrix often assumes the form

$$\mathbf{P}(\Theta) = \frac{1}{k^2 C_{sca}} \begin{pmatrix} S_{11} & S_{12} & 0 & 0 \\ S_{12} & S_{22} & 0 & 0 \\ 0 & 0 & S_{33} & S_{34} \\ 0 & 0 & -S_{34} & S_{44} \end{pmatrix} \tag{5.41}$$

where $S_{11} = 1/2[|S_2|^2 + |S_1|^2]$, $S_{12} = 1/2[|S_2|^2 - |S_1|^2]$, $S_{33} = 1/2[S_2^* S_1 + S_2 S_1^*]$ and $S_{34} = i/2[S_2^* S_1 - S_2 S_1^*]$. Matrices of this type are valid for spherical particles and also valid for nonspherical particles given certain scenarios of particle symmetry and orientation that are frequently relevant to the atmosphere. We will not discuss these symmetries here in any more detail but note that matrix (5.41) applies for:

1. spherical particles $S_{11} = S_{22}$ and $S_{33} = S_{44}$ and the additional property $S_{12} = 0$ applies for Rayleigh scatterers
2. randomly oriented particles that each possess a plane of symmetry (i.e., oblate raindrops or needlelike ice crystals that are randomly oriented in the horizontal plane)
3. randomly oriented asymmetric particles if half of the particles are mirror images of the others.

For spherical particles, the relationship between incident and scattered Stokes parameters follows from (5.33), (5.39), and (5.41) as

$$\begin{pmatrix} I_{sca} \\ Q_{sca} \\ U_{sca} \\ V_{sca} \end{pmatrix} = \frac{1}{k^2 R^2} \begin{pmatrix} S_{11} & S_{12} & 0 & 0 \\ S_{12} & S_{11} & 0 & 0 \\ 0 & 0 & S_{33} & S_{34} \\ 0 & 0 & -S_{34} & S_{33} \end{pmatrix} \begin{pmatrix} I_o \\ Q_o \\ U_o \\ V_o \end{pmatrix} \tag{5.42}$$

Examples of measured scattering matrix elements are shown in Fig. 5.21a for small spheres and Fig. 5.21b for irregularly shaped quartz particles. The experimental set up involved detection of He–Ne laser light scattered by a small volume of particles as a function of Θ. The solid curves in Fig. 5.21a correspond to Lorenz–Mie calculations assuming a log–normal size distribution with $r_e = 0.75~\mu m$ and a variance of 0.45. The points correspond to the measured phase matrix normalized in such a way that

$$P_{11} = \frac{1}{4\pi} \int \frac{S_{11}}{k^2 C_{sca}} d\Omega = 1$$

Values of P_{11} agree well with theory but the ratios $-P_{12}/P_{11}$, P_{34}/P_{11}, and P_{33}/P_{11} exhibit small deviations near $\Theta = 155°$ and toward the forward direction. For spherical particles, the ratio P_{22}/P_{11} should be exactly unity for all Θ but the results show departures that exceed the estimated experimental error of ± 0.05 for this ratio. These errors are thought to be due to slight effects of multiple scattering within the scattering volume.

Elements of the phase matrices for irregular quartz particles with a distribution characterized by an equivalent volume effective radius of 15 μm and a dispersion of 0.8 are also shown in Fig. 5.21b as a function of Θ. The angular variation of P_{11} is flatter than that for the spherical droplets of Fig. 5.21a due to the greater width of the quartz distribution and the irregularities in particle shape that wash out distinctive structures in the angular scattering patterns. The ratio P_{22}/P_{11} differs substantially from the example of water spheres as do P_{34}/P_{11} and P_{44}/P_{11} particularly in the back scattering directions $\Theta > 150°$. We will describe how polarization of backscattered radiation is used qualitatively to diagnose irregularities of particle shape in both Section 5.7 and later in Chapter 8.

Excursus: A Polynomial Representation of the Phase Function

A convenient way to represent the scattering phase function for radiative transfer applications is to express it as a polynomial in scattering angle. The following polynomial series,

$$P(\cos \Theta) = \sum_{\ell=0}^{N} \chi_\ell P_\ell(\cos \Theta) \qquad (5.43)$$

(a)

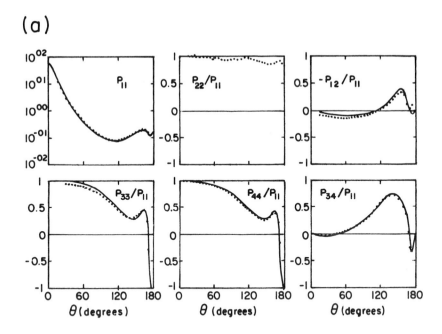

Figure 5.21 (a) Six scattering matrix elements of an ensemble of water droplets as a function of scattering angle. Measurements are denoted by symbols and the solid curve is from Lorenz–Mie theory. The wavelength is 632.8 nm and the refractive index m=1.332 (after Kuik et al., 1991).

is most commonly used since it has special advantages when applied to radiative transfer, but these need not concern us here. In this series, P_ℓ is the ℓth order Legendre polynomial and χ_ℓ are the associated expansion coefficients, defined as

$$\chi_\ell = \frac{(2\ell + 1)}{2} \int_{-1}^{1} P(\cos\Theta)P_\ell(\cos\Theta)d\cos\Theta \qquad (5.44)$$

A general guideline is that the larger the particle, and hence the more forward the scattering, the more polynomial terms are required to represent the true phase function. Simpler, analytic functions require fewer polynomial terms. Table 5.1 presents a few phase functions commonly encountered in different scattering problems, together with their first four expansion coefficients.

(b)

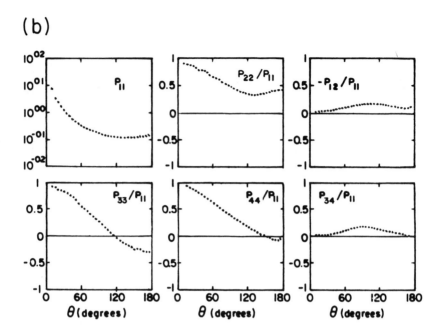

Figure 5.21 (Cont.) (b) Six scattering matrix elements of an ensemble of randomly oriented irregularly shaped quartz particles as a function of scattering angle (after Kuik et al, 1991).

The expansion coefficients for the two analytic functions given in Table 5.1 are expressed in terms of the *asymmetry parameter g*. This is a parameter of some importance to particle scattering problems and is defined as

$$g = \frac{1}{3}\chi_1 \tag{5.45}$$

An interpretation of this parameter is:
- $g = 1$ for complete forward scattering.
- $g = -1$ for complete backward scattering.
- $g = 0$ for isotropic or symmetric scattering (e.g., Rayleigh scattering).

For solar wavelengths and water droplet clouds, g is quasi–constant with an approximate value 0.85. Values of this parameter for irregular particles, like ice crystals, at solar wavelengths are not well known, but they are thought to be significantly different from the values associated with spherical particles.

Table 5.1 Legendre coefficients of selected phase functions

Scatter type	Formula for $P(\Theta)$	χ_0	χ_1	χ_2	g
Isotropic	1	1	0	0	0
Rayleigh	$\frac{3}{4}(1 + \cos^2 \Theta)$	1	0	$\frac{1}{2}$	0
Henyey–Greenstein	$\frac{1-g^2}{(1+g^2-2g\cos\Theta)^{3/2}}$	1	$3g$	$5g^2$	g
Forward[†] plus backward	$(1+g)\delta^+$ $+(1-g)\delta^-$	1	$3g$	$5g^2$	g

[†]$\delta^+ = 1$ when $\theta = 0°$ and zero otherwise. Similarly, $\delta^- = 1$ when $\theta = 180°$ and zero otherwise; g is the asymmetry parameter.

5.4.3 An Example of the Remote Sensing of Ice Crystal Phase Functions

A method for measuring the scattering phase function for nonspherical ice particles was introduced by Platt and Dilley (1984). The technique involves the measurement of solar radiation transmitted through the cloud (Fig. 5.22a) and collected by a receiver at the ground. Monitoring this radiation throughout the day as the sun passes across the sky provides a way of determining the scattering phase function for different values of the scattering angle. A number of assumptions are necessary for the analysis scheme of Platt and Dilley to work. The cloud must exist in a quasi–steady state so that the ice crystal properties remain unchanged with time. The cloud must also be spatially homogeneous so as not to change as it advects over the receiver. The cloud must also be optically thin so that smearing effects of multiple scattering are minimized. These conditions rarely apply in the real atmosphere, but they supposedly existed during the experiment reported by Platt and Dilley.

For the experimental configuration given in Fig. 5.22a, the energy E per unit area and per unit time received at the detector which has a field of view $\Delta\Omega$, and a spectral bandwidth $\Delta\lambda$, is

$$E \approx E_d + E_s + E_a \approx (I_d + I_s + I_a)\Delta\Omega\Delta\lambda \qquad (5.46)$$

where I_d is the intensity due to scattering of the direct solar radiation by clouds, I_s is the component of solar radiation reflected upwards from the surface and back to the receiver from the cloud, and I_a is the solar radiation scattered by the atmosphere below the cloud into the receiver. The amount of solar radiation, inclined at an angle θ_o from the vertical, that reaches some level in the cloud is $F_o \exp(-\tau/\mu_o)$ where $\mu_o = \cos\theta_o$ (see Chapter 6 for further discussion and the derivation of this expression). The rate of change of I_d with τ then follows from radiative transfer theory (Section 6.3),

$$\frac{dI_d}{d\tau} = -F_o \exp(-\tau/\mu_o)P(\Theta)/4\pi \qquad (5.47)$$

where Θ is the scattering angle. The intensity at some level corresponding to τ^* is

$$I_d(\tau^*) = F_o \frac{P(\Theta)}{4\pi}\mu_o[1 - \exp(-\tau^*/\mu_o)] \qquad (5.48)$$

which specifies the intensity of sunlight scattered from cloud base toward the receiver along the scattering angle Θ in terms of τ^*, the optical depth of the cloud. According to (5.46), the energy received at the detector due to the scattering of direct sunlight is

$$E_d = I_d\Delta\Omega\Delta\lambda = F_o\frac{P(\Theta)}{4\pi}\mu_o[1 - \exp(-\tau^*/\mu_o)]\mathcal{T}_a\Delta\Omega\Delta\lambda \quad (5.49)$$

where $\mathcal{T}_a$ is the transmission function for the atmosphere below the cloud.

The phase function follows if experimentally derived values of all other factors in (5.49), together with estimates of the remaining two contribution terms of (5.46), are determined. The interested reader can refer to the reference for a discussion of how these different factors were estimated. Values of $P(\Theta)/4\pi$ retrieved from (5.49) are shown in Fig 5.22b as a function of scattering angle. The theoretical reference curve shown on this diagram applies to a hexagonal columnar crystal $30\mu m$ wide and $120\mu m$ long, based on the ray tracing calculations of Wendling et al. (1982). Certain remarkable

(a)   (b)

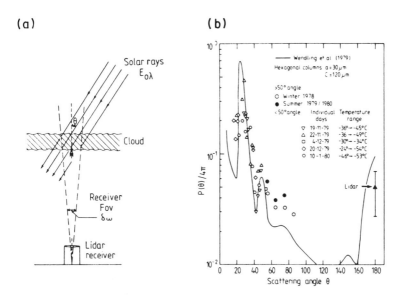

Figure 5.22 (a) The geometric configuration of the Platt and Dilley experiment. (b) The retrieved, normalized scattering phase function at various scattering angles compared to a theoretical curve for hexagonal ice crystals (from Platt and Dilley, 1984).

features appear in the retrieved phase function that match the theoretical curve for hexagonal columns, including the 22 degree and 46 degree haloes. There is also a suggestion that the actual scattering differs from the theoretical scattering curve in the range $\Theta > 60°$.

5.5 A Simple Diffraction Theory of Particle Extinction

The earlier remark that extinction arises more from interference than from a blocking of the incident wave is the basis for an approximate theory for particle extinction and absorption. This approximate theory owes its origins to van de Hulst (1957) and is known as the *anomalous diffraction theory* (ADT). One of the major virtues of this theory, and one exploited in different remote sensing problems, is the inherent simplicity of its mathematical form.

The anomalous diffraction theory models forward scattering in the following way. Consider a plane wave[6] that passes through a

---

[6] We consider a spherical particle here only for simplicity. The ideas described apply to any shape of particle and the approach requires only the specification of the path length of a ray within the particle.

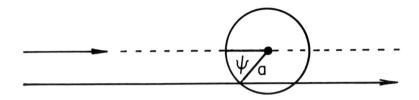

Figure 5.23 ADT geometry for a spherical particle of radius a.

spherical particle of radius r. Extinction can be defined in terms of
the characteristics of the wave projected on a reference screen some
distance from the particle. The geometry relevant to this discussion
is given in Fig. 5.23. Two basic assumptions are now introduced:
• The particle is assumed to be large relative to the wavelength of
the wave such that $x \gg 1$. Under these circumstances, we can ignore
the individual waves within the particle and trace the passage of the
wave as a ray.
• The refractive index of the particle is close to that of the background (i.e., m≈ 1) so that a ray passes through the particle without
suffering any significant refraction.
    The resultant wave on the screen is

$$\mathcal{E}_{res} = \mathcal{E}_{inc} + \mathcal{E}_{sca} \tag{5.50}$$

The ADT supposes that this wave may be expressed entirely in terms
of the phase difference between the ray that penetrates the particle
to one that is diffracted by the particle's edge. Thus we write

$$\mathcal{E}_{res} = \mathcal{E}_{inc}[1 + e^{-i\Delta\phi}]e^{i\phi_o} \tag{5.51}$$

where from simple geometric considerations the relative phase difference for the ray shown in Fig. 5.23 is

$$\Delta\phi = 2x \sin\psi(m - 1) = \rho \sin\psi \tag{5.52}$$

where ρ is the relative phase lag experienced by the ray passing along
the diameter of the sphere relative to a ray outside the particle and

ψ is the angle defined in Fig. 5.23. Making use of (5.32), we arrive at

$$S(\Theta = 0) = \frac{k^2}{2\pi} \int_G \frac{\mathcal{E}_{sca}}{\mathcal{E}_{inc}} dA \qquad (5.53)$$

where the integration is performed over the geometric shadow area G of the particle. Substitution of (5.51) and (5.52) into (5.53) leads to

$$S(\Theta = 0) = x^2 \mathcal{K}(i\rho) \qquad (5.54)$$

where

$$\mathcal{K}(w) = \frac{1}{2} + \frac{e^{-w}}{w} + \frac{e^{-w} - 1}{w^2}$$

and thus

$$Q_{ext} = 4\Re\{\mathcal{K}(i\rho)\} \qquad (5.55)$$

follows from the fundamental extinction formula (5.34).

Figure 5.24, taken from the original work of van de Hulst, compares an example of $Q_{ext} = Q_{sca}$ calculated using this simple formula for extinction with that from Lorenz–Mie theory. The approximate formula clearly has some shortcomings, lacking the very fine structure observed in the Lorenz–Mie theory, but the large maxima and minima of Q_{ext} and the position of these defined relative to the central phase shift ρ are well represented. This simple theory allows us to interpret these pronounced extrema as interference features. The maxima occur when the term $-e^{-i\Delta\phi}$ enhances the term 1 in the integrand of (5.53) from which Q_{ext} was computed. Since one term represents the transmitted light and the other term represents the diffracted light, we can deduce that the maxima are due to favorable interference of the transmitted and diffracted waves and that the minima are a result of unfavorable interference. By plotting Q_{ext} as a function of ρ, we are able to present the extinction for different values of m on a common scale.

Excursus: Particle Absorption in the ADT Approximation

The derivation given earlier applies to nonabsorbing particles. For an absorbing particle with m $= n - i\kappa$, the requirement that $|$ m $-1| \ll 1$ means that both $| n - 1 |$ and $| \kappa - 1 |$ be $\ll 1$. It is also convenient to express the absorption in terms of an angle ζ such that

$$\tan\zeta = \frac{\kappa}{n-1} \qquad (5.56)$$

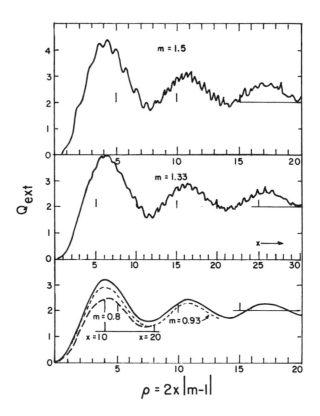

Figure 5.24 Extinction curves computed from Lorenz–Mie formulas for m =1.5,1.33,0.93,0.8. The abscissa is $\rho = 2x(m-1)$ and is common to the upper two Lorenz–Mie curves as well as to the bottom ADT curves (van de Hulst, 1957).

assumes values varying from 0 (no absorption) to ∞. The real phase shift parameter $\rho = 2x(n-1)$ was introduced earlier. The complex phase shift parameter for a path through the center of the sphere is

$$\rho^* = 2x(m-1) = \rho(1 - i\tan\zeta)$$

where the real part is the actual phase shift and the imaginary part is associated with the decay of the wave amplitude. The derivation of Q_{ext} and Q_{abs} will not be described further as the interested reader can find these in van de Hulst (1957, Chapter 11, p. 179). The results of the derivations are:

$$Q_{ext} = 4\Re e\{\mathcal{K}(i\rho + \rho\tan\zeta)\} \qquad (5.57a)$$

$$Q_{abs} = 2\mathcal{K}(4v) \qquad (5.57b)$$

where the absorption by the particle is determined as simply the integrated attenuation of all rays that penetrate the particle. In the expression for Q_{abs}, $v = 2x\kappa$ and $x = 2\pi a/\lambda$ is the size parameter of a sphere of radius a. Bohren and Nevitt (1983) provide a slight improvement to Q_{abs} which is cast in terms of $\mathcal{K}(v)$ by Flatau (1992) as

$$Q_{abs} = c[2\mathcal{K}(4v) - h^2\mathcal{K}(4av)] \qquad (5.58)$$

where

$$h = \frac{(n^2 - 1)^{1/2}}{n}$$

and

$$c \approx n^2$$

for n close to unity. Equation (5.58) predicts that Q_{abs} increases systematically with increasing v in the manner illustrated by the solid curve in Fig. 5.25. The symbols represent calculations of Q_{abs} obtained from Lorenz–Mie theory and the solid curve is (5.58) derived for a 5 μm water droplet for wavelengths spanning from 0.7 to 3 μm. The value of (5.58) is that it shows how the behavior of Q_{abs} uniquely depends on v. Two particles of different radius a, and composition κ, absorb the same amount of radiation when the respective values of v match each other.

5.6 Scattering by Spheres: A Brief Outline of Lorenz–Mie Theory

The theory for scattering by dielectric spheres was developed independently by Ludwig Lorenz in 1890 and Gustav Mie in 1908 (refer to the discussion of these developments in the bibliographical notes at the end of this chapter). The derivation of the solution is a straight forward application of classical electromagnetic theory and only the resulting formulas are given here. The solutions are expressed as infinite series and the rates of convergence of these series depend on the value of the size parameter x. The two scattering amplitude functions have the form

$$S_1(\Theta) = \sum_{n=1}^{\infty} \frac{2n+1}{n(n+1)} [a_n \pi_n(\cos\Theta) + b_n \tau_n(\cos\Theta)]$$

$$S_2(\Theta) = \sum_{n=1}^{\infty} \frac{2n+1}{n(n+1)} [a_n \tau_n(\cos\Theta) + b_n \pi_n(\cos\Theta)] \qquad (5.59)$$

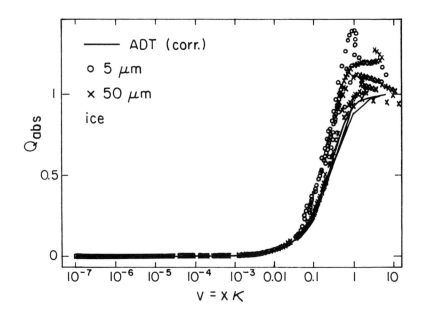

Fig. 5.25 The absorption efficiencies derived from Lorenz–Mie theory (symbols) and from (5.58) (solid curve) as a function of the absorption similarity parameter v.

where

$$\pi_n(\cos\Theta) = \frac{1}{\sin\Theta} P_n^1(\cos\Theta)$$

$$\tau_n(\cos\Theta) = \frac{d}{d\Theta} P_n^1(\cos\Theta)$$

(5.60)

and where P_n^1 is the associated Legendre polynomial (e.g., Abromowitz and Stegun 1971). The coefficients a_n and b_n are referred to as Mie scattering coefficients and are functions of refractive index m and size parameter x. The mathematical forms of these coefficients are given as ratios of Ricatti–Bessel functions and can be found in the references on Lorenz–Mie scattering cited at the end of this chapter. The extinction and scattering efficiencies are also given by the series

$$Q_{ext} = \frac{2}{x^2} \sum_{n=1}^{\infty} (2n+1)\, Re\,(a_n + b_n)$$

(5.61a)

$$Q_{sca} = \frac{2}{x^2} \sum_{n=1}^{\infty} (2n+1)\left(|\,a_n\,|^2 + |\,b_n\,|^2\right)$$

(5.61b)

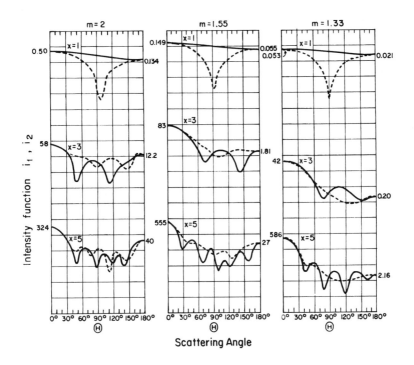

Figure 5.26 Scattering diagrams derived from Lorenz–Mie theory for a single particle. The solid curves are for i_1 and the broken curves are for i_2. Numerical values of i_1 and i_2 are specified at $\Theta = 0$ and 180 degrees (after van de Hulst, 1957).

Example calculations of the intensity functions $i_{1,2} = |\, S_{1,2}\,|^2$ are shown in Fig. 5.26 for a given particle of fixed size. These intensity functions reveal how the number of fringes increases over the range $0° \leq \Theta \leq 180°$ as the size parameter increases (compare these to Fig. 5.17). It is also noteworthy how the scattering in the forward direction increases as x increases according to our earlier expectation.

Figure 5.27 presents the scattering phase function $P = \frac{1}{2}(P_1 + P_2)$ for unpolarized radiation[7] versus scattering angle Θ (upper panels) and the degree of linear polarization (lower panels), defined as

$$LP = \frac{P_1 - P_2}{P_1 + P_2}$$

also plotted as a function of scattering angle. The calculations are shown for m= 1.33 and 1.5 and three size distributions characterized

<hr>

[7] We use P_1 and P_2 for $S_{11}/k^2 C_{sca}$ and $S_{22}/k^2 C_{sca}$, respectively.

by three different values of the effective size parameter $x_m = 2\pi a / \lambda$ where a is the parameter of the size distribution given by (5.22). The scattering phase function becomes more peaked in the forward direction as x_m increases in accordance with our expectations. A sharp increase in scattering also occurs in the backscattering direction ($\Theta = 180°$) for the large spheres. This strong enhancement, also noted in reference to Fig. 5.17, is a characteristic of spherical particles and is referred to as the *glory*. One of the major contributors to this effect is the tangential ray that just grazes the surface, setting up surface waves that travel around the sphere. The interference of these surface waves with those that penetrate the particle produces this effect. Also noteworthy is the way the polarization patterns alter as x_m changes. A comparison of Fig. 5.27 with Fig. 5.4a illustrates how the magnitude of the maximum degree of polarization is considerably reduced from the value of 100% that we associate with pure Rayleigh scattering at $\Theta = 90°$.

The formal Lorenz–Mie solution does not, at first sight, seem to offer much in the way of a physically intuitive picture of the mechanism of the scattering process. However, some qualitative insights can be obtained from the theory. We know that the scattered wave arises from oscillations of the electrons in the particle that are excited by the incident wave. Distributions of electric charges may be represented by a superposition of electric multipoles with arbitrary moments located at some origin point. If, as in the case of scattering, the distributions of charges and currents oscillate synchronously with the exciting wave, the scattered radiation arises from the corresponding oscillating multipoles. The oscillating electric multipoles produce partial electric waves and the oscillating magnetic multipoles produce partial magnetic waves. The amplitude and phase of each wavelet associated with a particular electric multipole is given by a_n and b_n is associated with the corresponding magnetic multipole. The electric dipole moment is therefore proportional to a_1 and the magnetic dipole moment is proportional to b_1 and the higher order a_n and b_n coefficients are each related to corresponding higher order multipole radiation.

An important example of the application of Lorenz–Mie theory in remote sensing is in the interpretation of the polarization of sunlight that results from the reflection by clouds in the atmosphere of Venus. Even up to the 1920s it was debatable whether sunlight reflected by Venus was in fact polarized. However, with the development of precision polarimeters capable of accuracies of about 0.1

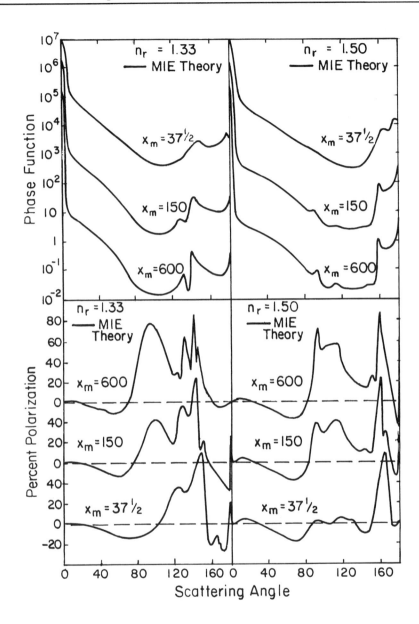

Figure 5.27 The scattering phase function and degree of polarization calculated from Lorenz–Mie theory as a function of Θ. The results are shown for two real refractive indices and different values of the effective size parameter x_m (adapted from Hansen and Travis, 1974).

%, it was revealed that reflected light from Venus exhibits distinct polarization features. Hansen and Hovenier (1974) used these polarization properties to deduce information about the size of the cloud particles. An example of the measured variation of polarization as a function of phase angle (which is defined as $180° - \Theta$) is presented in Fig. 5.28. The model results of Hansen and Hovenier (1974) are superimposed on these observations. From the comparisons between the calculated reflectivities and observed quantities, Hansen and Houvenier deduced that:

• the cloud particles at cloud top were spherical.
• the effective radius was $\sim 1\mu$m.
• the particle size distribution was very narrow with an effective variance, $b \sim 0.07$.
• the particles were deduced to be sulphate spheres based on the refractive index $n = 1.44$ that provided a match to the observations.

The successful analysis of the polarization of Venutian clouds, however, has not been repeated for terrestrial clouds. The success of this kind of analysis for Venus depends largely on the fact that the cloud particles are spheres. Irregularly shaped ice particles that occur in the upper portions of deep convective clouds or in the cirrus clouds that veil the Earth introduce effects that greatly complicate polarization patterns making unambiguous interpretation of polarization in terms of particle size more difficult.

5.7   Particle Backscattering

Understanding which properties of particles govern the amount of radiation that is backscattered from them is crucial to many remote sensing applications and especially to applications involving active sensing. We now discuss particular properties of particle backscattering relevant to both radar and lidar backscattering measurements. It is convenient to introduce the following definitions of cross–sectional areas:

• The differential cross–section

$$C_d(\Theta) = \frac{C_{sca}}{4\pi}P(\Theta) \qquad (5.62a)$$

where C_d is a measure of the amount of incident radiation scattered into the direction Θ per unit solid angle.

• Bistatic (radar) cross–section

$$C_{bi} = 4\pi C_d(\Theta) \qquad (5.62b)$$

Plates

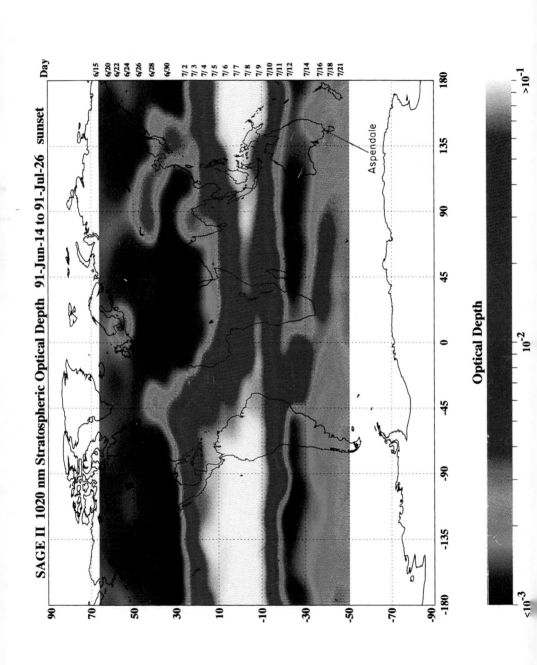

SAGE II 1020 nm Stratospheric Optical Depth 91-Jun-14 to 91-Jul-26   sunset

Day

6/15
6/20
6/22
6/24
6/26
6/28
6/30
7/2
7/3
7/4
7/5
7/6
7/7
7/8
7/9
7/10
7/11
7/12
7/14
7/16
7/18
7/21

Aspendale

Optical Depth

$<10^{-3}$ 10^{-2} $>10^{-1}$

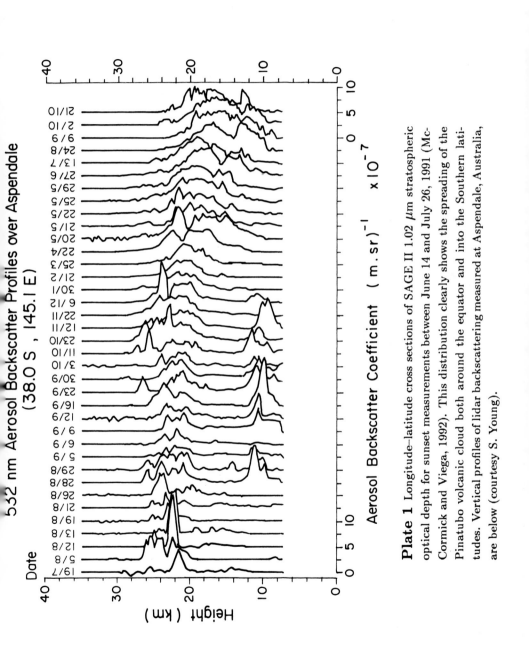

Plate 1 Longitude–latitude cross sections of SAGE II 1.02 μm stratospheric optical depth for sunset measurements between June 14 and July 26, 1991 (McCormick and Viega, 1992). This distribution clearly shows the spreading of the Pinatubo volcanic cloud both around the equator and into the Southern latitudes. Vertical profiles of lidar backscattering measured at Aspendale, Australia, are below (courtesy S. Young).

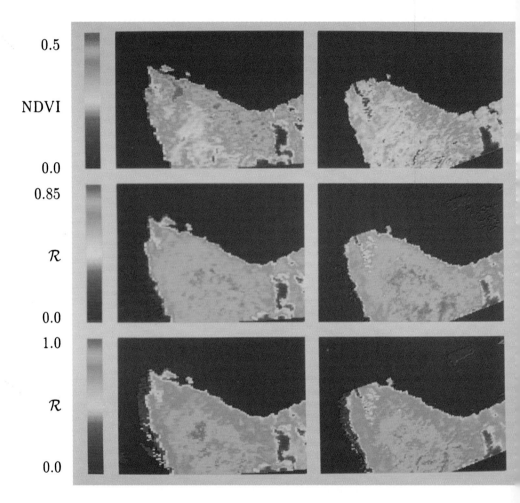

Plate 2 Correction of NDVIs for the effects of atmospheric scattering. The top two panels are the NDVIs derived from uncorrected intensity data as seen from two successive orbits by the AVHRR on NOAA 11. The center two panels show the NDVI corrected for molecular scattering, while the lower panel shows the effect of adding a thin aerosol layer at 2 km. The left panels correspond to the view looking east (toward the sun) and the right panels are the westward view (Mitchell and O'Brien, 1993).

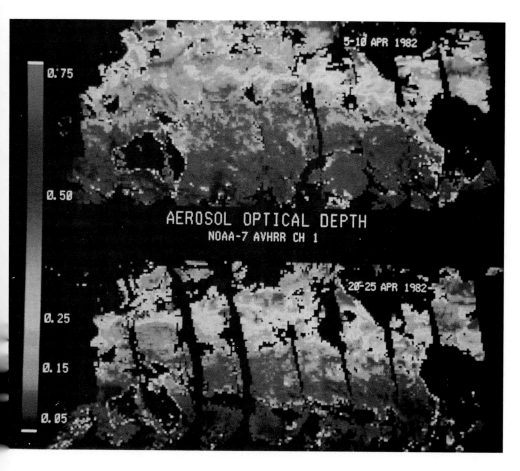

Plate 3 Averaged optical depth derived from AVHRR channel 1 for April 1982 (Durkee et al., 1991).

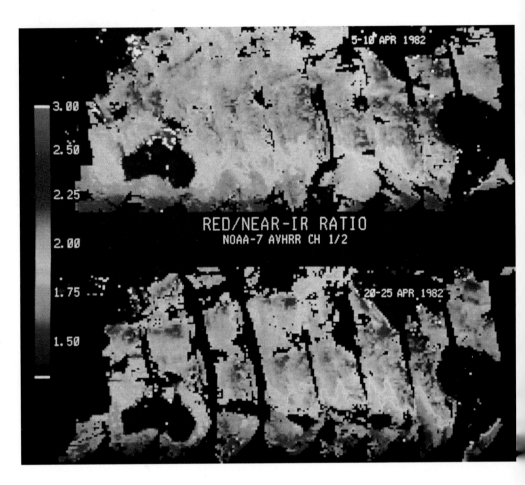

Plate 4 Averaged S_{12} values derived from AVHRR channels 1 and 2 for April 1982 (Durkee et al., 1991).

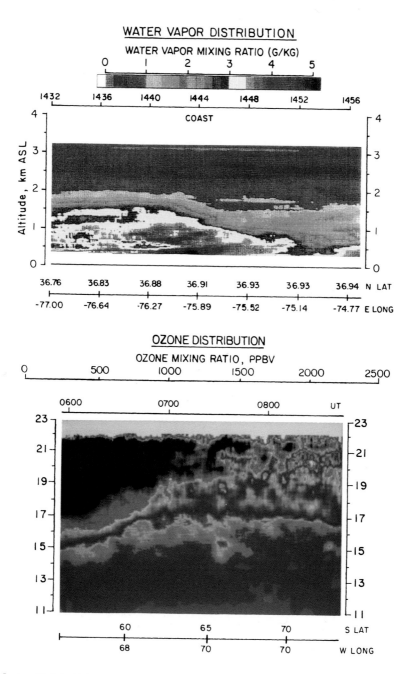

Plate 5 (a) Airborne DIAL measurements of daytime water vapor along an easterly flight track from Emporia (Virginia) to 120 km off the shore over the Atlantic ocean (Browell, 1993). (b) A vertical cross section of O_3 distribution across the edge of the ozone hole over Antartica obtained by an airborne DIAL system from an aircraft at an altitude of 9 km ASL on September 26, 1987 (Browell, 1989).

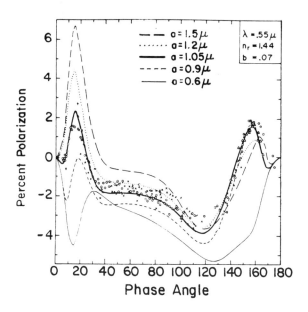

Figure 5.28 Polarization of Venus. Observational points and calculations centered at $\lambda = 0.55$ μm. All curves are for spheres with a
refractive index of 1.44. The different curves show the effect of different values of a in (5.23). All size distributions assume $b = 0.07$ (after
Hansen and Hovenier 1974).

which is the total scattering cross–sectional for a particle that
scatters isotropically by an amount $C_d(\Theta)$.

- Backscattering cross–section

$$C_b = 4\pi C_d(\Theta = 180^o) \qquad (5.62c)$$

which has a similar interpretation to (5.62b) except that the
scattering direction is now specified to be directly opposite the
incident direction.

5.7.1 Backscattering by Small Spheres

Since scattering by particles is described in its most general form by
the phase matrix $\mathbf{P}(\Theta)$, then the scattering cross–sections defined
earlier are also matrix quantities as described shortly. For now,
however, the much simpler situation of backscattering of unpolarized
light by spherical particles is treated. Lorenz–Mie theory provides

us with an expression of C_b for a sphere of radius r

$$C_b = \frac{\pi r^2}{x^2} \mid \sum_{n=1}^{\infty} (-1)^n (2n+1)(a_n - b_n) \mid^2 \qquad (5.63)$$

which follows from (5.62c) and (5.59) assuming $\Theta = 180°$.

When applying (5.63) to radar backscattering problems, it is useful to consider the behavior of C_b in the limit as $x \rightarrow 0$. In this limit it is possible to carry out a small argument expansion of the Bessel functions that define these coefficients and express them as polynomials of x (refer to Bohren and Huffman, Chapter 5). If we neglect all terms of higher order than x^5, then

$$b_1 \approx -\frac{i}{45}(m^2 - 1) x^5 + O(x^7)$$

$$a_1 \approx -\frac{2i}{3}\left(\frac{m^2-1}{m^2+2}\right)x^3 \left[1 + \frac{3}{5}\left(\frac{m^2-2}{m^2+2}\right)x^2\right] + O(x^6) \qquad (5.64)$$

$$a_2 \approx -\frac{i}{15}\left(\frac{m^2-1}{2m^2+3}\right)x^5 + O(x^7)$$

where all other coefficients ($a_3 \dots; b_2, b_3, \dots$) are neglected. Retaining x^3 terms leaves only the first term of a_1, the electric dipole term, which then yields on substitution in (5.63)

$$C_b = \frac{\lambda^2}{\pi}x^6 \mid \frac{m^2-1}{m^2+2} \mid^2 = \frac{\pi^5}{\lambda^4}\mid K \mid^2 D^6 \qquad (5.65)$$

where K is used for $(m^2 - 1)/(m^2 + 2)$ and D is the particle diameter. The path from Lorenz–Mie theory to (5.65) serves to emphasize how the Rayleigh scattering limit is a special case of Lorenz–Mie scattering.

Figure 5.29a provides an example of particle backscattering calculated using (5.63) as a function of the size parameter. The backscattering is expressed in terms of the *backscattering efficiency* $Q_b = C_b/\pi r^2$ and is shown for water and ice spheres assuming a wavelength of 10 cm which is commonly used by weather radar. Also shown for comparison are the efficiencies derived from Rayleigh scattering using (5.65). We note that the backscattering by water spheres in the range $x < 2$ is considerably larger than backscattering by ice spheres of the same size. As x increases, the comparative roles reverse with backscattering by ice spheres exceeds that of water spheres. Large water droplets significantly absorb 10 cm radiation, and backscattering is small compared to large ice particles. Figure 5.29b also shows how the Rayleigh approximation to backscattering deviates significantly from the Lorenz–Mie solution as x is increased beyond about $x = 1$.

The range of x over which the Rayleigh approximation applies for wavelengths typical of radars has been examined in detail by several investigators. Figure 5.29b also presents the variation of the ratio $C_{b,L-M}/C_{b,Ray}$ as a function of x for various radar wavelengths. It may be concluded from this diagram that scattering of 3cm wavelength radiation by precipitation drops smaller than about 2mm can be approximated by Rayleigh scattering to within 25% of Lorenz–Mie scattering and that virtually all precipitation, except perhaps hail, may be regarded as Rayleigh scatterers at wavelengths of about 10 cm.

5.7.2 Backscattering by Nonspherical Particles

It was mentioned in section 5.3 how scattering, especially in the backward direction, is sensitive to particle shape. Backscattering by irregularly shaped particles is now discussed.

In the most general terms, the scattering matrix introduced in Section 5.5 has 16 independent parameters. For many situations applicable to particles in the atmosphere, the phase matrix has a simpler form like that given by (5.41). We will now consider the backscattering by particles governed by this particular phase matrix. In doing so, we find it convenient to discuss polarization by backscattering in terms of ratio quantities that are used in lidar and radar studies of clouds. These quantities are introduced here to demonstrate their relationship to the scattering phase matrix.

Lidar Depolarization Ratio

The depolarization ratio is a quantity often used in lidar studies of particle scattering. It is convenient to think of this lidar parameter in terms of a receiver such as that described in our simple, earlier experiments discussed in Section 2.3. Most lidar systems transmit a laser pulse with a linear polarization, and receive the returned pulse using the optical elements as arranged in Fig. 2.13. We denote the intensity of the returned pulse by I_ℓ for the case when the polarizer is aligned along the direction parallel to the polarization of the transmitted pulse. When perpendicular to the polarization direction of this transmission, the detected intensity on transmission through

(a)

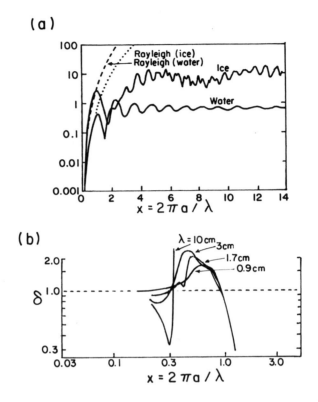

Figure 5.29 (a) Backscattering efficiency for ice and water at $0^\circ C$ as a function of size parameter. The Rayleigh approximation is indicated for both phases. (b) Ratio of the Lorenz–Mie backscattering to Rayleigh backscattering for water spheres as a function of x (adapted from Gunn and East, 1954).

the polarizer is I_r. The ratio of these measured intensities, namely

$$\delta = \frac{I_r}{I_\ell}$$

is the depolarization ratio. The physical interpretation of δ is clear when we consider the properties of I_r and I_ℓ in more detail. To do so, consider a linearly polarized transmitted pulse which we represent by the Stokes vector

$$\mathbf{I}_o = \begin{pmatrix} 1 \\ 1 \\ 0 \\ 0 \end{pmatrix} \qquad (5.66)$$

Suppose this pulse, backscattered by particles with properties that enable us to use (5.41), is passed through an ideal polarizing filter placed in front of the detector. In matrix form, the detected Stokes vector is

$$I = M_{\ell,r} P I_o \qquad (5.67)$$

where $M_{\ell,r}$ are the Mueller matrices as discussed in Section 2.3 relevant to the parallel [equation (2.40)] and perpendicular orientation of the polarizer, P is given by (5.41) and I_o by (5.66). With simple matrix algebra, it can be shown that

$$\delta = \frac{S_{11} - S_{22}}{S_{11} + S_{22}} \qquad (5.68)$$

for the depolarization ratio. Since $S_{11} = S_{22}$ for spherical particles, $\delta = 0$. Thus δ may be thought of as a measure of the departure of the scatterer from sphericity.

A number of studies attempt to quantity relationships between δ and the size and shape of ice crystals, and the ratio of water mass to ice mass. An example of the use of δ in the study of cirrus clouds is provided in Fig. 5.30. This diagram is taken from the work of Sassen et al. (1985). It is a height–time display of δ derived from a vertically pointing lidar. The near–zero values of δ at the bottom of the cirrostratus cloud is indicative of supercooled liquid water spheres, and the presence of these droplets was confirmed by in situ measurements.

Z_{DR} Ratios

The backscattering ratio Z_{DR} is used in the analyses of radar reflectivities to study the microphysics of precipitating clouds. Examples of these applications are presented in more detail in Chapter 8. Here we will try to show how Z_{DR} is related to the phase matrix. In the operation of a dual polarization radar, as opposed to lidar, two linearly polarized pulses are transmitted with the directions of the polarization orthogonal to each other. One direction is referred to as the vertical although this is not necessarily the local vertical[8]. The

---

[8] Using the nomenclature adopted for our lidar example, both I_{VV} and I_{HH} are equivalent to the parallel component I_ℓ as the polarization of the detected signal is parallel to the polarization of the transmitted signal in both cases. For radar, the scattering plane is taken to be the vertical plane and the direction parallel to this plane is referred to as the vertical component (see footnote, p 192).

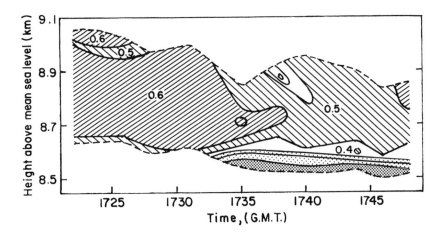

Figure 5.30 Height versus time display of the depolarization ratio from vertically pointing lidar measurements. A gradual water–to–ice cloud transformation is shown by the decreasing stippling (increasing values of δ) above cloud base. The cloud boundaries (dashed lines) derived from the δ analysis often do not correspond to the actual cloud boundaries (from Sassen et al., 1985).

second direction is referred to as the horizontal polarization. We use I_{VV} for vertically polarized transmitted and received intensities and I_{HH} for horizontally polarized intensities that are also transmitted and received. Using arguments similar to those presented earlier,

$$\mathbf{I_{VV,HH}} = \mathbf{M_{V,H}} \mathbf{P} \mathbf{I}_{o,V,H} \tag{5.69}$$

where V is synonymous with ℓ and H with r. The Stokes vectors of the transmitted pulses are

$$\mathbf{I}_{o,V} = \begin{pmatrix} 1 \\ 1 \\ 0 \\ 0 \end{pmatrix}, \quad \text{and} \quad \mathbf{I}_{o,H} = \begin{pmatrix} 1 \\ -1 \\ 0 \\ 0 \end{pmatrix} \tag{5.70}$$

Defining

$$Z_{DR} = 10 \log \frac{I_{HH}}{I_{VV}} \tag{5.71}$$

and using (5.41), the Mueller matrices specified by (2.40) and the matrix expression immediately above (2.40) with $\psi = \pi/2$, then

$$Z_{DR} = 10\log[\frac{S_{11} - 2S_{12} + S_{22}}{S_{11} + 2S_{12} + S_{22}}] \tag{5.72}$$

which is expressed in units of dB. For spherical particles, or for randomly oriented nonspherical particles (tumbling hail for example), $I_{HH} \approx I_{VV}$ and $Z_{DR} = 0$. In precipitating regions of clouds, elongated raindrops tend to be preferentially orientated. We have tacitly assumed that dipoles are spherical in shape. Raindrops illuminated by centimeter wavelength microwave radiation may be thought of as nonspherical dipoles with a larger dipole moment parallel to the long axis of the drop (and aligned horizontally) than parallel to the short axis of the drop (and aligned vertically). This results in larger amounts of backscattering for horizontally polarized radiation than for vertically polarized radiation so that $I_{HH} > I_{VV}$. Thus $Z_{DR} > 0$ for rain drops. The larger the drop impling heavier rainfall, the more asymmetric it becomes. The more asymmetric the geometry of the drop, the larger is the value of Z_{DR} (Fig. 5.31).

Circular Depolarization Ratios

Another example of how we may use the polarization properties of particle backscatter applies to radars that transmit circularly polarized pulses. Suppose that the radar transmits a left–circularly polarized pulse. Backscattering by spheres returns a right circularly polarized pulse. Since any given polarization (including linear polarization, see Section 2.4) can be thought of as a mixture of left– and right–hand circular polarization, the scattering by irregular particles produces a component that has some degree of left–hand polarization as well as right–hand polarization. We can conveniently represent this mixture in terms of the circular depolarization ratio (CDR, again in units of dB)

$$CDR = 10\log(\frac{I_{LH}}{I_{RH}})$$

where I_{LH} and I_{RH} are the detected return powers for left–hand (LH) and right–hand (RH) circular polarization assuming the transmitted pulse is left–circularly polarized. For water droplet clouds, I_{LH} is several orders of magnitude less than I_{RH} yielding corresponding values of CDR that are less than about –25 dB. Ice clouds

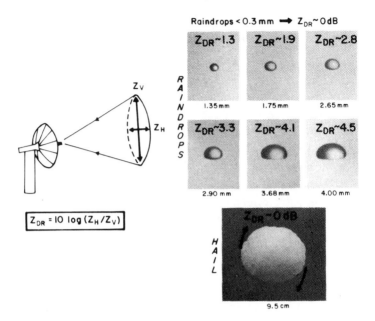

Figure 5.31 Summary of typical Z_{DR} values for raindrops of various sizes and hail. The black arrows on the hail represent its tumbling motions as it falls. Sizes are the median volume diameter (from Wakimoto and Bringi, 1988).

and clouds composed of larger, more oblate droplets are characterized by CDR values in the range between −15 to −25 dB whereas melting aggregate particles have values of CDR around −10 to −5 dB. Rarely does CDR exceed about −5 dB for naturally occurring scatterers at radar wavelengths. By contrast, the backscattering of circularly polarized radiation from an axial particle, like a piece of chaff, becomes linearly polarized along the direction of the long axis of the chaff. A linearly polarized beam can be decomposed into RH and LH circularly polarized components of equal amplitude in which case the value of CDR is 0 dB. CDR for chaff, therefore, differs significantly from the CDRs associated with scattering by cloud particles. Because of this difference, the chaff backscattering signal can be readily distinguished from the cloud particle scattering. This property provides a way of using chaff as a tracer of air motions inside clouds and thus offers potential for studying venting processes of boundary–layer pollutants by clouds or for studying the entrainment of surrounding air into clouds. Figure 5.32 is a visualization of the three–dimensional distribution of cloud volumes characterized

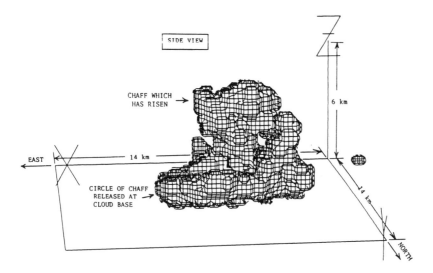

Figure 5.32 A vertical cross–section of the distribution of chaff in a convective cloud. The chaff was released at cloud base in a circle and rose into the cloud by convection (Martner 1990, private communication).

by values of $CDR > -5\text{dB}$ after chaff was released from an aircraft at the base of a convective cloud. This diagram shows the surface of chaff in the cloud and indicates how it rose in towers primarily on the southwest side of the cloud in strong and persistent updrafts.

5.8   Notes and Comments

5.1. A discussion of polarized light in the atmosphere, and in a Rayleigh atmosphere specifically, is given in Coulson (1988).

An account of Haidinger's Brush, and an experiment to observe it, is contained in the book of Minneart (1954) as well as in Kliger et al. (1990).

5.2. The discussion of scattering by a two dipole particle follows Bohren (1987).

The discrete dipole approximation (DDA) as a method for calculating the scattering by nonspherical particles was first proposed by Purcell and Pennypacker (1973). A number of studies since then have appeared in the literature, including the work of Singham and

Bohren (1988), Draine (1988), Flatau et al. (1990), and Goedecke and O'Brien (1988), among others. Tests of the method against analytical solutions have also been reported (e.g., Goodman et al., 1991; Evans and Vivekanandrum, 1990). Computational constraints have limited its application to particles characterized by small size parameters ($x \approx 5$, say). Thus the approach is well suited to study particle scattering for microwave radiation. The studies of Evans and Vivekanandan (1990) and Dungey and Bohren (1992) use DDA to examine the backscattering properties of nonspherical particles illuminated by microwave radiation. Goedke and O'Brien (1988) also use the method to model the scattering by irregular ice particles at microwave radiation. Evans and Stephens (1992) apply the DDA to examine the feasibility of using microwave scattering for remotely sensing ice clouds.

An important extension to the method is provided in the work of Goodman et al. (1991) who employ fast Fourier transform procedures to extend the applicability of the DDA out to size parameters of about 15.

A somewhat related, but nonetheless different, method of calculating scattering by particles of complex geometry is that of Fuller and Kattawar (1987) who developed a solution for light scattering by clusters of spheres.

5.3. An interesting piece of history is associated with the Poisson spot. In 1818, when Augustin Fresnel submitted his wave theory of diffraction, Simeon Poisson attempted to contradict Fresnel's explanation. Poisson deduced from the wave theory that a maximum should be observed on the axis directly behind a circular stop. He believed that his deduction would disprove the wave theory. At a later time Arago performed an experiment and found the spot; instead of dismissing a theory for diffraction, therefore, Poisson's calculations ironically reinforced it.

5.4. The method of ray tracing provides a way of calculating the scattering properties of irregular particles that are much larger in size than the wavelength of illumination. This method is not described in this book and the interested reader is referred to the work of Takano and Liou (1989) for examples of the application of this approach to the scattering by hexagonal ice crystals.

5.5. The ADT formulas (5.55) and (5.58) are indeed most convenient. van de Hulst (1957) qualifies his ADT results as "..one of the most useful formulae in the whole domain of Mie theory, because it describes the salient features of the extinction curve, not only for m close to 1 but even for m values as large as 2."

5.6. The historical perspective of the solution to scattering by spheres has recently been revisited by Logan (1990). Other historical perspectives and anecdotes are contained in selected articles in a special issue of Applied Optics (November 20, 1991). Logan recalls the significant contributions to this problem by Clebsch in a remarkable paper published in 1863 and the solution of Lorenz reported in 1890 which is identical to that presented nearly 20 years later by Mie and Debye after Mie.

Clebsch sought to derive the response of a vector wave at the surface of a curved reflector. Maxwell's famous papers in 1864 and 1873 and Hertz's vindication of Maxwell's theory were still in the future when Clebsch began his study. Clebsch constructed the solutions of the scalar wave equation as an infinite series of terms involving spherical Bessel functions which he had invented for this purpose before Bessel. The solution to the vector wave equation was also developed. It was his wish that his analysis would enable him to deduce the laws of reflection from a spherical mirror by both longitudinal and transverse elastic waves from the solution of the boundary–value problem for elastic waves. The series were so complex that he was unable to derive any information from them except in the case of a sphere that was small compared to the wavelength. We now call this Rayleigh scattering.

It is noteworthy that the influence of Clebsch's 1863 paper is found in the 1890 work of Lorenz. In contrast to Clebsch, Lorenz noted that longitudinal waves should be excluded in studies of the propagation of light. The theory of light employed by Lorenz turns out to be mathematically identical to the Maxwell equations and he arrived at the identical solution to that presented by Mie. Within this historical perspective, Kragh (1991) suggests that the basic theory of plane wave scattering by spheres be called "Lorenz–Mie" theory and this is the terminology adopted in this book.

The numerical features of the Lorenz–Mie solutions are well described by Dave (1968) and Wiscombe (1980), and useful references are cited therein.

5.7. A review of particle backscattering, with an emphasis on the calculation of this backscatter, is provided by Bohren (1991). The method of DDA is mentioned in that paper and studies that employ this method to study backscatter were mentioned under 5.3.

The recent theoretical developments of Fuller and Kattawar (1987) for clusters of spheres were used by Muinonen and Lumme (1991) in which they propose a mechanism associated with coherent scattering between the spheres of the cluster to explain the observed polarization signatures of scattered starlight.

5.9   Problems

5.1. Explain or interpret the following:
    a. Unpolarized radiation scattered by spherical particles much smaller than the wavelength of incident radiation becomes polarized.
    b. A wind profiler antenna of 135 elements is capable of measuring wind to a greater range than an antenna of 35 elements.
    c. On rare occasions the moon appears blue. Assuming that this is due to selective scattering by spheres of index of refraction 1.33, define an upper limit to the radius of the sphere? (You may use Fig. 5.16 to answer this question).
    d. Under steady haze conditions, mountains viewed to the west during the morning hours appear clearer than the same mountains viewed in the afternoon.
    e. Smoke from an automobile appears blue against a dark background but yellow against bright objects.
    f. Why is sunlight observed at 90 degrees to the direction of the sun highly polarized? Why is the amount of polarization not 100%?
    g. Under most circumstances, the single scatter albedo (the ratio Q_{sca}/Q_{ext}) for large particles approaches 0.5. Under which circumstances do you think this albedo is smaller than 0.5 for large particles?

5.2. Rayleigh scattering problem:
    a. From (5.33) and (5.39), show that the scattered intensity at angle Θ is

$$I(\Theta) = I_o \frac{C_{sca}}{R^2} \frac{P(\Theta)}{4\pi}$$

and that the cross–sectional area per molecule is

$$C_{sca} = \frac{|\alpha|^2 128\pi^5}{3\lambda^4}$$

for Rayleigh scattering (note ϵ_o is replaced by the factor $1/4\pi$).

b. Using the Clausius–Mosotti relationship, calculate the scattering cross–section of molecules at 0.3, 0.5 and 0.7 μm assuming $N = 2.55 \times 10^{19}$ molecules per cubic centimeter, which is a typical value at sea level and

$$(m-1) \times 10^8 = 6.4326 \times 10^3 + \frac{2.94981 \times 10^6}{146 - \lambda^{-2}} + \frac{2.554 \times 10^4}{41 - \lambda^{-2}}$$

for air where λ is in μm.

c. If you assume that the variation of the density of air with altitude follows the variation of pressure with altitude $p = p_o \exp(-z/H)$, show that the Rayleigh optical depth, measured from the top of the atmosphere to level z is directly proportional to the pressure at z. [*For later problems and for purposes of comparison, a parameterization of the Rayleigh optical depth as a function of wavelength (in microns) and altitude z (in kilometers) is*

$$\tau_{Ray}(\lambda, z) = 0.0088\lambda^{(-4.15+0.2\lambda)}[e^{-0.1188z-0.00116z^2}]$$

(after Maggraf and Griggs (1969).]

5.3. A plane wave electric field may be expressed as

$$\mathcal{E} = \mathcal{E}_o \exp\{-i(kr - \omega t)\}.$$

a. In the following diagram, the surface of constant phase emerging from two dipoles arrives at location 1 at a different time than at location 2. Convert this time difference into a phase difference. Express your answer in terms of the separation of the dipoles d and angle of incidence γ.

b. Combine the proceeding answer with a similar expression for the difference in the phase between waves scattered from 1 and 2 arriving a distant detector at an angle Θ to the incoming plane wave. Show that the wave fronts are in phase whenever $\Theta = 0°$ (i.e., for the geometry of forward scattering).

5.4. Assume that a particle of diameter d is approximated in a simple way as for Problem 5.3 and that dipoles are non–interacting. For a wave incident on this particle along a direction normal to a line that separates these dipoles (i.e., for $\gamma = 0$), derive an

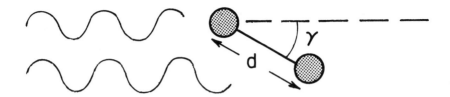

Figure 5.33 Geometry for Problem 5.3.

expression for the scattered intensity as a function of scattering angle.

   a. What is the smallest scattering angle at which you would expect to see a minimum in scattered intensity?

   b. Suppose that an instrument measures $0.5\mu m$ radiation and observes rings around the moon. If the location of the first bright ring is 10 degrees from the center of the moon's disc, what is the size d of the particle?

   c. Under what real cloud conditions would you expect to see such a scattering feature?

5.5. Phase function problem:

   a. Show that

$$\mathbf{S}(\Theta) = k^3 \alpha \begin{pmatrix} \cos \Theta & 0 \\ 0 & 1 \end{pmatrix}$$

for Rayleigh scattering, and that the phase function for unpolarized radiation is

$$P(\Theta) = \frac{3}{4}(1 + \cos^2 \Theta)$$

   b. Calculate the first three expansion coefficients χ_ℓ of this phase function.

5.6. This problem is not for the faint of heart, but it is nevertheless rewarding for those game enough to try it. It is designed to illustrate the concept of the discrete dipole method further and follows the approach of Draine (1988) closely. We begin with

(5.19) which can be rearranged in matrix form

$$\mathbf{Ap} = \mathbf{E}_{in} \tag{5.73}$$

where, for N dipoles, the $3N \times 3N$ matrix $\mathbf{A}$ has the 3×3 block matrix form represented using dyadic notation

$$a_{jk}p_k = \frac{exp(ikr_{jk})}{r_{jk}^3}\{k^2 \vec{r}_{jk} \times (\vec{r}_{jk} \times \vec{p}_k)+$$

$$\frac{(1 - ikr_{jk})}{r_{kj}^2}\left[r_{jk}^2\vec{p}_k - 3\vec{r}_{jk}(\vec{r}_{jk} \cdot \vec{p}_k)\right]\}$$

for the off diagonal blocks and

$$a_{jj} = 1/\alpha_j$$

for the 3×3 block matrix along the diagonal. In this expression $\vec{r}_{jk} \equiv \vec{r}_j - \vec{r}_k$ and p_k is a three–element column vector.

We now wish to solve (5.73) for a simple two dipole particle. Consider this particle to be composed of touching spherical dipoles of diameter d located at the point $(0, d/2, 0)$ and $(0, -d/2, 0)$. We will approximate the volume of this particle to be $2d^3$ and consider the incident plane wave propagating along the $+x$ direction. The $x - y$ plane is taken to be the scattering plane. Consider two polarizations of the incident field, one linear and parallel to the scattering plane (also parallel to the line connecting the dipoles) and one linear but perpendicular to the scattering plane.

a. Show that the 6×6 matrix $\mathbf{A}$ has the following 3×3 block structure

$$a_{12} = a_{21} = -\frac{k^2 e^{ikd}}{d}\begin{pmatrix} 1 & 0 & 0 \\ 0 & 0 & 0 \\ 0 & 0 & 1 \end{pmatrix} +$$

$$\frac{e^{ikd}}{d^3}(1 - ikd)\begin{pmatrix} 1 & 0 & 0 \\ 0 & -2 & 0 \\ 0 & 0 & 1 \end{pmatrix}$$

and

$$a_{11} = a_{22} = \frac{1}{\alpha}\begin{pmatrix} 1 & 0 & 0 \\ 0 & 1 & 0 \\ 0 & 0 & 1 \end{pmatrix}$$

where we assume that the material is isotropic and further-more that the dipoles are homogeneous. [*Hint: In formulat-ing these matrices, you need to specify the dipole moment vector $\vec{p}$ at each dipole location as a column vector with elements $p_x, p_y,$ and p_z.*]

b. Solve (5.74) for **p** assuming the following six element col-umn vectors for the incident electric field

$$\mathbf{E}_{in,\ell/r} = \begin{pmatrix} e_{\ell/r} \\ e_{\ell/r} \end{pmatrix}$$

where

$$e_\ell = e^{-i\omega t} \mathcal{E}_{o,\ell} \begin{pmatrix} 0 \\ 1 \\ 0 \end{pmatrix}$$

and where

$$e_r = e^{-i\omega t} \mathcal{E}_{o,r} \begin{pmatrix} 0 \\ 0 \\ 1 \end{pmatrix}$$

and set $\mathcal{E}_{o,\ell/r} = 1$. [*Do this using symbolic algebra if pos-sible; otherwise, solve numerically using the values given below in part (c)*].

c. Specifying the polarizability α of each dipole is not a trivial issue as Draine (1988) describes. Here we will adopt his values for α according to

$$\alpha = \alpha_o \left[1 - \frac{2i}{3N}(ka)^3 \frac{m^2 - 1}{m^2 + 2} \right]^{-1}$$

where

$$\alpha_o = \frac{3}{4\pi n} \frac{m^2 - 1}{m^2 + 2},$$

n is the number density of dipoles (in this case we assume $n = 1/d^3$), $N = 2$, $a = (3N/4\pi n)^{1/3}$, m is the refractive index set as m= 1.31, $\lambda = 0.7$ μm so that $d \approx 0.028$ μm. With these values, calculate the 2×2 complex amplitude matrices

$$S_2(\Theta) = -ik^3 \sum_{j=1}^{N} \vec{p}_{j,\ell} \cdot \vec{\xi}_\ell \exp(-ik\hat{n} \cdot \vec{r}_j)$$

$$S_3(\Theta) = ik^3 \sum_{j=1}^{N} \vec{p}_{j,r} \cdot \vec{\xi}_\ell \exp(-ik\hat{n} \cdot \vec{r}_j)$$

$$S_4(\Theta) = ik^3 \sum_{j=1}^{N} \vec{p}_{j,\ell} \cdot \vec{\xi}_r \exp(-ik\hat{n} \cdot \vec{r}_j)$$

$$S_1(\Theta) = -ik^3 \sum_{j=1}^{N} \vec{p}_{j,r} \cdot \vec{\xi}_r \exp(-ik\hat{n} \cdot \vec{r}_j)$$

where $\hat{n}$ is the unit vector that defines the direction of scatter (note that $\hat{n} \cdot \vec{r}_j = \pm d \sin \Theta/2$, $\vec{p}_{j,\ell/r}$ is the vector dipole moment of the jth dipole for parallel (ℓ) and perpendicular (r) incident polarization, $\vec{\xi}_{\ell/r}$ is a unit vector defining the direction of polarization after scattering $\vec{\xi}_r = (0,0,1)$ and $\vec{\xi}_\ell = (\sin \Theta, \cos \Theta, 0)$ respectively, and Θ is the scattering angle). Plot the intensity functions $i_1 = |S_1|^2$ and $i_2 = |S_2|^2$ as a function of Θ.

d. The more advanced student wishing to explore this technique further may want to repeat these calculations for a four dipole particle with the additional particle positioned at $(0, 3d/2, 0)$ and $(0, -3d/2, 0)$ and compare results. An interesting application of these calculations is to evaluate the functions S_{11}, S_{12}, and S_{22} from the formulas (e.g., Bohren and Huffman, p. 65)

$$S_{11} = \frac{1}{2}(|S_1|^2 + |S_2|^2 + |S_3|^2 + |S_4|^2)$$

$$S_{12} = \frac{1}{2}(|S_2|^2 - |S_1|^2 + |S_4|^2 - |S_3|^2)$$

$$S_{21} = \frac{1}{2}(|S_2|^2 - |S_1|^2 - |S_4|^2 + |S_3|^2)$$

$$S_{22} = \frac{1}{2}(|S_1|^2 + |S_2|^2 - |S_3|^2 - |S_4|^2)$$

and use the values of these at $\Theta = 180°$ to estimate Z_{DR} using

$$Z_{DR} = 10\log\frac{[S_{11} - S_{12} - S_{21} + S_{22}]}{[S_{11} + S_{12} + S_{21} + S_{22}]}$$

Contrast the Z_{DR} values calculated for a two and four dipole particle.

5.7. For the hypothetical phase function expressed by

$$P(\Theta) = f\delta_{\Theta-0°} + b\delta_{\Theta-180°}$$

where $\delta_{\Theta-\Theta'} = 1$ when $\Theta = \Theta'$ and $\delta_{\Theta-\Theta'} = 0$ when $\Theta \neq \Theta'$, show that

$$g = f - b,$$

and that the forward and backscattering fractions are

$$f = \frac{1}{2}(1 + g)$$

$$b = \frac{1}{2}(1 - g)$$

respectively.

5.8. Extinction and optical depth:
   a. The extinction cross–section at a wavelength λ of a droplet of radius r is approximately $2\pi r^2$ when $2\pi r/\lambda > 1$. Find the optical depth of a cloud 0.5km thick containing 150 drops cm^{-3} having $r = 5\mu$m.
   b. Repeat (a), but for a droplet $r = 10~\mu$m assuming the liquid water content is the same as in (a).
   c. For water, m=(5.46,-2.94) at 19 GHz, what is the optical depth of the cloud defined in (a) at this frequency?
   d. Repeat (c), but assume the particles are ice spheres. The relevant value of the refractive index is m=(1.79,-0.003).

5.9. According to anomalous diffraction theory, the scattering amplitude function is given by the expression

$$S(\Theta = 0) = \frac{k^2}{2\pi}\int\int_G [1 - e^{i\rho(x,y)}]dxdy$$

where $\rho(x, y)$ is the phase lag difference between a ray outside the particle and a ray that enters the particle at the position

(x, y) on its shadow projection. G is the cross–sectional area of the particle's shadow. Using this expression and the fundamental extinction formula (5.34),

   a. derive the extinction efficiency for a cube of length l illuminated normally on its upper face. Plot this efficiency as a function of the size parameter $2\pi l/\lambda$ and compare the spectrum of Q_{ext} with that predicted by the ADT for spheres (assume m=1.33).

   b. Derive the extinction efficiency per unit length of an infinite circular cylinder illuminated by radiation normal to its long axis. [*Hint:*

$$\frac{2}{\pi} \int_0^{\pi/2} [1 - \cos(\rho \cos \gamma)] \cos \gamma d\gamma = H_1(\rho)$$

where $H_1(\rho)$ is a first order *Struve function* (*refer to Abromowitz and Stegun, Ch 12*).]. From the limiting forms of this function, derive Q_{ext} as $\rho \to 0$ and $\rho \to \infty$.

5.10. Anomalous diffraction theory:

   a. For the first four wavelengths and the corresponding refractive indices of water tabulated in problem 4.7, calculate the values of the angle ζ and v as defined by (5.56) and in the text prior to (5.58) for a sphere of radius 5 μm. Estimate $Q_{ext,abs}$ from anomalous diffraction theory (5.57) and the modification provided by (5.58)

   b. Repeat your calculations of Q_{abs} for ice spheres at $\lambda = 10.8$ μm and $\lambda = 12$ μm (the refractive indices are given in the caption to Fig. 7.24). Vary your particle radius from 5 to 20 μm and refer to Fig. 7.24 for comparison and discussion.

5.11. Circular polarization has been observed in light from noctilucent clouds. One explanation for this observation is that these clouds are composed of aligned, nonspherical particles whose matrix is expressed by (5.41). If the incident light on these particles has the form

$$\begin{pmatrix} I \\ Q \\ U \\ V \end{pmatrix} = \begin{pmatrix} I_o \\ 0 \\ 0 \\ 0 \end{pmatrix} + \begin{pmatrix} I_{ms} \\ Q_{ms} \\ U_{ms} \\ V_{ms} \end{pmatrix}$$

where the first term of the right–hand side is the unscattered sunlight and the second term is the contribution by multiply scattered sunlight. Derive the degree of circular polarization of this sunlight due to single scattering by these particles.

5.12. Derive (5.72).

5.13. The volume backscattering coefficient β (this has units of inverse length and the relation between this coefficient and the radar reflectivity is explored in Chapter 8) for a volume filled with N_o identical cloud drops is defined as [see (5.65)]

$$\beta = N_o C_b.$$

a. Show that for spherical Rayleigh scatterers of radius a,

$$\beta = N_o \frac{64\pi^5}{\lambda^4} a^6 \mid K \mid^2$$

b. Assuming that the number density and radius of cloud droplets are 100 cm^{-3} and 20 μm, respectively, calculate β for the following two wavelengths (see table) where m is the refractive index of water. Repeat these calculations for raindrops that are 1 mm in size and with a concentration of $N_o = 1$ liter^{-1}.

Wavelength (cm)	10	3.21
m= (n, κ)	(3.99,1.47)	(7.14,2.89)

6
Passive Sensing — Extinction and Scattering

The preceding four chapters describe different aspects of how electromagnetic radiation interacts with matter. Most practical applications, however, rely on measurements of the accumulated effect of many interactions not a single interaction. The actual path along which these interactions occur might be just a few meters in some cases or hundreds of kilometers for instruments on meteorological satellites. Some way of describing the accumulated effects of all processes as radiation is transferred from one volume of atmosphere to another along these paths is needed.

The theory of radiative transfer provides such a description and two special forms of radiative transfer are introduced in this chapter. One treats only extinction and examples of remote sensing of path integrated quantities based on measurements of this extinction are discussed in the following two sections. The problem of radiative transfer in an opaque scattering atmosphere is unfortunately not described by this extinction equation. In dense media, like clouds, photons reappear along a given direction because of multiple scattering. Accounting for these scattered photons complicates the radiative transfer and ways of simplifying this complexity are discussed in the context of the remote sensing of aerosol, ozone, and clouds.

6.1   Beer's Law and the Remote Sensing of Aerosol

Lambert's law of extinction, introduced in Chapter 3, states that the change in the intensity of radiation emerging from the end of a path of length ds is proportional to the incoming intensity, namely

$$dI_\lambda = -\sigma_{ext,\lambda}(s)I_\lambda ds \qquad (6.1)$$

where the proportionality constant is the volume extinction coefficient $\sigma_{ext,\lambda} = \sigma_{sca,\lambda} + k_{\lambda,v}$, where $\sigma_{sca,\lambda}$ is the volume scattering coefficient, and $k_{\lambda,v}$ is the volume absorption coefficient. This equation only accounts for radiation lost from incidence by extinction and does not include radiation accrued along the path by either scattering into the volume (which is described later in this chapter) or by emission from the volume (see Section 7.1). Equation (6.1) is readily cast into a radiative transfer equation

$$\frac{dI_\lambda}{ds} = -\sigma_{ext,\lambda}I_\lambda \qquad (6.2)$$

which has a solution of the form

$$I_\lambda(s'') = I_\lambda(s')\exp(-\tau_\lambda) \qquad (6.3)$$

where $\tau_\lambda = \int_{s'}^{s''} \sigma_{ext,\lambda}(s)ds$ is the optical thickness. This solution is referred to as *Beer's law* and serves as the basis for the retrieval of selected atmospheric constituents integrated along the path from s' to s''. For example, consider the measurement of direct sunlight. If the sun is inclined at an angle θ_o from the vertical (the solar zenith angle), then (6.3) becomes

$$I_\lambda(\tau_\lambda^*) = I_\lambda(\tau_\lambda = 0)\exp(-\tau_\lambda^*/\cos\theta_o) \qquad (6.4)$$

where τ_λ^* is the optical depth. The logarithmic form of (6.4) is

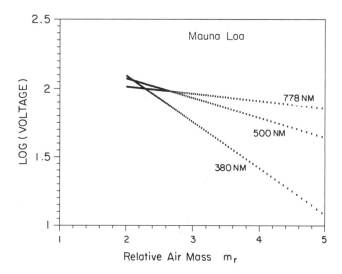

Figure 6.1 An example of a Lambertian plot for three wavelengths (in nm): the logarithm of solar intensity is plotted as a function of optical air mass for clear, stable atmospheric conditions (this air mass is $\sec\theta_o$, from Dutton, private communication).

$$\ln I_\lambda(\tau_\lambda^*) = \ln I_\lambda(\tau_\lambda = 0) - \tau_\lambda^*/\cos\theta_o \qquad (6.5)$$

Figure 6.1 is an example of this type of relationship derived from radiometer measurements obtained at the Manua Loa Observatory. The data are from a spectral radiometer pointed toward the sun and measurements are recorded as the sun moves across the sky throughout the course of a day. If the logarithms of these measured intensities are plotted as a function of $\sec\theta_o$, then the optical depth τ_λ^* is the slope of the line and the incident intensity $I_\lambda(\tau_\lambda = 0)$ is given by the intercept determined by extrapolating $\sec\theta_o$ to zero.

The data presented in Fig. 6.1 indeed obey the kind of straight line relation predicted by (6.5) and this approach offers a simple and seemingly reliable way of deriving aerosol optical depths provided it is reasonable to neglect multiple scattering of sunlight and provided it is possible to remove the contributions to τ_λ^* by other constituents. The type of diagram shown in Fig. 6.1 is referred to as a *Lambertian* or *Langley* plot.

6.1.1 Aerosol Turbidity

The clear sky optical depth τ^* derived from Fig. 6.1 generally has three components. One is due to molecular scattering (Rayleigh scattering), another arises from aerosol scattering, and, depending on the wavelength in question, a third component is the absorption by certain trace gases such as ozone. The aerosol contribution is described in terms of a quantity referred to as the atmospheric *turbidity*. Measurements of this quantity have been made for about seventy years starting with Linke and Boda (1922) and Ångström (1929), and are now routinely performed worldwide as part of the World Meteorological Organization's (WMO's) solar monitoring network. The purposes of such a network include:
• determination of the "clean air" or background turbidity and the geographical, seasonal, and long–term variations of it
• detection of any unusual air pollution occurrence
• provision of information about the optical quality of the atmosphere as it relates to aerosol and gaseous pollution thereby establishing the effects of industrialization on turbidity.

There are three common definitions of turbidity. One is the turbidity index of Linke (T) which relates the total extinction in the real atmosphere to the extinction in a pure Rayleigh atmosphere by

$$I/I_o = 10^{-T\tau_{d,Ray}m_r} \qquad (6.6)$$

so that

$$T = \frac{\log(I_o/I)}{\tau_{d,Ray} m_r} \tag{6.7}$$

where reference to wavelength dependences on all quantities is omitted. One way of interpreting this index is that it represents the number of Rayleigh atmospheres that must be stacked one on top of the other to produce the measured attenuation of sunlight. This index thus contains the effects of all forms of attenuation.

A second index is that of Volz (1959) which follows from (6.4) expressed as

$$I/I_o = 10^{-(\tau_{d,Ray} + \tau_{d,O_3} + B)m_r} \tag{6.8}$$

where I is the observed solar radiation at a wavelength of 0.50 μm adjusted to the mean sun–Earth distance, I_o is the solar radiation at the same wavelength outside the Earth's atmosphere at mean sun–Earth distance, $\tau_{d,Ray}$ and τ_{d,O_3} are, respectively, the decadic optical depths for Rayleigh scattering by O_2 and N_2 molecules and for absorption by ozone. The quantity B is the turbidity factor which is a measure of the aerosol optical depth. The factor m_r is referred to as the optical air mass and accounts for the slant path of the sun. This factor is approximately[1] equal to $\sec\theta_o$ for $\theta_o < 80°$.

The third turbidity index is Ångström's *turbidity coefficient* β, which is related to the aerosol optical depth according to

$$\tau_{a,\lambda} = \beta\lambda^{-\alpha} \tag{6.9}$$

where λ is the wavelength in microns and α, the wavelength exponent, is related to the aerosol particle size distribution in the manner described shortly. Rayleigh scatterers are characterized by $\alpha \approx 4$ whereas $\alpha \approx 0$ for large (relative to the wavelength) scatterers like cloud particles illuminated by sunlight. A typical value of α for atmospheric aerosol is 1.3.

6.1.2   Measurement of Turbidity

An instrument designed to measure the spectral intensity of direct sunlight at 0.50 μm and estimate B at this wavelength is the Volz sun–photometer. Another instrument used in turbidity analyses is the pyrheliometer. The first instrument of this kind dates back to

---

[1] The optical airmass differs significantly from $\sec\theta_o$ only when the solar elevation is low because of refraction and curvature effects on the path.

the early nineteenth century, and much of the motivation for the development of these instruments in the early part of the twentieth century stemmed from the desire to estimate the solar constant (by extrapolation of the Langley plot) and time variations of that quantity. Since then, pyrheliometers have greatly improved in their design and sensitivity, and they are currently used as a standard instrument in the WMO's global solar network. A schematic of the basic design of a pyrheliometer is shown in Fig. 6.2. The instrument consists of a thermopile [2] imbedded in an absorbing disc at the bottom of a tube with diaphragms and a limiting aperture that restricts the instrument's field of view. A filter wheel is located at the entry to the aperture. Modern pyrheliometers are often mounted on an automated sun tracker. A more detailed discussion of the operation of various types of radiometers, including pyrheliometers, is provided by Coulson (1975).

Figure 6.3 presents a time series of approximately 80 years of the Linke turbidity derived from data collected at a high altitude site far removed from the effects of urban areas. Wu et al. (1990) correlate these turbidity measurements with global and hemispheric sea surface temperatures (SSTs) obtained for the same period of time. The turbidity data suggest a slight increase in the transmittance of the atmosphere (i.e., a reduction in turbidity) throughout much of this century due to the apparent decrease in volcanic activity from the early 1900s to the 1980s. Wu et al. claim this decrease in turbidity significantly correlates with the rise in sea surface temperature over the same period of time.

Michalsky et al. (1990) analyzed a 10 year time series of turbidity data from measurements at a single site and removed the seasonal component of τ^* from this time series arguing that this variable component is representative of the tropospheric aerosol optical depth. The residual, plotted in Fig. 6.4a for the wavelength of 0.785 μm, is thus taken to be the stratospheric component of the aerosol optical depth. The various turbidity peaks correlate with

---

[2] Thermopiles are combinations of thermocouples configured together in some way. A thermocouple is made of two dissimilar metals that are joined. Absorption of radiation heats these metals in a dissimilar way producing a difference in temperature between the junction of the metals and a reference junction that in turn produces an electromotive force. The voltage across a single junction is small and is built up by joining several of these in series. The measured voltage is proportional to radiation incident on the thermopile.

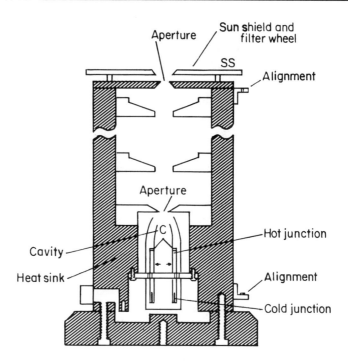

Figure 6.2 An illustration of the components of a modern pyrheliometer. A typical configuration of the filter wheel includes three filters with one aperture left open for measurement of the total spectrum.

times of volcanic activity evidenced by the prominent effects of the El Chichon eruption on the stratospheric aerosol during the early 1980s.

6.1.3   Retrieval of Particle Size

We know from the earlier discussion in Chapter 5 that the properties of extinction depend on the distribution of particle sizes within the observed volume of atmosphere. Differences in spectral measurements of particle extinction can be used to infer information about particle size distributions. To do so, however, requires some a priori information about the form of the size distribution. The simplest example of this approach is based on estimating the wavelength exponent α from measurements of τ using (6.9) and then relating this to a simple power law size distribution. To examine the method,

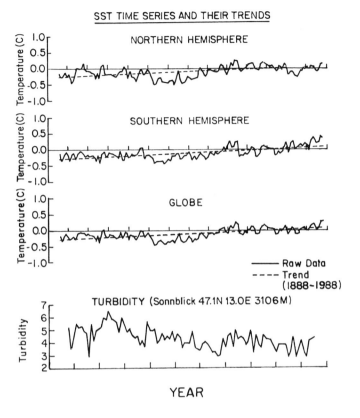

Figure 6.3 SST anomalies and trends for the northern and southern hemispheres and for the globe contrasted with the time series of Linke turbidity for a high mountain site (from Wu et al., 1990).

recall that

$$\sigma_{ext} = \int_0^\infty n\,(r)\,\pi r^2 Q_{ext} dr \qquad (6.10)$$

is the volume extinction coefficient. The observations of Junge (1955) suggest that aerosol size distributions commonly follow a simple power law of the form

$$n(r) = \text{const}\ \ r^{-(\gamma+1)} \qquad (6.11)$$

which, upon substitution into (6.10), produces

$$\sigma_{ext} = \text{const}\ \int_0^\infty Q_{ext} r^{1-\gamma} dr \qquad (6.12)$$

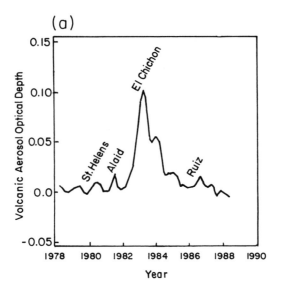

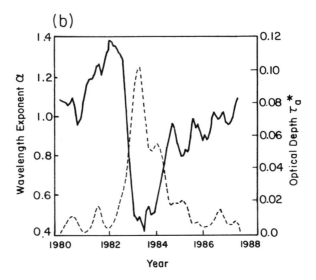

Figure 6.4 (a) Smoothed time series of volcanic aerosol optical depth perturbation at 0.785 μm. The observations were collected at a Mountain Observatory
at 46.4°, 119.6°W. Volcanic events are indicated. (b) Time series of the wavelength exponent, α (solid curve) and the 0.785 μm optical depth of (a) for
comparison (from Michalsky et al., 1990).

Changing the integration over aerosol size r to an integration over size parameter $x = 2\pi r/\lambda$ leads to

$$\sigma_{ext} = \text{const} \left(\frac{\lambda}{2\pi}\right)^{2-\gamma} \int_0^\infty Q_{ext} x^{1-\gamma} dx \qquad (6.13)$$

If x is large and the refractive index constant with λ over the spectral range of the instrument, then Q_{ext} is approximately constant with respect to x. It follows that

$$\sigma_{ext} = \text{const } \lambda^{2-\gamma} \qquad (6.14)$$

where, according to (6.9), $\alpha = \gamma - 2$.

We now introduce a basic assumption, partly for convenience, and partly out of necessity. This assumption is that the shape of the aerosol size distribution remains unchanged with height in the atmosphere so that

$$\tau_a^*(\lambda) = \text{const } H\lambda^{2-\gamma} \qquad (6.15)$$

where H is a constant obtained from

$$H = \int_0^\infty h(z) dz$$

and where $h(z)$ is a normalized profile function describing the relative variation of total aerosol number density along the vertical.

The procedure for estimating α is therefore straightforward once the wavelength exponent $\alpha = \gamma - 2$ is obtained from measurements of τ_a^* at, say, two different wavelengths. The problem with this approach, as we shall see later, is that the wavelength dependence of τ_a^* is often much more complicated than predicted by the simple power law expressed by (6.15). Under these circumstances, interpretation of α is highly ambiguous. Nevertheless, this simple type of analysis offers useful qualitative information about aerosol size distributions. An example of the type of information provided by such an analysis is given in Fig. 6.4b, which shows a time series of α (solid curve) corresponding to the time series of turbidity shown in Fig. 6.4a. The aerosol optical depth shown in Fig. 6.4a is also superimposed on Fig. 6.4b for comparison. The results point to the dramatic impact of El Chichon on the mean particle size of stratospheric aerosol. At peak loading, α falls to about 0.4 implying a substantial increase in large particles. As the El Chichon layer decays in time, the size distribution of the aerosol gradually returns to the more typical pre–eruption values of $\alpha \approx 1.3$.

Excursus: Size Distributions from Anomalous Diffraction Theory

The mathematical problem of retrieving aerosol size distributions from multispectral measurements of light extinction reduces to the problem of inverting a Fredholm integral equation

$$\tau_a^*(\lambda) = \pi H \int_0^\infty r^2 Q_{ext}(m, \lambda, r) n(r) dr \qquad (6.16)$$

for the unknown function $n(r)$. $\tau_a^*(\lambda)$ is obtained from measurements of sunlight using the method described earlier in relation to (6.5) and the function $r^2 Q_{ext}$ is specified in some way. Several approximate analytical solutions to (6.16) exist for the expressions of Q_{ext}, derived from the anomalous diffraction theory formulas introduced in Section 5.5. We start with the ADT expression for a nonabsorbing wavelength written as

$$Q_{ext} = 2 - 4\frac{\sin \rho}{\rho} + 4\frac{1 - \cos \rho}{\rho^2} \qquad (6.17)$$

where $\rho = k'r, k' = 4\pi(m - 1)/\lambda$. The inversion of this equation proposed by Fymat (1978) has the form

$$\pi r^2 n(r) = -\frac{1}{2\pi} \int_0^\infty [\cos \rho + \rho \sin \rho](\tau(k') - 2A) dk' \qquad (6.18)$$

where A, the area of the distribution, is the necessary a priori information required to invert (6.16). It is not the intention to dwell on specific methods for evaluating this equation and reference to a number of studies that address this problem are given in the notes at the end of this chapter. All methods, in principle, require extinction measurements for a large number of wavelengths (i.e., for a large number of k's). Klett (1984) introduced a general inversion to (6.16) based on the Laplace transform theory and provides a framework for coping with more practical situations that occur when measurements are made only for a limited number of wavelengths. An example of the Klett method, verified against synthetic

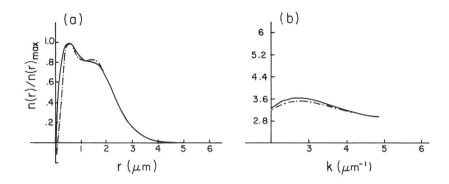

Figure 6.5 (a) Inversion of anomalous diffraction theory extinction data. The input distribution used to simulate particle extinction is plotted as a continuous line and the retrieved distribution is represented by the dashed line. (b) Anomalous diffraction, with a correction factor included (dashed), and Lorenz–Mie extinction (solid) as a function of wavenumber for the bimodal distribution in (a) (after Klett, 1984).

distributions, is given in Fig. 6.5a by the dashed line. Ten extinction values were selected to resolve the $\tau - k'$ spectrum which is also shown in Fig. 6.5b for reference. This is a more complex spectrum than typically predicted by Ångström's turbidity formula (6.9) and is a result of the more complicated bimodal size distribution used to generate the spectrum. The Klett retrieval appears at least capable of reproducing this bimodality.

6.2 More on Extinction–Based Methods

6.2.1 Total Ozone from UV Extinction Measurements

During the 1970s a surge of public concern surfaced about partial destruction of the ozone layer due to anthropogenically produced pollutants. Since then, the substantial depletion of ozone over the Antarctic during the southern hemisphere springtime months was discovered. This has led a significant push into research on ozone chemistry. In recognition of the pressing need for a coordinated international ozone monitoring network, in the 1970s the WMO initiated a special effort to enhance the existing global ozone network. The Dobson ozone spectrometer was selected as the instrument for this worldwide network. Not only do these measurements serve the important purpose of determining hemispheric and global ozone trends, they also provide a means for "ground–truthing" the satellite ozone observations discussed later in this chapter.

The operational principles of the Dobson spectrometer are discussed in Chapter 3. The spectrometer is used to measure the relative logarithmic attenuation of radiation for pairs of wavelengths in the Hartley–Huggins ultraviolet (UV) ozone bands; one where UV absorption is strong and a second where absorption is weak. These attenuation measurements may be made by observing the direct sun, zenith sky under clear conditions or by observing the zenith sky under cloudy conditions. It is beyond the scope of this book to consider the many different ways of estimating ozone from these spectral measurements. A method based on measurement of direct sunlight is described here, although the basis for the methods associated with zenith sky intensity measurements is developed later in this chapter.

The intensity of the direct sun sensed by an instrument obeys Beer's Law (6.3), which we write in the following way [3]

$$\Delta = \log_{10}(I/I') - \log_{10}(I_o/I'_o) =$$

$$- [\Delta\tau_{d,Ray} + \Delta\tau_{d,0_3} + \Delta\tau_{d,a}] \, m_r \qquad (6.19)$$

obtained from differences of the two adjacent wavelengths (the primes identify quantities at the second wavelength). Of the terms in (6.19), the differential optical depth due to aerosol scattering, namely $\Delta\tau_{d,a}$, is the most uncertain. To overcome this uncertainty two pairs of wavelengths, designated by the superscripts A and D, are used. Then from the subtraction of the logarithm differences of pair D from pair A, we obtain

$$\Delta^{AD} = \Delta^A - \Delta^D$$

and

$$\Delta^{AD} \approx - \left[\Delta\tau_{d,O_3}^A - \Delta\tau_{d,O_3}^D\right] m_r - \left[\Delta\tau_{d,Ray}^A - \Delta\tau_{d,Ray}^D\right] m_r. \quad (6.20)$$

Because of the more slowly varying wavelength dependence of the aerosol optical thickness [this dependence is predicted, for example,

---

[3] In the basic reference to the theory of the Dobson measurement, a different slant path factor is used for each attenuating species. In each case, this factor only differs significantly from $\sec\theta_o$ at very low solar elevations. In our discussion of the Dobson measurements, the factor m_r is used for each species only to simplify matters.

by (6.15)], the difference terms for aerosol are taken to be the same for each wavelength pair so aerosol effects subtract out of (6.20). Various combinations of pairs of wavelengths could be used in this approach but the following were proposed as a standard procedure by WMO: A – 0.3055, 0.3254 μm; D – 0.3176, 0.3398 μm. If we take the decadic optical depth of ozone to be simply the product of the decadic absorption coefficient k_d of ozone and the column integrated ozone amount[4] X_o, then inversion of (6.20) yields

$$X_o = \left[\frac{\Delta^{AD}}{m_r} - (\Delta\tau_{d,Ray}^{D} - \Delta\tau_{d,Ray}^{A})\right] / \left[\Delta k_d^{A} - \Delta k_d^{D}\right] \qquad (6.21)$$

Inserting known values for the absorption coefficient and Rayleigh optical depth for the wavelength pairs A and D leads to the simplification

$$X_o = \frac{\Delta^{AD}}{1.388 m_r} - 0.009 p_o \qquad (6.22)$$

where the dependence of X_o on surface pressure p_o (expressed in units of atmosphere) arises through the dependence of Rayleigh optical depth on pressure. The factor 1.388 is the decadic ozone absorption coefficient difference for the A and D double pair in units of atm–cm^{-1}. The total ozone amount defined by (6.22) has units of milli atm–cm or, alternatively known as a Dobson unit (DU). A DU is the equivalent vertical thickness of atmosphere, in thousandths of a cm, that is occupied by the ozone when concentrated into a uniform layer of pure gas at S.T.P. Values of X_o range from about 200DU to about 400DU.

Total ozone derived from a Dobson spectrometer at the South Pole is shown in Fig. 6.6. These ozone amounts are actually supplemented by ozonesonde observations during the twilight months of March to mid–April and September to mid–October, when it is not possible to make Dobson measurements. Shown are the annual variations of total ozone at the South Pole for 1987, 1986 and portions of the 1978 total ozone record. These results show the marked springtime (November) increase of ozone and the significant depletion of

---

[4] The symbol Ω is most commonly used to represent column ozone and the symbol X is taken to represent the ozone profile in the literature. To avoid confusion with later usage, X_o here refers to total column ozone and $X(p)$ or $X(z)$ refers to the ozone path integrated from the top of the atmosphere to level p or z.

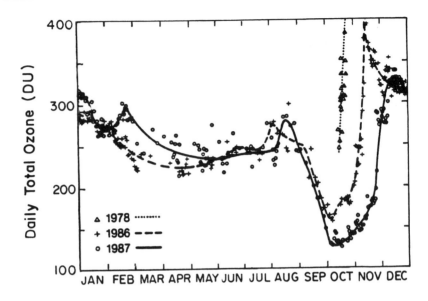

Figure 6.6 Daily total ozone at the south pole in 1986 and 1987 and a portion of the 1978 total ozone record showing how the minimum in October has deepened in the 1980s and how the springtime ozone subsequently increases during (Komhyr et al., 1989b).

total ozone (the so–called ozone hole; Solomon, 1988) observed during the 1980s just prior to the springtime increase of ozone.

6.2.2   Limb Profiling by Extinction—SAGE

One method of obtaining vertical profiles of atmospheric constituents is to measure the radiation emerging from the atmosphere along a horizontal direction. Radiation from the horizon is referred to as *limb* radiation, and the general approach for obtaining profiles of atmospheric constituents derived from limb measurements is referred to as limb sounding. A series of satellite experiments conducted over the past decade have adopted this approach for studying the composition of the stratosphere and limb sounders are now used operationally for stratospheric profiling.

The Stratospheric Aerosol Measurement (SAM) experiment was conducted in the middle to late 1970s and demonstrated the feasibility of solar extinction measurements obtained during satellite sunrise and sunset events observed by a satellite. This experiment served as a prototype for the Stratospheric Aerosol and Gas Experiment (SAGE I, and now SAGE II).

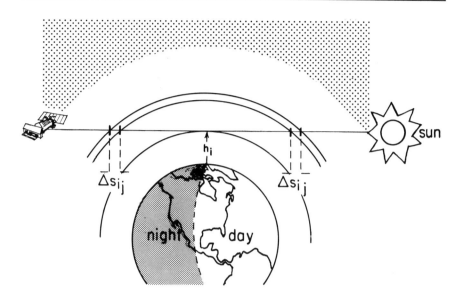

Figure 6.7 The geometry for limb–extinction measurements.

The SAGE I instrument is described in Chapter 3. The SAGE II instrument is similar but has seven channels centered at 1.02, 0.94, 0.6, 0.525, 0.453, 0.448, and $0.385\mu m$. The $0.94\mu m$ measurements are used to obtain water vapor amounts, $0.6\mu m$ is used for ozone, the difference between 0.452 and $0.448\mu m$ used for NO_2 concentrations, and aerosol extinction is deduced from measurements at all wavelengths.

Figure 6.7 illustrates the limb viewing geometry of relevance to this experiment. The instrument on the satellite scans across the solar disc as the sun rises and sets relative to the motion of the satellite. The extinction is taken to be 0 for the highest altitude scan. The ratio of the intensity, I_o, at the center of this scan to the intensity $I(h_i)$ obtained from a scan at a lower altitude h_i provides a measure of the transmittance

$$\mathcal{T}_\lambda = I(h_i)/I_o = \exp(-\tau_\lambda^i) \qquad (6.23)$$

where in this instance,

$$\tau_\lambda^i = \int_0^\infty \sigma_{ext,\lambda}(s)ds$$

is the optical thickness along the tangent path defined by h_i at one of the instrument wavelengths. The extinction at any point s along

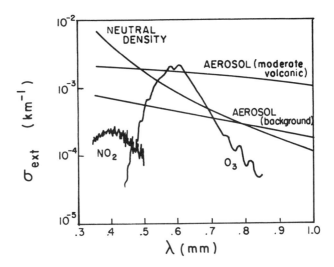

Figure 6.8 The volume extinction at an altitude of 18 km as a function of wavelength (Chu and McCormick, 1979).

this path typically includes several contributions such as

$$\sigma_{ext,\lambda}(s) = \sigma_{\lambda}^{Ray}(s) + \sigma_{\lambda}^{O_3}(s) + \sigma_{\lambda}^{NO_2}(s) + \sigma_{\lambda}^{a}(s) \qquad (6.24)$$

including contributions from Rayleigh scattering, ozone, NO_2, and aerosol, respectively. An example of the relative contributions to the total extinction by each of these attenuations at an altitude of 18 km is illustrated in Fig. 6.8 for the range of wavelengths relevant to the SAGE II spectrometer.

The basic principle of SAGE is to make measurements at several wavelengths and at many altitudes to arrive at profiles of the attenuators. In the retrieval scheme of Chu and McCormick (1979), the atmosphere is divided into 80 layers. The optical thickness for the kth channel then follows as

$$\tau_k^i = 2 \sum_{j=i}^{N=80} \sigma_{ext,k,j} \Delta s_{ij}, \qquad (6.25)$$

where $\sigma_{ext,k,j}$ is the averaged extinction coefficient in the jth layer, and Δs_{ij} is the path length in the jth layer of the ray of the sun passing through the tangent height at the base of the ith layer. The procedure is to determine τ_k^i from the inversion of (6.23) using the measured intensities. The extinction coefficient for each layer

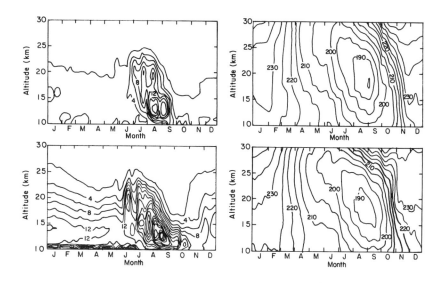

Figure 6.9 Contour plots (left) of the weekly averaged extinction ratio at 1 μm for 1982 (upper) and 1984 (lower) and the corresponding plots of averaged temperature (in K and to the right) for the same years (McCormick and Trepte, 1986).

then follows from the inversion of (6.25) and the extinction by each constituent inferred from either known properties, as in the case of Rayleigh scattering, or from the wavelength dependence of the derived extinction. Details of the inversion algorithms for SAGE I and SAGE II are described by Chu and McCormick (1979) and Chu et al. (1989), respectively. Based on an analysis of the experimental error for typical constituent amounts, Chu and McCormick estimate that the aerosol and ozone profiles can be retrieved to an accuracy of about 10% and the retrieved nitrogen dioxide has an accuracy of about 25% near its peak concentration which occurs between about 25 and 38 km.

SAGE I and II have provided important and unique information about the extinction properties of the atmosphere. For example, we have learned much about the gross properties of polar stratospheric clouds which play a major role in the chemical reactions that produce the Antarctic ozone hole. Some general properties of these clouds may be inferred from the ratio $\sigma_\lambda^a/\sigma_{ext,\lambda}$ derived from SAGE I data over the South Pole. Figure 6.9 shows the annual variation of weekly averages of this extinction ratio at the South Pole. The aerosol extinction maxima that occur during the winter months indicate the presence of stratospheric clouds between about 10–17 km. These episodes of high extinction coincide with stratospheric temperatures colder than about 190 K, a factor considered important to the formation of polar stratospheric clouds and, in turn, to the chemistry responsible for the destruction of atmospheric ozone.

Another demonstration of the value of these data is given in Fig. 6.10 in the form of a six–year time series of weekly averaged aerosol optical depths integrated from 200 mb upward over the Arctic (dashed) and Antarctic (solid) regions. This time series indicates the regular appearance of polar stratospheric clouds in the Antarctic and the lack of these clouds over the Arctic. Also evident is the general increase of stratospheric aerosol over both poles after the El Chichon volcanic eruption.

Not only can SAGE detect volcanic clouds, but observations from SAGE also allow us to map their movement as they are dispersed by the stratospheric winds. Plate 1 (refer to the front of the book) shows a map of the optical depth derived by integrating the 1.02 μm SAGE extinction profiles from 2 km above the tropopause to the highest level of observation for the period in 1991 from mid–June to late July. This map highlights regions where optical depths exceed 0.1 and shows how the Pinatubo volcanic cloud has spread around the equatorial region into both hemispheres (McCormick and Viega, 1992). The line profiles below the SAGE map are the vertical profiles of lidar backscatter (a topic discussed more in Chapter 8) which were measured at Aspendale, Australia, during the same time as SAGE observations. These shows accentuated levels of backscatter between 20 and 24 km associated with the Pinatubo ash cloud.

6.3   Scattering as a Source of Radiation

So far we have only considered scattering as a process that removes photons from a particular direction such as along an instrument's line–of–sight. The radiative transfer equation used in extinction–based retrievals described earlier assumes only single scattering which is reasonable for tenuous scattering media. In dense clouds or under highly turbid skies, photons from the sun can actually reappear again along the direction of the sun when scattered

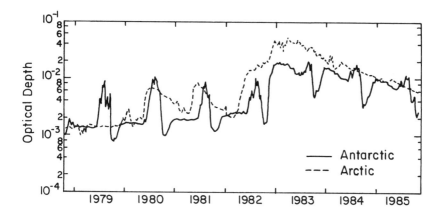

Figure 6.10 Weekly averaged optical depths integrated upward from 200 mb obtained from SAM II observations. The solid line represents Antarctic measurements; the dashed represents Arctic measurements (McCormick and Trepte, 1987).

a multiple of times. In fact, many scattering problems of interest to the atmospheric sciences have to deal with multiple scattering. Multiple scattering of sunlight, for instance, gives rise to many observable phenomena that cannot be explained from single scattering arguments alone. Single scattering predicts a sky that is of uniform brightness and color contrary to what we observe. The whiteness and brightness of clouds is also a result of multiple scattering. Reflection of visible and microwave radiation from various surfaces is largely a product of multiple scattering. Multiple scattering is thus relevant to many topics of remote sensing, and especially to methods based on reflected sunlight.

To account for multiple scattering, we need to introduce a mathematical expression for the reappearance of photons along a specified direction. Unfortunately, to do so requires that we pay more attention to what we mean by direction so that we can talk about changes in direction that occur after a scattering event. We do this using a frame of reference as a basis and describe all scattering processes relative to this frame of reference. Discussion of a particular type of frame of reference used here, and the general definition of a direction vector set on this framework, is relegated to Appendix 1. A few exercises to familiarize the reader with these geometric concepts are

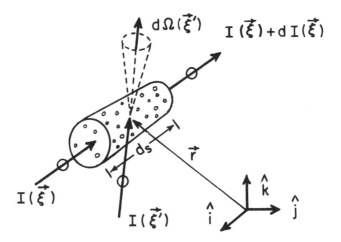

Figure 6.11 Geometry for scattering of diffuse light. $\vec{\xi}$ is the unit vector that defines the direction of the flow and $\vec{r}$ is the vector that specifies the position of the volume element relative to an origin point.

also included in the problem section at the end of this chapter. In brief, we use the vector $\vec{\xi}$ to denote the unit direction. It is relevant to note how this vector may be expressed as a function of the zenith angle θ and azimuth angle ϕ.

Consider a small volume of scatterers illuminated by two beams of monochromatic radiation flowing along directions defined by $\vec{\xi}$ and $\vec{\xi}'$. The volume is centered at $\vec{r}$ and is of length ds (Fig. 6.11). This volume is taken to be small enough, and the distribution of particles sparse enough, that only single scattered photons emerge from it. The incremental increase in intensity along the direction specified by $\vec{\xi}$ due to the scattering of this incident beam is

$$\delta I(\vec{r},\vec{\xi}) = \sigma_{sca} ds \frac{P(\vec{r},\vec{\xi},\vec{\xi}')}{4\pi} I(\vec{r},\vec{\xi}') d\Omega(\vec{\xi}') \qquad (6.26)$$

by virtue of the definition of the phase function given in Chapter 5. Here σ_{sca} is the volume scattering coefficient (the wavelength dependence on all quantities is understood as are the position dependences of σ_{sca} and σ_{ext}). The total contribution to $I(\vec{r},\vec{\xi})$ by scattering of the complete diffuse field surrounding the volume is given by the integral of (6.26), namely

$$dI(\vec{r},\vec{\xi}) = \sigma_{sca} ds \int_{4\pi} \frac{P(\vec{r},\vec{\xi},\vec{\xi}')}{4\pi} I(\vec{r},\vec{\xi}') d\Omega(\vec{\xi}') \qquad (6.27)$$

which prompts the following definition

$$J(\vec{r},\vec{\xi}) = \tilde{\omega}_o \int_{4\pi} \frac{P(\vec{r},\vec{\xi},\vec{\xi}')}{4\pi} I(\vec{r},\vec{\xi}') d\Omega(\vec{\xi}') \qquad (6.28)$$

such that

$$dI(\vec{r},\vec{\xi}) = \sigma_{ext} ds J(\vec{r},\vec{\xi}) \qquad (6.29)$$

The quantity $J(\vec{r},\vec{\xi})$ is the source of radiation due to scattering of diffuse light (sometimes this source is referred to as *virtual emission*) and $\tilde{\omega}_o$ is the single scatter albedo described previously in Chapter 5 and defined here as

$$\tilde{\omega}_o = \frac{\sigma_{sca}}{\sigma_{ext}}$$

this ratio varies between zero for pure absorption and unity for pure scattering (the latter condition is referred to as *conservative* scattering). The quantity $1 - \tilde{\omega}_o$ is the fraction of the incident radiation that is absorbed by the small volume.

The monochromatic radiative transfer equation expresses the net change in intensity of a beam traversing as set path. The intensity change due to both absorption and scattering along ds is

$$dI = dI(extinction) + dI(scattering) \qquad (6.30)$$

or

$$\frac{dI(\vec{r},\xi)}{ds} = -\sigma_{ext}[I(\vec{r},\xi) - J(\vec{r},\xi)] \qquad (6.31)$$

after collecting (6.1) for extinction and (6.29) for the change induced by diffuse scattering. It is relevant to note the similarity of this equation to (7.3)[5] where the scattering source in (6.31) is replaced by the Planck function in (7.3).

Excursus: Single Scatter Albedo of Water and Ice Spheres

We shall soon appreciate the fundamental importance of the single scattering albedo to multiple scattering. In anticipation of this, it is helpful to consider some general properties of $\tilde{\omega}_o$ for spherical particles of a size typically found in clouds. Example spectra of $\tilde{\omega}_o$ for 5 and 50μm water and ice spheres are shown in Fig 6.12a for wavelengths between 0.3 and 50μm. Spectra of $\tilde{\omega}_o$ for longer wavelengths from 100μm to 5mm are presented in Fig. 6.12b, also for ice and water spheres, but of radii of 100μm and 2mm. Notable features of these spectra are:

• Values of $\tilde{\omega}_o$ ≥ 0.99 are typical of wavelengths less than about 1.5 μm (this is shown more clearly in Fig. 6.13). The spectra of $\tilde{\omega}_o$ in the near infrared region is actually complex and contains several weak absorption bands of liquid and solid water.

• The minima of $\tilde{\omega}_o$ are the absorption features of ice and water mentioned earlier and these align with the corresponding maxima of the complex part of the refractive index κ. This is consistent with our expectations from the simple model of particle absorption described in section 5.5 based on anomalous diffraction theory (ADT). This simple theory predicts that Q_{abs} for a particle of radius r increases in a systematic way when the parameter $v = 2x\kappa$ increases, where $x = 2\pi r/\lambda$.

• The single scatter albedo is also sensitive to the size of the particle. An example of this sensitivity is shown in Fig. 6.12b for longer wavelengths where $\tilde{\omega}_o \to 0$ as the size parameter $x \to 0$ when $\lambda \to \infty$. The approach to this regime, the Rayleigh regime, occurs at shorter wavelengths for smaller spheres (as illustrated in the example of the 100μm sphere compared to the 2mm sphere). Ice and water

---

[5] We can also follow the procedure of Section 7.1 to obtain an integral equation of transfer analogous to (7.5). This is not, however, a solution to (6.31) as there is a fundamental difference between (7.5) and the equivalent integral equation that follows from (6.31). In (7.5) the source function appearing in the integrand is known a priori (assuming that the temperature distribution along the path is known) and the solution requires a straightforward integration of known functions. For scattering, the source function appearing in the integrand unfortunately contains the desired intensity and cannot be evaluated a priori without approximation. The presence of the intensity in the definition of $\mathcal{J}$ is what complicates the problem of multiple scattering and why a host of different approximations exist to overcome it.

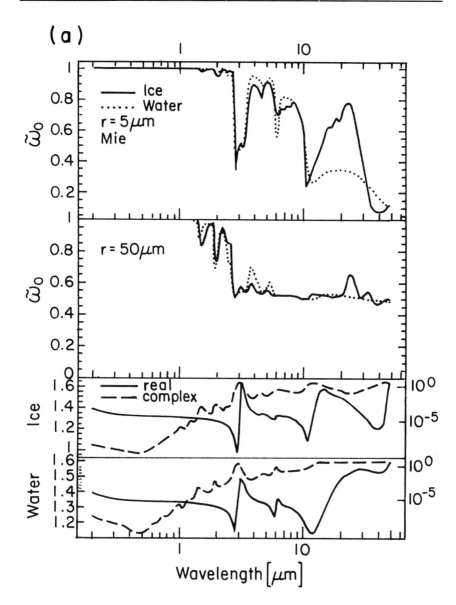

Figure 6.12 (a) The single scatter albedo as a function of λ for 5 and 50 μm water and ice spheres.

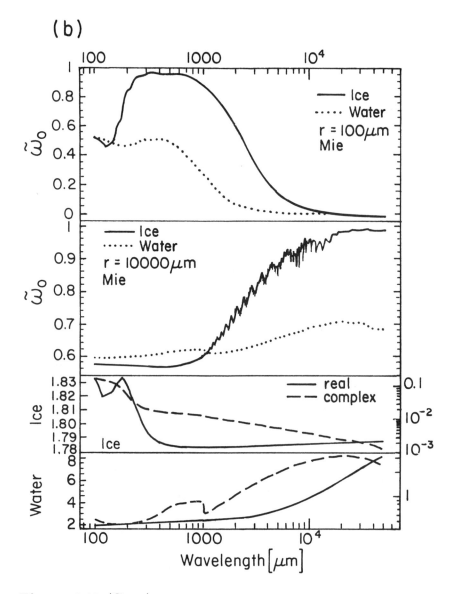

Figure 6.12 (Cont.) (b) Same as (a) but for 100 μm and 2 mm ice and water spheres for $100\mu m < \lambda < 5$ mm.

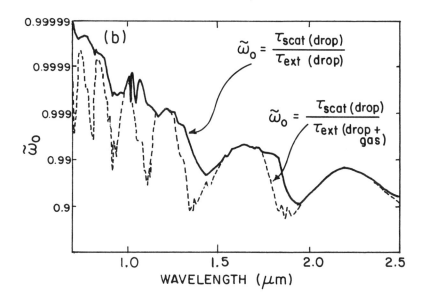

Figure 6.13 The single scatter albedo of a model cloud as a function of wavelength. The solid line refers to droplet absorption alone whereas the dashed line refers to droplet plus vapor absorption (from Twomey and Seton, 1980).

spheres differ at these long wavelengths. Ice particles primarily scatter radiation (i.e., $\tilde{\omega}_o \to 1$) whereas water spheres absorb radiation strongly at these wavelengths (i.e., $\tilde{\omega}_o < 0.5$). These characteristics are crucial elements in the remote sensing of rainfall from passive measurements of microwave radiation.

• There are a number of other spectral regions where the differences between ice and water spheres are large. Notable differences occur in the near infrared region (such as near 1.6 μm) although these are not apparent in Fig. 6.12a. Another difference occurs around $\lambda = 10$ μm where $\tilde{\omega}_o$ for small ice spheres decreases more steeply with increasing λ than does $\tilde{\omega}_o$ for an equivalent sized water sphere. This is fundamental to the change in spectral emission at these wavelengths as discussed in Chapter 7.

The single scatter albedo $\tilde{\omega}_o$ is a volumetric quantity defined as the ratio of the scattering properties of the volume to the properties that define the total extinction by the volume. For most wavelengths of interest, this extinction is a result of both absorption and scattering by cloud particles as well as absorption by the minor gases in the volume, especially water vapor. We see the effects of water vapor absorption on $\tilde{\omega}_o$ over the wavelength range 0.5 μm to 2.5 μm in Fig 6.13. What makes the problem of multiple scattering particularly troublesome at these wavelengths is the fact that both liquid water and bands containing thousands of water vapor lines overlap in the same spectral region producing a very complicated spectrum of $\tilde{\omega}_o$.

6.4 Multiple Scattering — A Natural Solution

A natural solution to problems of multiple scattering in the atmosphere is one that decomposes the light field into components that can be identified with the number of times a photon is scattered. This is more generally known as the method of *orders of scattering*. Figure 6.14a provides the general geometric setting for reference. Light of intensity I_o enters the medium along the direction $\vec{\xi}$ at the point $\vec{r}_o$. The amount of radiation leaving the distant point $\vec{r}$ along $\vec{\xi}$ is

$$I^0(\vec{r},\vec{\xi}) = I_o(\vec{r}_o,\vec{\xi})T(\vec{r}_o,\vec{r},\vec{\xi}) \tag{6.32}$$

where $T(\vec{r}_o,\vec{r},\vec{\xi})$ is the transmission function defined by the path $\vec{r}_o \rightarrow \vec{r}$ along $\vec{\xi}$ as shown in Fig. 6.14a. We refer to I^0 as the *reduced* or *unscattered* intensity.

When some light $I^0(\vec{r}',\vec{\xi}')$ at an intermediate point $\vec{r}'$ undergoes a scattering event, a first order or primary scattered intensity is generated at that point. The intensity gain per unit length of path at $\vec{r}'$ by such a process is

$$I_*^1(\vec{r}',\vec{\xi}) = \mathcal{J}^1(\vec{r}',\vec{\xi})\sigma_{ext},$$

where, according to (6.28), $\mathcal{J}^1$ may be considered as the source associated with the primary scattering of $I^0(\vec{r}',\vec{\xi}')$ integrated over all $\vec{\xi}'$ directions. It therefore follows that the radiation from primary scattering of light from all directions is

$$I_*^1(\vec{r}',\vec{\xi}) = \sigma_{sca}\int_{4\pi}\frac{P(\vec{r}',\vec{\xi},\vec{\xi}')}{4\pi}I^0(\vec{r}',\vec{\xi}')d\Omega(\vec{\xi}') \tag{6.33}$$

The amount of this primary scattered radiation accumulated along the path from $\vec{r}_o \rightarrow \vec{r}$ is

$$I^1(\vec{r},\vec{\xi}) = \int_{\vec{r}_o}^{\vec{r}} I_*^1(\vec{r}',\vec{\xi})T(\vec{r},\vec{r}',\vec{\xi})d\vec{r}' \tag{6.34}$$

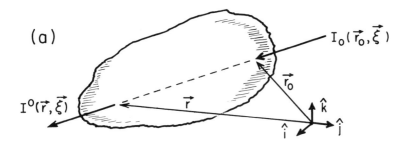

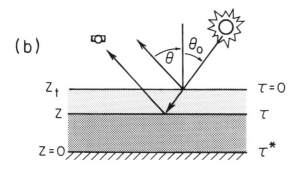

Figure 6.14 (a) Geometry for orders of scattering. (b) Geometry of a plane parallel atmosphere used to compute the primary scattered intensity induced by scattering of a collimated source of solar radiation of intensity I_o.

It is a simple and somewhat intuitive matter to show that construction of the intensities associated with higher order scattering then follows from the repeated application of (6.33) and (6.34) such that

$$I_*^{n+1}(\vec{r},\vec{\xi}) = \sigma_{sca} \int_{4\pi} \frac{P(\vec{r},\vec{\xi},\vec{\xi'})}{4\pi} I^n(\vec{r},\vec{\xi'})d\Omega(\vec{\xi'}) \qquad (6.35a)$$

$$I^{n+1}(\vec{r},\xi) = \int_{\vec{r}_o}^{\vec{r}} I_*^{n+1}(\vec{r'},\xi)T(\vec{r},\vec{r'},\xi)d\vec{r'} \qquad (6.35b)$$

for each integer order $n = 0, 1, \cdots$ of scattering. The total intensity is therefore the sum of all orders of scattering, namely

$$I = I^0 + I^1 + I^2 + \cdots I^n + \cdots = \sum_n I^n \qquad (6.36a)$$

which may be written as

$$I = I^0 + I^* \qquad (6.36b)$$

where I^* is the total diffuse intensity $I^* = \sum_{n=1}^{\infty} I^n$.

Excursus: How Many Times Does a Photon Get Scattered?

The answer to this question largely depends on how many particles there are in the volume, the size of the volume, and on how efficiently the particles scatter the radiation. The latter is characterized by $\tilde{\omega}_o$, and a rough idea of the effect of the single scatter albedo on multiple scattering is given by the following arguments. Suppose $\bar{I}^0$ is an upper bound on I^0. Then from (6.33)

$$I^1_*(\vec{r}',\vec{\xi}) \le \bar{I}^0 \sigma_{sca} \frac{1}{4\pi} \int P(\vec{r}',\vec{\xi},\vec{\xi}') d\Omega(\vec{\xi}') = \bar{I}^0 \sigma_{sca} \qquad (6.37)$$

by virtue of the phase function renormalization condition (Section 5.4). Equation (6.34) can be expressed as

$$I^1(\vec{r},\vec{\xi}) = \int_{\vec{r}_o}^{\vec{r}} I^1_*(\vec{r}',\vec{\xi}) e^{-\sigma_{ext}|\vec{r}-\vec{r}'|} d\vec{r}'$$

for an exponential transmission function defined for the path of length $d = |\vec{r} - \vec{r}'|$. From the condition on I^1_* given by (6.37), it follows that

$$I^1(\vec{r},\vec{\xi}) \le \bar{I}^0 \tilde{\omega}_o (1 - e^{-\sigma_{ext}d}) \le \bar{I}^0 (\eta \tilde{\omega}_o),$$

where we set $\eta = (1 - e^{-\sigma_{ext}d})$. Repeating this procedure for the next order of scattering leads to

$$I^2(\vec{r},\vec{\xi}) \le \bar{I}^0 (\eta \tilde{\omega}_o)^2$$

and

$$I^n(\vec{r},\vec{\xi}) \le \bar{I}^0 (\eta \tilde{\omega}_o)^n \qquad (6.38)$$

for every scattering order n. With the following notation

$$I^{(k)}(\vec{r},\vec{\xi}) \quad \text{for} \quad \sum_{n=0}^{k} I^n(\vec{r},\vec{\xi})$$

it follows that the difference $\Delta = I(\vec{r},\vec{\xi}) - I^{(k)}(\vec{r},\vec{\xi})$ is

$$\Delta = \sum_{j=k+1} I^j(\vec{r},\vec{\xi}) \le \bar{I}^0 \sum_{j=k+1} (\eta \tilde{\omega}_o)^j$$

or

$$\Delta \leq \bar{I}^0(\eta\tilde{\omega}_o)^{k+1} \sum_{j=0} \tilde{\omega}_o^j = \bar{I}^0 \frac{(\eta\tilde{\omega}_o)^{k+1}}{(1 - \eta\tilde{\omega}_o)} \tag{6.39}$$

Consider two examples, one with $\tilde{\omega}_o = 0.5$ and with $\tau = \sigma_{ext}d = 0.1$ and a second with $\tilde{\omega}_o = 0.5$ and $\tau \gg 1$. Suppose that we require I^k (which is the intensity defined for the summation of the first k orders of scattering) to differ from the intensity derived from all orders of scattering by an amount that is not to exceed 1% of $\bar{I}^0$. It follows that $\Delta/\bar{I}^0 \leq 0.01$ and that for the first case

$$0.01 \leq \frac{0.05^{k+1}}{0.95}$$

which implies that $k = 1$ for the nearest integer. Thus single scattering is sufficient to model the intensity field in an optically thin ($\tau = 0.1$) medium for the given value of $\tilde{\omega}_o$. The second case requires that

$$0.01 \leq \frac{0.5^{k+1}}{0.5}$$

where we assume that $\eta \to 1$. In this case $k = 7$ for the nearest integer value. Thus seven orders of scattering are required to model the diffuse intensity with a one percent accuracy when $\tilde{\omega}_o = 0.5$ in an optically thick medium.

This simple exercise provides a clear illustration of the significance of both τ and $\tilde{\omega}_o$ to multiple scattering. We generally infer that the number of scatterings required to represent the total intensity decreases as the absorption by the particle increases (or as $\tilde{\omega}_o \to 0$). For example, many orders of scattering contribute to the total radiation field in optically thick clouds at solar wavelengths where $\tilde{\omega}_o > 0.9$; however, we infer that relatively few scatterings contribute at the infrared wavelengths where $\tilde{\omega}_o < 0.5$ (e.g., Fig. 6.13a and b).

Excursus: Primary Scattering in a Simple Plane Parallel Atmosphere

We now illustrate how (6.35a), (6.35b), and (6.32) can be used to calculate the intensity of primary and higher order scattered light in a plane parallel atmosphere. Consider the geometry shown in Fig. 6.14b. In this simple atmosphere, the distribution of the scattering and extinction properties are taken to be vertically uniform up to a level z_t where sunlight enters the atmosphere along the direction defined by the zenith angle θ_o. In this case, (6.32) is the familiar statement of Beer's law

$$I^0(z,\vec{\xi}_o) = I_o e^{-\sigma_{ext}(z_t-z)/\mu_o} \tag{6.40}$$

where $\vec{\xi}_o$ is used to refer to the direction of the sunlight [note that $\vec{\xi}_o = (\theta_o, \phi_o)$ and $\mu_o = \cos\theta_o$ as discussed in the Appendix]. From (6.35a) it follows that

$$I_*^1(z,\vec{\xi}) = I_o e^{-\sigma_{ext}(z_t-z)/\cos\theta_o} \sigma_{sca} \int_{\Omega_o} \frac{P(\vec{\xi}',\vec{\xi})}{4\pi} d\Omega(\vec{\xi}') \tag{6.41}$$

where the integral is over the small solid angle Ω_o of the sun and where $\vec{\xi}'$ is confined to this solid angle. Since Ω_o is small, we approximate (6.41) as

$$I_*^1(z,\vec{\xi}) = \frac{F_o}{4\pi} e^{-\sigma_{ext}(z_t-z)/\cos\theta_o} \sigma_{sca} P(\vec{\xi}_o,\vec{\xi}) \tag{6.42}$$

where $F_o = I_o\Omega_o$ is the flux of solar radiation incident at the top of the atmosphere through a surface perpendicular to $\vec{\xi}_o$.

The primary scattered intensity that accumulates along the path from z_t down to z is therefore

$$I^1(z,\vec{\xi}) = \int_z^{z_t} I_*^1(z',\vec{\xi}) e^{-\sigma_{ext}(z'-z)/\cos\theta} \frac{dz'}{\cos\theta} \tag{6.43}$$

where θ is the zenith angle that defines $\vec{\xi}$.

Suppose that we are interested in modeling the radiation scattered to an instrument on a satellite. We write (6.43) as

$$I^1(z,\vec{\xi}) = \int_0^z I_*^1(z',\vec{\xi}) e^{-\sigma_{ext}(z-z')/\cos\theta} \frac{dz'}{\cos\theta} \tag{6.44a}$$

for radiation accumulated from z' up to z and after integrating this equation,

$$I^1(z,\vec{\xi}) = \frac{\tilde{\omega}_o F_o}{4\pi\mu m} P(\vec{\xi}_o,\vec{\xi}) \left[e^{-\sigma_{ext}(z_t-z)/\mu} - e^{-\sigma_{ext}(z_t/\mu_o+z/\mu)} \right]$$

where $\mu = \cos\theta$. For $z = z_t$, it follows that the primary scattered radiation observed by a satellite is

$$I^1(z,\vec{\xi}) = \frac{\tilde{\omega}_o F_o}{4\pi\mu m} P(\vec{\xi}_o, \vec{\xi})[1 - e^{-\tau^* m}] \qquad (6.44b)$$

where $m = \sec\theta_o + \sec\theta$ and $\tau^* = \sigma_{ext} z_t$ is the optical depth of the atmosphere. In deriving (6.44b), it is assumed that both $\mu, \mu_o \geq 1$ and it is relevant to recall that this equation is derived assuming that all scattering properties are uniformly distributed throughout the atmosphere.

6.4.1 An example: An Atmospheric Correction for NDVI

Remote sensing of surface properties from satellites using measurements of reflected sunlight often requires some type of correction for attenuation of sunlight within the atmosphere. A simple order of scattering approach applied specifically to AVHRR measurements was implemented by Mitchell and O'Brien (1993) to remove the specific effects of molecular scattering from calculations of the normalized vegetation index (NDVI, refer to Section 4.5 for the definition of this index). In their method, they separate the satellite measured intensity

$$I_{meas} = I_{surf} + I^*$$

into a surface term I_{surf} and a correction or atmospheric term I^*. The latter, they propose, contains three main contributions: One is the primary scattered radiation from the atmosphere back to the satellite, another includes this single scatter forward to the surface followed by a surface reflection, and the third accounts for a surface reflection followed by an atmospheric scattering forward to the satellite. Derivation of the latter two terms is not given here, but the primary scattering term in their study is given by (6.44b).

Mitchell and O'Brien (1993) estimate I^* in this way and subtract this from the intensity measured by the AVHRR. This corrected intensity is assumed to approximate the intensity reflected directly from the surface more closely and they use this corrected quantity in the definition of the NDVI. Examples of the NDVI derived with and without this correction are given in Plate 2 (refer to the front of the book). The upper two panels show the NDVI for two viewing geometries derived from the uncorrected, raw AVHRR intensities. The target area of north western Tasmania is shown for two successive orbits of the NOAA 11 satellite. If we suppose that the surface is a Lambertian reflector and that the surface properties remain the same between these satellite passes, then the NDVI should be independent of the viewing geometry. There are large differences between the two uncorrected NDVI distributions. The NDVIs derived from the westward view (upper right panel and away from the sun) are consistently lower than those for the eastward view (toward the sun). The NDVI derived from the intensities corrected for the effects of molecular scattering are much less sensitive to viewing geometry (middle panels of Plate 2) although slight view angle anisotropies remain. Contributions to this asymmetry may arise from a number of other factors including scattering processes by aerosol as well as by non–Lambertian surface reflection effects. The bottom two panels include an additional correction for aerosol scattering.

6.4.2 A Second Example: Aerosol From AVHRR

The retrieval of aerosol properties from satellite measurements of reflected sunlight is a difficult task. These difficulties stem from the relatively small influence of aerosol on reflected sunlight necessitating accurate measurements of intensities. This is further compounded by the variable effects of the surface below and still further by the influence of aerosol microphysics on the reflection.

One way to consider these problems is to suppose that the intensities reflected by an aerosol layer of optical thickness τ^* may be approximated as

$$I(0, \mu) = \frac{F_o \tilde{\omega}_o}{4\pi \mu m} P(\vec{\xi}_o, \vec{\xi})[1 - e^{-\tau^* m}]$$

and for $\tau^* < 1$,

$$I(0, \mu) \approx \frac{F_o \tilde{\omega}_o}{4\pi \mu} P(\vec{\xi}_o, \vec{\xi}) \tau^* \qquad (6.45)$$

For wavelengths of red light and longer, the additional contribution to the reflected intensity by Rayleigh scattering is relatively small, although it is not difficult to calculate this contribution, as Mitchell and O'Brien (1993) did in their study described earlier. Ignoring Rayleigh scattering, the aerosol optical depth follows from (6.45) as

$$\tau^* \approx \frac{4\pi \mu}{\tilde{\omega}_o F_o P(\vec{\xi}_o, \vec{\xi})} I(0, \mu) \qquad (6.46)$$

This demonstrates how the retrieval of optical depth depends on both the scattering phase function and $\tilde{\omega}_o$. Both depend on the microphysical properties of the aerosol in a manner not known a priori.

Under some circumstances, it may be assumed that the aerosol particles are nonabsorbing with $\tilde{\omega}_o = 1$ so that the effects of particle size on the retrieval of optical thickness expressed by (6.46) enter only through the phase function. Since the size distribution is not known, some assumption about the form of $P(\vec{\xi}, \vec{\xi}_o)$ and its sensitivity to particle size is required. Durkee et al. (1991) propose that the following ratio of intensities measured at two wavelengths

$$S_{12} = \frac{I_1}{I_2} \approx \frac{[\tilde{\omega}_o P(\vec{\xi}_o, \vec{\xi})\tau^*]_1}{[\tilde{\omega}_o P(\vec{\xi}_o, \vec{\xi})\tau^*]_2}$$

could be used to parameterize the scattering phase function. The subscripts 1 and 2 refer to channels 1 (red) and 2 (near infrared) of the AVHRR, respectively. The idea is that any changes in the aerosol size distribution result in spectral variations in the measured intensities both by variations in optical depths and variations in phase functions. We note from earlier discussion how gross changes in size distributions of aerosol affect the wavelength variation of aerosol optical depth. The viewing geometry dictates that scattering is always along the backward direction and the ratio of forward to backward scattering increases as the size of the scatterer increases (Chapter 5). Thus the phase function is sensitive to the slope of the size distribution – as this slope increases (i.e., as the distribution shifts to smaller particles) then the proportion of backscattered light increases. The combined effects of optical depth and phase function lead to an increase in S_{12} with increasing slope. Durkee et al. (1991) introduce an aerosol phase function parameterized solely in terms of S_{12} and tune this parameterization to values typical of marine aerosol (they propose $S_{12} \approx 1.2$ is a typical value) and to values typical of rural aerosol ($S_{12} \approx 1.8$).

In retrieving the aerosol optical depth from (6.46), Durkee et al. developed criteria to identify cloud–free pixels (which is approximately 15–20% of all pixels) from other channels of the AVHRR. For the month of April, 1983, they gridded these cleared pixels into $1° \times 1°$ regions of intensity and maps of these gridded data, converted to τ^* and S_{12}, are presented in Plates 3 and 4 (refer to the front of the book). They find smaller values of optical depth in the Southern Hemisphere (SH), typically 0.1 or less, than in the Northern Hemisphere (NH) where values exceed 0.15 or more. A small area with high values is found along the western coast of Central America during this month is perhaps indicative of higher aerosol concentrations associated with agricultural burning in this region. High values of optical depth are also apparent over the East China Sea as well as surrounding the island of Borneo and large values of optical depth extend out of Western Africa into the North Atlantic with values exceeding $\tau^* > 0.5$ near the coast. This is also consistent with observations of the transport of Saharan dust from Africa into the western Atlantic basin and into the Caribbean (e.g., Prosperso, 1982).

Plate 4 presents the distribution of S_{12} for April 1983. The general pattern is for a decrease from values $S_{12} \approx 2$ in equatorial regions to smaller values over the higher latitude oceans with $S_{12} \approx 1.5$ over the southern oceans. High values of S_{12} over tropical oceans are consistent with aerosol production from biogenic sources of oceanic Dimethyl Sulphide (DMS) which tends to have highest concentrations in upwelled tropical waters (Andreae, 1985). Durkee et al. lend credence to this proposition by showing a high correlation between ship–measured DMS concentrations and S_{12}.

6.5   The Remote Sensing of Ozone From Scattered UV Radiation

Ozone has been observed from the ground for many years using Dobson spectrometers. The operation of this spectrometer is described in Section 3.6, and the basis for the retrieval of total ozone from measurements made with this instrument is presented in Section 6.2. The need for global measurements, together with the added advantages gained in using a single instrument to make these measurements, stimulated the development of satellite ozone measuring techniques.

The general principle for measuring ozone from backscattered UV radiation lies in the relationship between ultraviolet sunlight reflected from the atmosphere and the amount of ozone within the atmosphere. This relationship may be in a form that either relates the total column ozone to reflected sunlight or one that relates the vertical profile of ozone to this radiation. The primary scattering model described earlier provides a convenient way of examining these relationships. For our purposes, we treat the atmosphere in the highly idealized way as shown in Fig. 6.15. Ozone absorbs UV sunlight as it passes downward into the atmosphere and again after being reflected by the atmosphere below. It is assumed that the ozone layer does not scatter any radiation and that sunlight is scattered by only a relatively thin layer adjacent to the ground and well below the bulk of the absorbing ozone layer. Suppose also that the majority of the radiation reflected by the atmosphere to the satellite arises from single scattering in this lower layer and let us ignore the contribution by reflection from the ground for now. We adapt (6.44a) in the form

$$I(\tau = 0, \mu) = \frac{F_o P_{Ray}(\vec{\xi}_o, \vec{\xi})}{4\pi} \int_0^{\tau_*} \tilde{\omega}_o(t) e^{-tm} \frac{dt}{\mu} \qquad (6.47)$$

where the scattering by molecules is defined by the Rayleigh phase function $P_{Ray}(\vec{\xi}_o, \vec{\xi})$ and where we now treat $\tilde{\omega}_o$ as variable throughout the atmosphere. The optical depth of a layer extending from the top of the atmosphere at $\tau = 0$ to some level defined by $\tau = t$ is

$$t(p) = \tau_{Ray}(p) + k_\lambda X(p) \qquad (6.48)$$

where p is the pressure level corresponding to t, $X(p)$ is the ozone path measured from the top of the atmosphere to p, and τ_{Ray} is the contribution to the optical depth at p due to Rayleigh scattering. Combining (6.48) with (6.47) leads to

$$I(\tau = 0, \mu) = \frac{\tilde{\omega}_o F_o}{4\pi\mu} P_{Ray}(\vec{\xi}, \vec{\xi}) \beta_o \int_0^{p_o} e^{-[\tau_{Ray}(p) + k_\lambda X(p)]m} dp \qquad (6.49)$$

This equation exploits the dependence of optical depth on pressure (refer to Problem 5.2c),

$$\tau_{Ray}(p) = \beta_o p$$

where $\beta_o = \tau_{Ray}(p_o)/p_o$ is defined at the surface pressure p_o. Another simplification can be introduced here as long as we assume the ozone layer sits above the scattering layer and approximate the pressure dependence of optical depth as

$$t(p) = \beta_o p + k_\lambda X_o$$

where X_o is the total column ozone. This equation in (6.47) followed by integration yields

$$I(\tau = 0, \mu) = \frac{\tilde{\omega}_o F_o}{4\pi\mu} \frac{P_R(\vec{\xi}_o, \vec{\xi})}{m} [1 - e^{-\beta_o p_o m}] e^{-k_\lambda X_o m} \qquad (6.50)$$

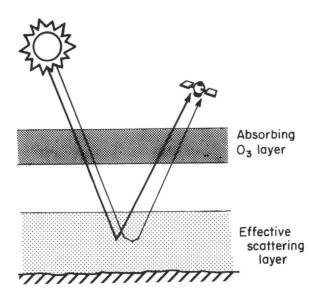

Figure 6.15 A simplified view of UV radiation in the atmosphere. This radiation is absorbed by the ozone layer and scattered by the atmosphere below this layer.

The factors that multiply the ozone amount X_o in the exponent can be adjusted by varying either wavelength (and thus the absorption coefficient k_λ), or changing the viewing direction — the factor μ in the definition of m or making measurements at different solar zenith angles — the μ_o factor, also in the definition of m. For all of these situations, the basic procedure for estimating ozone is the same; namely, ozone amount is retrieved from the inversion of (6.50) given a set of observations $I(\tau = 0, \mu)$ at the top of the atmosphere. An entirely equivalent procedure can also be developed for a similar set of measurements obtained at the surface (Problem 6.7).

For the more general situation in which the optical depth varies with pressure in the manner defined by (6.48), we can write (6.49) in the following way

$$I(\tau = 0, \mu) = \text{constant} \int_{-\infty}^{\ln p_o} \mathcal{W}(0,p) d\ln p$$

where all factors preceding the integral in (6.49) are constant for fixed viewing angle θ, and where

$$\mathcal{W}(0,p) = pe^{-[\beta_o p + k_\lambda X(p)]m}$$

The product $\mathcal{W}(0,p)\Delta p$, defined for a thin layer centered on p and of thickness Δp, is the contribution by scattering in this layer to the intensity observed at $p = 0$. The function $\mathcal{W}(0,p)$ is referred to as the *weighting function*. The simple model introduced here predicts a bell shape variation with pressure for this weighting function similar to that for other weighting functions described in the next chapter. This bell shape mathematically occurs as a result of the product of two factors, one increasing with increasing p and the exponential factor that decreases with increasing p.

Numerically derived weighting functions for scattered UV sunlight are given in Fig. 6.16 as dashed curves for the three wavelengths 0.3025, 0.3075, and 0.3125 μm. The results apply to realistic distributions of ozone and scatterers (as opposed to the simple disjoint two–layer example used earlier). The curves are arbitrarily normalized to unity at the level of their maxima. For $\lambda = 0.3025$ μm, the effective scattering layer actually occurs above the maximum concentration of ozone. At this wavelength, the model described earlier is unrealistic and only limited information about total ozone is available since very little of the diffusely reflected UV sunlight originates from below the ozone layer. The solid curves for each wavelength correspond to the weighting function derived for all orders of scattering. These indicate that multiple scattered photons contribute only 7% to the total backscattered energy at $\lambda = 0.3025$ μm, so the assumption of single scattering is reasonable at this wavelength. The situation for $\lambda = 0.3075$ μm is considerably different with the contribution by multiple scattered sunlight now 31% of the total backscattered energy. This wavelength clearly contains more information about total ozone, although an appreciable fraction of the emergent photons fail to penetrate the ozone layer even at this wavelength. The longer wavelength, $\lambda = 3125$ μm, with a weighting function peaking in the lower troposphere, is the best choice for total ozone measurements since most of the emergent photons pass through the ozone layer and the scattering emanates largely from the lower troposphere in a way similar to that of our two–layer atmosphere. Multiple scattering, however, is an essential ingredient in developing a relationship between reflected sunlight and total ozone at these wavelengths. Furthermore, since an appreciable portion of the incoming radiation at the longer wavelength reaches the bottom of the atmosphere, the reflectivity of the underlying surface is also important, although this reflection is taken to be 0 for this discussion. We shall now outline how surface reflection modifies these arguments.

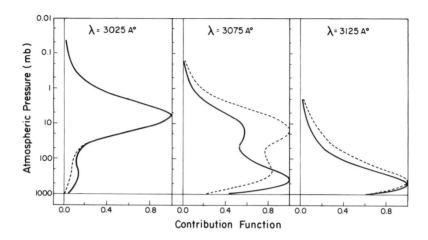

Figure 6.16 Normalized scattering weighting functions for three UV wavelengths as a function of altitude. For each wavelength, the contribution to the weighting function by single scattering (dashed curves) is compared to the weighting function corresponding to all orders of scattering with $\mu = 1$ and $\theta_o = 60°$ (Dave and Mateer, 1964).

6.5.1 Reflection from Below: The Principle of Interaction

Adding a reflecting surface below a scattering layer involves steps that are fundamental to solving the most general of multiple scattering problems. At the heart of all methods is one very important principle called *the principle of interaction*. This principle embodies the linear nature of the interactions between radiation and matter. Before introducing this principle here, and before solving our particular problem, we need to define properties that describe how a surface both reflects and transmits radiation in different directions.

Consider the thin slab of material illustrated in Fig. 6.17a. The surface may be the top of a cloud, the top of the atmosphere, the ground, or any other "surface" capable of both reflecting and transmitting radiation. If the surface is illuminated by radiation with an intensity I_i along the direction $\vec{\xi}$ contained in an element of solid angle of $d\Omega(\vec{\xi})$, then the

$$\text{incident flux density} = I_i(\vec{r}, \vec{\xi}')d\Omega(\vec{\xi}')$$

The flux density reflected from the surface into the set of directions

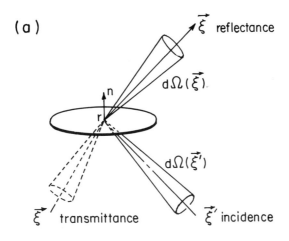

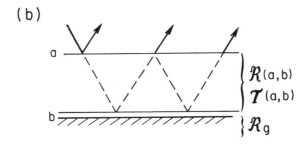

Figure 6.17 (a) A thin slab of a surface showing the diffuse reflection and transmission separately. (b) A single layer of atmosphere overlying a reflecting surface.

$d\Omega$ along the direction $\vec{\xi}$ produces a

$$\text{response flux} = I_r(\vec{r}, \vec{\xi})d\Omega(\vec{\xi})$$

and the ratio of these fluxes defines a generalized reflection function

$$r(\vec{r}, \vec{\xi}, \vec{\xi}') = \frac{I_r(\vec{r}, \vec{\xi})d\Omega(\vec{\xi})}{I_i(\vec{r}, \vec{\xi}')d\Omega(\vec{\xi}')}$$

Suppose that the flux density emerging from the surface is measured by an instrument looking at the surface. If this instrument collects radiation within a field of view defined by a small solid angle $d\Omega_{fov}$, then it follows from (6.47) that the measured intensity is

$$I_r(\vec{r}, \vec{\xi}) = \frac{1}{d\Omega_{fov}} r(\vec{r}, \vec{\xi}, \vec{\xi}') I_i(\vec{r}, \vec{\xi}') d\Omega(\vec{\xi}')$$

If the flow is such that it opposes the incoming flow of radiation, namely if $\vec{\xi} \cdot \vec{\xi}' < 0$, then the radiation is said to be reflected, and we define

$$\mathcal{R}(\vec{r}, \vec{\xi}, \vec{\xi}') = \frac{1}{d\Omega_{fov}} r(\vec{r}, \vec{\xi}, \vec{\xi}')$$

as the bidirectional reflection function such that the reflected intensity is

$$I(\vec{r}, \vec{\xi}) = \mathcal{R}(\vec{r}, \vec{\xi}, \vec{\xi}') I(\vec{r}, \vec{\xi}') d\Omega(\vec{\xi}')$$

This reflection function applies equally to a surface that is infinitely deep, as we assumed in the example of a land surface in Section 4.5, or to a thin slab of material in a vacuum such as illustrated in Fig. 6.17a. Using entirely analogous arguments, when the flow from the surface is observed in the same sense as the flow onto the surface $\vec{\xi} \cdot \vec{\xi}' > 0$, then the radiation is said to be transmitted through the slab and

$$T(\vec{r}, \vec{\xi}, \vec{\xi}') = \frac{1}{d\Omega_{fov}} \frac{I_t(\vec{r}, \vec{\xi}) d\Omega(\vec{\xi})}{I_i(\vec{r}, \vec{\xi}') d\Omega(\vec{\xi}')}$$

is the bidirectional transmittance function. The intensity of radiation transmitted through the slab is therefore

$$I(\vec{r}, \vec{\xi}) = T(\vec{r}, \vec{\xi}, \vec{\xi}') I(\vec{r}, \vec{\xi}') d\Omega(\vec{\xi}')$$

For a surface illuminated by multiple sources of radiation where $I(\vec{r}, \vec{\xi}_k')$ is the intensity of the kth source below and $I(\vec{r}, \vec{\xi}_j')$ is the intensity of the jth source above, then the radiation emerging from the surface along the direction $\vec{\xi}$ is

$$I(\vec{r}, \vec{\xi}) = \sum_{\text{no of j sources}} \mathcal{R}(\vec{\xi}, \vec{\xi}_j') I_i(\vec{\xi}_j') d\Omega(\vec{\xi}_j')$$

$$+ \sum_{\text{no of k sources}} T(\vec{\xi}, \vec{\xi}_k') I_i(\vec{\xi}_k') d\Omega(\vec{\xi}_k') \qquad (6.51)$$

This is a statement of the principle of interaction — the resultant intensity observed leaving the surface is a simple superposition of reflected and transmitted intensities. This expression can be written in a convenient way if we consider a set of n discrete incident and n distinct response directions $(\pm\vec{\xi}_1 \pm \vec{\xi}_2 \ldots \pm \vec{\xi}_n)$ and write

$$I^{\pm} = \left[I(\pm\vec{\xi}_1), I(\pm\vec{\xi}_2) \ldots I(\pm\vec{\xi}_n) \right]^t$$

as a column vector of intensities such that I^+ is the set of n intensities flowing along the n directions $\vec{\xi}_1 \ldots \vec{\xi}_n$ and I^- is the set of n intensities flowing along the n directions $-\vec{\xi}_1 \ldots -\vec{\xi}_n$.

Using the interaction principle expressed in the form of (6.51), together with the vectorial representation of the intensities, we can now determine the intensities emerging from a scattering layer overlying a reflecting surface as shown in Fig. 6.17b. For convenience, the scattering layer extends from a to b and is illuminated from above by a diffuse source of intensity characterized by the vector $I^-(a)$. R_g is the matrix of bidirectional reflection functions of the surface. The following $R(a, b), R(b, a), T(a, b)$ and $T(b, a)$ represent matrices of bidirectional reflection and transmission functions of the atmosphere such that

$$I^-(b) = T(a, b)I^-(a) + R(b, a)I^+(b)$$

$$I^+(a) = R(a, b)I^-(a) + T(b, a)I^+(b) \tag{6.52}$$

is an alternative way of writing (6.51). It is also possible to write an interaction statement for surface reflection as

$$I^+(b) = R_g I^-(b)$$

Re–arrangement of these equations yields

$$I^+(a) = [R(a, b) + T(b, a)R_g[1 - R(b, a)R_g]^{-1}T(a, b)]I^-(a) \tag{6.53}$$

where the first term of the right–hand side is the reflection of the incident light by scattering from the layer $a \to b$. The second term is the contribution to the reflected light by multiple scattering between $a \to b$ and the surface.

We can now deal with our problem of UV radiation reflected from both the surface and atmosphere. For a source of UV radiation of intensity I_o confined to the direction $\vec{\xi}_o$, the reflected intensity measured by a radiometer on a satellite viewing along the direction $\vec{\xi}_{sat}$ follows from (6.53) as

$$I(0, \vec{\xi}_{sat}) = R(p_o, X_o, \vec{\xi}_{sat}, \vec{\xi}_o)I_o(\vec{\xi}_o) + I_{atm}(\vec{\xi}_{sat}, \vec{\xi}_o, p_o, R_g, X_o) \tag{6.54}$$

where $I_{atm}(\vec{\xi}_{sat}, \vec{\xi}_o, p_o, R_g, X_o)$ is the intensity generated by multiple scattering between the surface and the scattering layer [this is just the second term of (6.53)] and is a function of the viewing direction

$\vec{\xi}_{sat}$, the direction of the collimated sunlight $\vec{\xi}_o$, the pressure of the reflecting surface p_o through the dependence of the Rayleigh optical depth on pressure, the column ozone amount X_o, and the surface reflection.

6.5.2   Total Ozone Algorithms from Backscattered UV

The Nimbus 7 satellite carries two ozone instruments. One is the Total Ozone Mapping Spectrometer (TOMS) which consists of a monochromator with a field of view of approximately $3° \times 3°$ at the subsatellite point which scans in the plane perpendicular to the orbital plane. The backscattered radiation is sampled at 6 wavelengths, 313, 318, 331, 340, 360, and 380 nm. The second instrument is the Solar Backscattering Ultraviolet (SBUV) nonscanning instrument which has a larger field of view than the TOMS (11.3° square). The wavelengths used for total ozone are 340, 331, 318, and 313 nm, which are also a subset of the wavelengths of the TOMS instrument. Both instruments are calibrated on board by viewing solar radiation reflected by a diffuser plate.

The principle for computing total ozone from measured UV radiation is similar to that described earlier for ground–based Dobson measurements. The procedure begins with a correction for surface reflection using precomputed tables of $R(p_o, X_o, \vec{\xi}_{sat}, \vec{\xi}_o)$, $T(b,a)$ and $R(a,b)$ for a full range of view angles $\vec{\xi}_{sat}$, solar angles $\vec{\xi}_o$, and for specified solar intensities $I_o(\vec{\xi}_o)$. Computations were carried out for 17 standard O_3 profiles appropriate to three different latitudes. Two sets of tables were computed, one for $p_o=1$ atm and another for $p_o=0.4$ atm. The surface reflection is determined from (6.53) based on measurements at longer UV wavelengths where the dependence on column ozone is taken to be negligible. For example, the TOMS algorithm uses measurements at 360 nm and 380 nm and the SBUV employs 340 nm measurements to estimate R_g. It is also assumed that R_g is independent of wavelength.

The first term of the right–hand side of (6.54) is evaluated once the surface reflections are taken into account. This term is the atmospheric reflection and its relation to column ozone follows from (6.50) if we assume only first–order scattering. Equation (6.50) can be written as

$$N_\lambda = k_\lambda^D X_o m = -100 \log_{10} \frac{I_\lambda}{S_\lambda}$$

N_λ is the logarithmic attenuation, I_λ is the intensity diffusely reflected along the nadir observing direction, and S_λ includes all other factors defined in (6.50). Measurements at two wavelength pairs thus yield

$$\Delta = X_o m \Delta k_\lambda^D = -100 \log_{10}[(\frac{I_1}{F_{o,1}})/(\frac{I_2}{F_{o,2}})]$$

which is analogous to (6.19) and where all other factors that define S_λ with the exception of F_o cancel from the intensity ratios. Δ is precomputed as a function of X_o for selected pairs of wavelengths. This provides a way of estimating X_o from measurements of Δ after correcting I_1 and I_2 for surface reflection. For TOMS there are 12 separate estimates of total ozone; three for each of the three wavelength pairs, the A–pair (313/331), the B–pair (318/331), and the C–pair (331/340); for each pair there are two values corresponding to the prescribed pressures (1 and 0.4 atm); and for each pressure two estimates are made from two sets of standard ozone profiles corresponding to the latitudes nearest the measurement latitude. The total ozone is then linearly interpolated between latitudes, except that between 0 and 15 degrees latitude only the 15 degree profile set is used and poleward of 75 degrees only the 75 degree profile set is used.

The total ozone value is then linearly interpolated between the two pressures assuming an effective pressure

$$\tilde{p} = wp_o + (1 - w)p_c$$

where p_o is the terrain pressure and p_c is the estimated cloud–top pressure. This cloud–top pressure may be estimated, for example, from a prescribed cloud–top pressure variation as a function of latitude or from co–located infrared measurements available from sensors on Nimbus 7. The weight w varies between zero and unity based on surface reflectivity. This procedure is then used for the three pairs and the final estimate is arrived at by a combination of all three.

6.5.3 Measured Trends in Total Ozone

In the 1970s, gloomy predictions about anthropogenic impacts on the depletion of the ozone layer surfaced. The search for evidence of downward trends in the thickness of the ozone layer at that time was inconclusive until the discovery of the Antarctic ozone hole reported by Farman et al. (1985). Ozone decreases during the Antarctic spring are now well documented, and ozone decreases outside the

Antarctic have also been reported. Estimates of these downward trends in ozone have been carried out using surface measurements obtained from the Dobson network since 1957 and using TOMS data collected by the Nimbus 7 data since November 1978. The problem in maintaining similar long–term ozone records derived from satellite measurements, among other factors, centers on maintaining the stability of the instrument over the time period in question. In the case of TOMS, the gradual degradation of optical components prompted the introduction of ad hoc calibration procedures to account for obvious instrument drifts imposed by this degradation. These procedures rely heavily on the surface observations which, in turn, are anchored to the highly calibrated world standard instrument (e.g., McPeters and Komhyr, 1991). With these corrections, it is estimated that trends predicted for the entire period of TOMS data up to 1989 are accurate to approximately ±1% relative to the Dobson measurements.

Decadal trends in the global distribution of total ozone obtained from TOMS data are presented in Fig. 6.18a and b. These results indicate that downward trends of column ozone are now being observed at all seasons in both the northern and southern hemisphere at middle and higher latitudes. Little trend is observed equatorward of about 25 degrees latitude. These results are also supported with the few Dobson stations that have long records of ozone and confirmed by the SAGE II ozone profile data. The latter show that the ozone decrease occur largely in the lower stratosphere. The decline in ozone amount measured by SAGE below about 25 km is about 10% per decade.

6.6   Sensing Clouds By Reflected Sunlight

The radiative transfer equation relevant to a cloud that scatters and absorbs radiation follows by considering the intensity change

$$dI = dI(extinction) + dI(scattering)$$

as a beam traverses the path element ds in cloud. Collecting appropriate expressions for these intensity changes yields the radiative transfer equation

$$\frac{dI(r,\vec{\xi})}{ds} = -\sigma_{ext}I(r,\vec{\xi}) + \frac{\sigma_{sca}}{4\pi}\int_{4\pi} P(r,\vec{\xi},\vec{\xi'})I(r,\vec{\xi'})d\Omega(\vec{\xi'}) \quad (6.55)$$

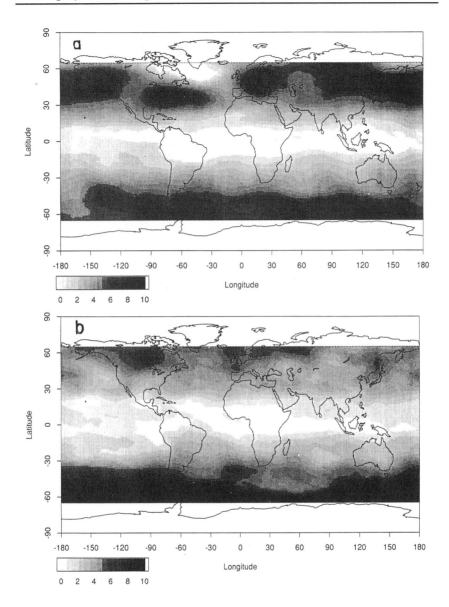

Figure 6.18 (a) Map of the TOMS averaged trends in percentages per decade for December through March over the period from November 1978 through March 1991. (b) As in (a) earlier except for the season from May through August (from Stolarski et al., 1992).

If it were not for the presence of the integral term, this equation would be a mere differential equation and the theory of multiple scattering would have been worked out and forgotten long ago. Formal solutions of (6.55) now exist for one–, two–, and three–dimensional geometries, but a thorough discussion of these are beyond the scope of this book. Instead, a simple conceptual approach is described shortly to offer a glimpse at the intricate way multiple scattered intensities depend on the scattering and absorbing properties of the cloud particles and, therefore, provide the basis for the remote sensing of these properties based on measurements of these intensities.

The integral term in (6.55) is the main stumbling block to the solution of this equation and the various published methods of solution primarily differ in the way this term is treated. Simpler methods, such as the one described shortly, make use of the property of the intensity illustrated in Figs. 6.19a and b. This diagram presents the intensity as a function of zenith angle on descent into a lake (Fig. 6.19a) and deep in a thick cloud (Fig. 6.19b). Apparently this structure approaches some sort of asymptotic form such that a fixed angular pattern is preserved with increasing depth into the "medium". This is the basic assumption of the simple methods — that a fixed and simple distribution of intensity exists so that a suitable approximation may be introduced for the integral.

6.6.1   The Two–Stream Approximation

The two–stream approximation is one method that makes use of this type of simplification. This approximation is described here in a conceptual way in the spirit of the pioneering work on radiative transfer by Schuster in 1905.[6]

Consider a parallel, horizontally uniform slab of cloud, and consider the fluxes flowing in two opposing directions.[7] The + superscript refers to quantities associated with flow in the upward direction and the − superscript refers to downward flow in the same sense as in (6.52). The two stream equations define the energy balance of this thin slab of thickness Δz in exactly the same way that (6.55)

---

[6] The equations developed here from conceptual arguments can also be derived directly from (6.55) if we introduce a mathematical simplification to the intensity field resembling the angular properties shown in Fig. 6.19. This alternative derivation is left as a problem for the interested reader (Problem 6.10).

[7] The relationship between radiative flux and intensity is also explored in Appendix 1.

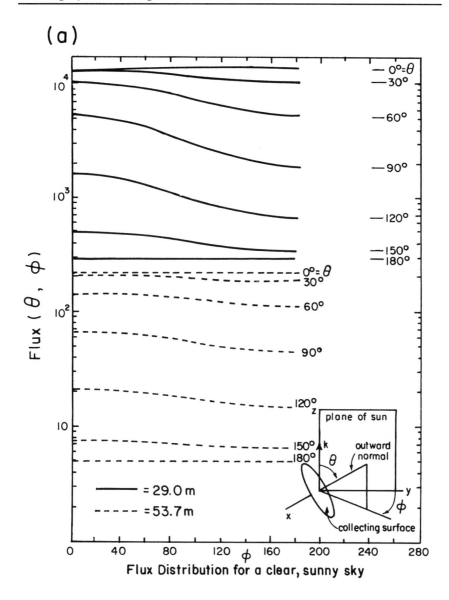

Figure 6.19 (a) The flux distribution on a clear sunny day at two indicated depths in Lake Pend Orielle, Idaho (adapted from Preisendorfer, 1976). These fluxes are defined for a collecting surface inclined at an angle θ as shown in the inset.

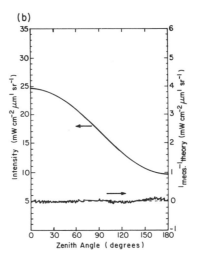

Figure 6.19 (Cont.) (b) Measured intensity as a function of zenith angle obtained from a scanning radiometer on an aircraft as it flew through the center of a deep stratiform cloud. The lower curve is the difference between measurement and a simple cosine of zenith angle variation (King et al., 1990).

describes an energy balance of a small volume of cloud. This balance requires the following optical properties:

• The proportion of the incident flux lost by absorption as the radiation flows through a layer of unit thickness is $k_{abs}D^{\pm}$ where $D^{\pm}$ is a measure of the "diffuseness" of the radiation field. This parameter more or less represents the mean extension of the path, relative to the vertical, that a diffuse radiation field travels as it penetrates the layer. It is a function of the angular properties of the intensity field (Problem 6.9) among other parameters and represents one of the simplifications mentioned earlier. If we suppose that the angular distribution of radiation that produces the flux is the same in both directions (the magnitudes may differ, however), then

$$D^+ = D^- = D.$$

Although this assumption is questionable, it is almost universally used in two–stream models.

• The proportional loss of flux by scattering is $\sigma_{ext}b^{\pm}$ per unit thickness. The process of scattering is treated differently here than is the process of absorption in that a measure of the path length is needed for estimating the latter. If we suppose that scattering is also the same for radiation flowing upward or downward, then

$$b^+ = b^- = b$$

and we can hereafter omit reference to the flow on these properties. Another parameter of relevance is the fraction of radiation f scattered in the forward direction. This fraction is defined such that

$$f + b = 1 \qquad (6.55)$$

(refer to Problem 5.4 for further insight regarding this relationship).
If the change in flux ΔF is defined as positive upwards, then

$$\Delta F^\pm = \mp (D \, k_{abs} + \sigma_{ext} \, b) \, F^\pm \Delta z \pm \sigma_{ext} \, b \, F^\mp \Delta z \quad (\pm S^\pm \Delta z) \qquad (6.56)$$

on transfer through the layer Δz where the last term in parentheses represents internal sources of $F^\pm$ in the layer Δz.[8] The first two terms on the right–hand side and enclosed in parentheses describe the losses of radiation through the processes of absorption and scatter respectively. The middle terms are the increases of flux by the backscatter of the opposing stream. We introduce the definition of optical depth as

$$\Delta \tau = -(k_{abs} + \sigma_{sca}) \Delta z$$

where the minus sign defines τ as increasing downwards from cloud top in the opposite sense to increasing z. In the limit $\Delta z \to 0$ and

---

[8] Two main sources of flux are usually considered in these models. One is the source of radiation due to thermal emission which, according to Kirchhoff's law, takes the form

$$S^\pm = k_{abs} \pi B(T)$$

for emitting cloud particles of temperature T. The second is the source of diffuse radiation that results from the single scattering of a collimated flux F_o of sunlight. This source has the form

$$S^\pm = F_o e^{-\tau/\mu_o} \sigma_{sca} \begin{pmatrix} b_o \\ f_o \end{pmatrix}$$

where f_o and b_o are the forward and backscattering fractions of the incident flux F_o and these are functions of the cosine of the solar zenith angle μ_o.

with $\tilde{\omega}_o = \sigma_{ext}/(\sigma_{ext} + k_{abs})$, the two flow radiative transfer equation becomes

$$\mp \frac{dF^\pm}{d\tau} = -\left[D(1 - \tilde{\omega}_o) + \tilde{\omega}_o \, b\right] F^\pm + \tilde{\omega}_o \, b \, F^\mp \quad (+S^\pm) \qquad (6.57)$$

All two–stream methods described in the literature essentially reduce to this equation. The only difference between the various methods lies in how D, b, and $S^\pm$ are specified. One example assumes the simple phase function introduced in Problem 5.4 such that

$$b = (1 - g)/2$$

where g is the phase function asymmetry factor. The radiative transfer equation then becomes

$$\mp \frac{dF^\pm}{d\tau} = -\left[D\left(1 - \tilde{\omega}_o\right) + \frac{\tilde{\omega}_o}{2}\left(1 - g\right)\right] F^\pm + \frac{\tilde{\omega}_o}{2}\left(1 - g\right) F^\mp \quad (+S^\pm)$$
$$(6.58)$$

The general solution to (6.57) for given sources is more complicated than we wish to discuss here. The source term is hereafter neglected and only solar radiation incident on cloud top in the form of purely diffuse radiation (as opposed to the more realistic case of a purely collimated incident flux) is considered. While the details of the solutions described below change with the addition of the source term for solar radiation, notably by introducing a solar zenith angle dependence to the solutions, the gross relationships between the optical properties of clouds ($\tau^*, \tilde{\omega}_o$, and g) and the diffuse reflectance and transmittances do not change.

For a vertically uniform cloud, the solutions become:
• For pure scattering, $\tilde{\omega}_o = 1, k_{abs} = 0$

$$F^\pm(\tau) = m_+ \mp m_- (1 + \tilde{\tau})$$

where m_+ and m_- are constants determined by boundary conditions and

$$\tilde{\tau} = (1 - g)\tau$$

is the optical depth scaled by the factor $(1 - g)$. The relevance of this scaled parameter is clear when considering an isolated scattering layer illuminated from above by flux F_o overlying a dark surface. Under these conditions, the albedo of the cloud layer is

$$\mathcal{R} = F^+(0)/F_o = \frac{\tilde{\tau}^*}{2 + \tilde{\tau}^*}, \qquad (6.59a)$$

and the transmittance is

$$T = F^-(\tau^*)/F_o = 1 - \mathcal{R} = 2/(2 + \tilde{\tau}^*), \qquad (6.59b)$$

where τ^* is the optical depth of the entire slab and $\tilde{\tau}^* = (1 - g)\tau^*$. Two cloud layers of different optical thicknesses τ^* and possessing different values of g at a nonabsorbing wavelength reflect precisely the same amount of radiation when the respective values of $\tilde{\tau}^*$ are the same. This is referred to as a similarity condition and implies that it is not possible to infer τ^* from a single reflection or a single transmission measurement without information about g. One of the problems associated with the remote sensing of ice crystal clouds is that g is neither well known nor is it well understood how g varies as the shape of the crystal changes. By contrast, g is better understood for water droplet clouds and is quasi–constant over most solar wavelengths with a typical value in the range 0.8–0.85.

• Nonconservative scattering $\tilde{\omega}_o < 1$, $k_{abs} > 0$: The solution to (6.55) for a sourceless, uniform medium has the form

$$F^\pm(\tau) = m_+\gamma_\pm e^{k\tau} + m_-\gamma_\mp e^{-k\tau} \qquad (6.60a)$$

where

$$k = \{(1 - \tilde{\omega}_o)\,D\,[(1 - \tilde{\omega}_o)\,D + 2\tilde{\omega}_o b]\}^{1/2} \qquad (6.60b)$$

and

$$\gamma_\pm = 1 \pm (1 - \tilde{\omega}_o)\,D/k \qquad (6.60c)$$

where again the coefficients $m\pm$ are determined from appropriate boundary conditions. For the boundary fluxes

$$F^+(\tau^*) = 0$$

$$F^-(0) = F_o$$

on an isolated layer of optical thickness τ^*, and after some manipulation of (6.60a), the albedo and transmittance of the layer can be written as

$$\mathcal{R} = \gamma_+\gamma_- \left[e^{k\tau*} - e^{-k\tau*}\right]/\Delta(\tau^*) \qquad (6.61a)$$

$$T = (\gamma_+^2 - \gamma_-^2)/\Delta(\tau^*) \qquad (6.61b)$$

where

$$\Delta(\tau^*) = \gamma_+^2 e^{k\tau^*} - \gamma_-^2 e^{-k\tau^*}$$

As $\tau^* \rightarrow \infty$, $\mathcal{R} \rightarrow \mathcal{R}_\infty = \gamma_-/\gamma_+$; a quantity referred to as the albedo of a semi–infinite cloud. This represents the upper limit to the albedo of a cloud and since $\mathcal{T} = 0$ and $\mathcal{A} = 1 - \mathcal{R}$, this is also the upper limit to the absorption within the cloud. This limit is denoted as $\mathcal{A}_\infty$. These upper limits are determined entirely by the optical properties $\tilde{\omega}_o, k, g$ and D of the cloud. From the substitution of (6.60b) in (6.60c), together with the definition of $\mathcal{R}_\infty$, it follows that

$$\mathcal{R}_\infty = \frac{[1 + 1/s^2]^{1/2} - \sqrt{2}}{[1 + 1/s^2]^{1/2} + \sqrt{2}} \qquad (6.62)$$

where

$$s = \left(\frac{1 - \tilde{\omega}_o}{1 - \tilde{\omega}_o g} \right)^{1/2}$$

is another similarity parameter. Equation (6.62) states that the reflections by two optically thick clouds defined by different sets of optical properties are identical when the similarity parameters for both are the same.

Excursus: Pollution Susceptible Clouds

The effect of ship stack effluents on cloud optical depth is discussed in Chapter 5. The simple two–stream model introduced earlier can now be used to consider the effects of pollution on cloud albedo and at the same time offer some insight into the effects of droplet size on the sunlight reflected by clouds. It was suggested in Chapter 5 that, with some assumption, the optical depth at wavelength λ may be expressed as

$$\tau^* \approx 2\pi N_o \bar{r}^2 h$$

for a cloud of depth h composed of N_o particles of a size $\bar{r}$ assuming $\bar{r} >> \lambda$. According to this formula, the optical depth may increase either by increasing the droplet concentration N_o, by increasing $\bar{r}$, or by a combination of both. Increased water content (occurring largely by increased $\bar{r}$), for instance, can increase the optical depth of clouds according to (5.29). For the case in which the liquid water content remains fixed, the sensitivity of optical thickness τ^* to N_o is given by

$$\frac{\Delta \tau^*}{\tau^*} = \frac{1}{3} \frac{\Delta N_o}{N_o}$$

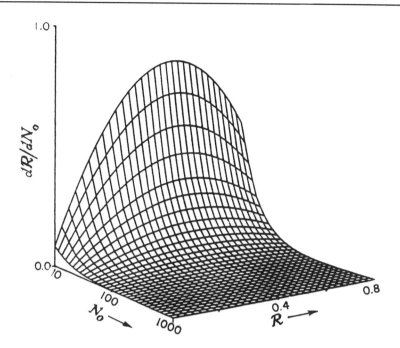

Figure 6.20 The susceptibility parameter $d\mathcal{R}/dN_o$ for different conditions. The vertical unit is the percentage of reflectance per additional droplet per cubic centimeter (from Twomey, 1991).

For cloud droplets under solar illumination, $g \approx 0.85$ and it follows from (6.59a) that

$$\mathcal{R} \cong \frac{\tau}{13 + \tau}$$

and that the sensitivity of $\mathcal{R}$ to droplet number N_o for fixed liquid water content w is

$$\left(\frac{d\mathcal{R}}{dN_o}\right)_w = \frac{\mathcal{R}(1 - \mathcal{R})}{3N_o}$$

which Twomey (1991) introduces as a measure of the susceptibility of clouds to changes in their optical depth due to the influence of pollution. For a given N_o, the most susceptible clouds are those with $\mathcal{R} \approx 1/2$, but the maximum of $\mathcal{R}$ is rather flat. For fixed $\mathcal{R}, (d\mathcal{R}/dN_o)_w$ is inversely related to N_o which, in the atmosphere, can vary by more than two orders of magnitude. The susceptibility parameter $d\mathcal{R}/dN_o$ (graphed in Fig. 6.20) reveals a considerable sensitivity of cloud optical depth to possible effects of pollution for those clouds formed in originally clean conditions — such as in oceanic and remote areas where N_o is low.

6.6.2 Cloud Droplet Radii from Reflected Sunlight

Plots of the similarity parameter s and the cloud reflectance are given in Fig. 6.21a as a function of wavelength for cloud droplet distributions characterized by different values of the effective radius r_e. The parameter s varies from zero for conservative scattering to unity for total absorption. The largest variation in s due to changes in r_e, and therefore the largest sensitivity between r_e and reflectivity (as shown in Fig. 6.21b), occurs in the 1.6 and 2.1 μm window regions. At these wavelengths, the reflection varies with both the optical depth and the similarity parameter s. The approach to the retrieval of r_e is to use the measurements at visible wavelengths corresponding to conservative scattering to estimate the optical depth of the cloud given an assumed value of g. It is then assumed that this value of τ^* is also representative of the window wavelengths where cloud droplets absorption is weak. The droplet size is estimated from theoretical relationships between r_e and reflectance at these wavelengths for a given value of τ^*.

An example of this approach is taken from the work of Nakajima and King (1990) and shown in Fig. 6.22a. This figure is a bispectral scatter plot of reflectivities $\mathcal{R}_1$ and $\mathcal{R}_2$ of a low stratiform cloud layer. The relationship between reflectivities measured at two wavelengths varies theoretically as a function of optical depth and r_e in the manner defined by the grid which is superimposed on the observations. This grid varies depending on the given solar geometry and the particular viewing geometry of the instrument. Thus a given observation of $\mathcal{R}_1$ and $\mathcal{R}_2$ translates to a specific value of τ^* and r_e.

Nakajima et al. (1991) provide perhaps what is the most detailed evaluation of this type of approach. In this evaluation, they compare estimates of r_e derived from measured droplet distributions obtained in the central portions of clouds from aircraft measurements and estimates of r_e obtained from simultaneous reflection measurements by a spectrometer flown on another aircraft above the cloud. The results of their study, summarized in Fig. 6.26b, indicate a general agreement in terms of the spatial variability of the remotely derived values of r_e compared to in situ measurements. However, a substantial bias exists in these values that persists even after some attempt is made to adjust the remote value down to the central portions of the cloud. This type of overestimate is also reported in other

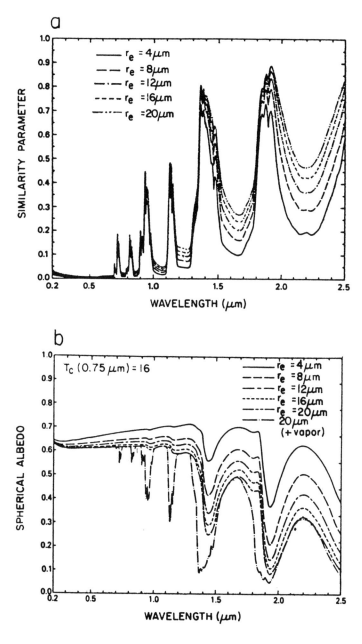

Figure 6.21 (a) The similarity parameter as a function of wavelength for selected values of r_e. Results apply to water clouds with a distribution defined by a modified gamma function with an effective variance of 0.111. $\tau^* = 16$ at 0.75 μm. (b) The albedo of clouds as a function of wavelength for selected values of effective radius.

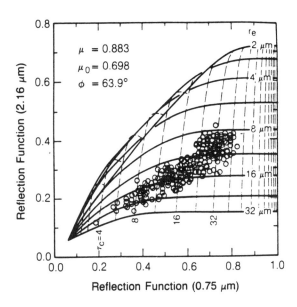

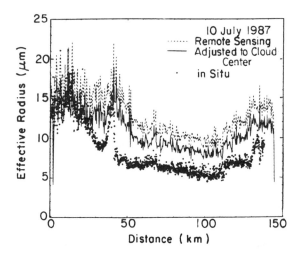

Figure 6.22 (a) Theoretical relationships between reflection at 0.75 and 2.16 μm for various values of τ^* and r_e for the specific case when θ_o =45.7°, θ =28°, and ϕ =63.9°. Data collected from aircraft overflying stratocumulus clouds and for the same geometry are also shown (Nakajima and King, 1990). (b) Comparison of r_e as a function of distance along the flight track from remote sensing and in situ measurements. The remote sensing values have also been arbitrarily adjusted to cloud center (Nakajima et al., 1991).

studies[9] and possible explanations for it are summarized in Stephens and Tsay (1990).

6.6.3 Cloud–top Pressure from Reflected Sunlight

Cloud height is a quantity important to many studies of the atmosphere, from assimilation of cloud winds to studies of the effect of clouds on the energy budget of the planet. Different methods for determining cloud–top pressure from satellite observations have been proposed over the years. These include:

• The brightness temperature technique where cloud–top height is estimated from a temperature profile and the cloud brightness temperature obtained from satellite measurements in a window channel.

• The CO_2 slicing method described in Chapter 7.

• Lidar measurements of backscattering, to be discussed in Chapter 8.

• Absorption by O_2 between 0.759 and 0.771 μm from analysis of reflected sunlight in the A band.

The latter technique exploits properties of the molecular absorption by O_2 that occurs primarily above a reflecting cloud. The difficulty of the approach is that the reflecting surface is not a hard surface and photons penetrate to different levels into the cloud. This depth of penetration depends on a number of factors, such as the vertical distribution of cloud droplets and the optical depth of the cloud. The fact that this penetration depth is not known creates some ambiguity in interpreting the reflection of sunlight in the oxygen A band, and methods developed to retrieve cloud–top pressure from measurements at these wavelengths require additional information to resolve these ambiguities.

The penetration depth depends on those factors that affect the multiple scattering and absorption of photons at the cloud–top. Wu (1985) systematically explored these factors using a model of the radiative transfer at 0.7609 and 0.7634μm located in the oxygen A band (Fig. 6.23a). In these calculations, the depth of the cloud was

---

[9] The remotely sensed values of droplet radii are ambiguous for many reasons. One cause of ambiguity is associated with just what level in the cloud the estimated particle size is supposed to represent. Intuitively, this level roughly corresponds to the upper regions of the cloud where the majority of the photons that exit cloud top are scattered. This ambiguity is also related to the photon penetration depth discussed in relation to cloud top pressure sensing and is a reason for the ambiguity of the latter.

varied to keep the scaled optical depth $\tilde{\tau}^* = 10$ fixed. The altitude of the cloud was also varied and the results of modeling calculations are presented in Fig 6.23b for different values of scaled volume extinction $\tilde{\sigma}_{ext}$. Shown on this diagram are the calculated bidirection reflection functions for a solar zenith angle $\theta_o = 30°$ and for an observation angle at nadir. These reflection functions are presented as a function of cloud–top pressure and are contrasted against the reflection derived for a hypothetical cloud in which there is no photon penetration (solid curve). The calculated reflections increase with altitude because the absorption by oxygen above the cloud decreases as the cloud top pressure decreases. Penetration of photons into a cloud reduces this reflection due to both absorption by oxygen within the cloud and by the forward scattering of cloud droplets. The effect of this photon penetration on the cloud reflection varies with the scaled extinction. As an example, consider the case illustrated in Fig. 6.23b with $\tilde{\tau}^* = 10$ and $\tilde{\sigma}_{ext} = 1$ km^{-1}. If the cloud has a reflection of 0.4 at 0.7069 μm, then the uncorrected cloud–top pressure is 480 mb compared to the pressure of approximately 280 mb derived after correcting for photon penetration.

The results of Wu (1985) quoted here serve to highlight the point that other information about the cloud, namely the scaled optical depth, is also needed in retrieving cloud top pressure. Wu proposes measurements at wavelengths adjacent to the A band for estimating $\tilde{\tau}^*$ and the combination of two measurements at wavelengths on the A band (as shown in Fig. 6.27a) to derive the cloud–top pressure. Fischer and Grassl (1991) propose an approach that, while philosophically similar, uses only two measurements in the oxygen A band. An example of their retrieval is shown in Fig. 6.27c where cloud–top height derived from A band measurements is shown as a function of distance along a research aircraft flight track and compared to lidar values of the cloud–top height.

6.7   Notes and Comments

6.1. A large body of literature exists on the topic of aerosol remote sensing using optical methods. Some of this literature is given in the review article of Reagan and Herman (1972). Other studies providing examples of the retrieval of aerosol distributions are Livingston and Russel (1989), Rizzi et al. (1982), and Shifrin and Perelman (1966) among others. Examples of the use of anomalous diffraction theory are Fymat (1978), Fymat and Smith (1979), Box

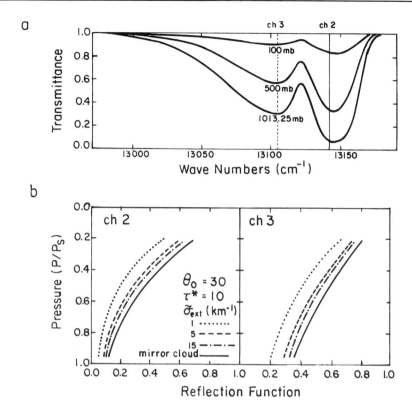

Figure 6.23 (a) The two–way transmission (from the top of the atmosphere to cloud–top and back) in the oxygen A band as a function of wavenumber for an airmass of 2 and for three different cloud–top pressures. (b) Simulated bidirection reflection functions for the two channels located in the A band as shown in (a). Parameter values discussed in the text are also given (Wu, 1985).

and McKellar (1976), Smith (1982), and Viera and Box (1985). A comprehensive description of radiometers that provide the kind of spectral measurements needed for these aerosol inversions is contained in Coulson (1975).

6.2. Ozone retrieval from Dobson measurements is outlined in the handbook of Dobson (1957). The Dobson spectrometer (number 83) was established in 1963 as a standard for total ozone measurement in the United States; it was designated by the WMO as a primary standard in 1980. This instrument and its calibration history is discussed by Komhyr et al. (1989).

The amount of literature on SAGE and the number of products derived from SAGE are steadily growing. Some of the earlier

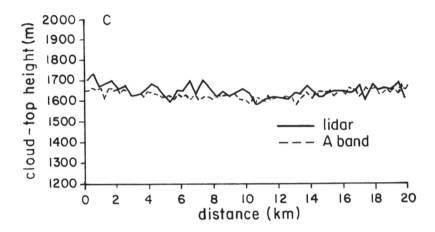

Figure 6.23 (Cont.) (c) Comparison of the cloud–top height derived from measurements in the oxygen A band and from a downward pointing lidar as a function of distance along the flight path of an aircraft (Fischer et al., 1991).

publications of McCormick and Trepte (1987), Chu and McCormick (1979) and Chu et al. (1989) describe the SAGE and its algorithm. Examples of the use of SAGE measurements of water vapor are found in Rind et al. (1991), and SAGE ozone data are reported by Mc-Cormick and Viega (1992).

6.3. A classic treatise on radiative transfer is that of Chandrasekhar (1960). The orders of scattering approach, as described in this section, was adapted from Preisendorfer (1976). Other examples of the orders of scattering calculation are also summarized in Lenoble (1985).

Direct measurements of the single scattering albedo of clouds are difficult, if not impossible, to make, so it is not surprising that none exist. King (1981) outlines a theoretical approach for indirectly deriving $\tilde{\omega}_o$ based on the angular properties of the intensity field as portrayed in Fig. 6.19b. This approach was then put to the test using aircraft radiometric data by King et al. (1990).

6.4. A cursory review of the literature, as reported in Lenoble (1985), suggests that there are a variety of different ways of solving radiative transfer problems similar to that posed in Section 6.3. Actually the number of truly different methods is small and all condense to the principle of interaction. From this principle, a few important statements about multiple scattering are established. These statements are referred to as the principles of invariance. The expressions derived for reflection by a surface underlying a multiple scattering layer are such statements. In fact, Equations (6.51) and (6.52) match, respectively, statements I and II as formulated by Chandrasekhar [(1960) pp. 162/3] except for the direct solar beam term which has been omitted to keep the discussion simple. (The interested student might note the connections between our R and T functions and Chandrasekhar's scattering functions S).

Problem 6.8 is a somewhat traditional problem in multiple scattering. It seeks to deduce the intensities emerging from the boundaries of a two layer atmosphere in terms of intensities incident on these boundaries, as well as to determine the intensities at the interface between the two layers. The interaction principle, in the form of statements like (6.51) and (6.52), provide the basis for the solutions to such problems.

In using the AVHRR data, as Durkee et al. (1991) do, it is important to consider calibration of the instrument. This is not a trivial problem as there is no onboard calibration procedure for the AVHRR. Other research programs that use these data (e.g., Rossow and Schiffer, 1991) attempt to calibrate the data using set targets on Earth, such as reflection from desert surfaces, as well as combining these satellite data with aircraft observations. This procedure is discussed in detail in Brest and Rossow (1989) as well as briefly in Section 7.8.

6.5. Dave and Mateer (1967) demonstrate the feasibility of determining atmospheric ozone from measurements of the backscattered UV sunlight near 310 nm. This served as a basis for an algorithm developed by Mateer et al (1971) and subsequently used to derive ozone from the Backscatter Ultraviolet (BUV) experiment flown on Nimbus 4. The BUV measured solar radiation in 12 wavelength bands 1.0 nm wide between 255.5 to 339.8 nm and discussion of the retrieval of total ozone from the BUV is given by Klenk et al. (1982).

The SBUV and TOMS instruments are calibrated on board by viewing solar radiation reflected by a diffuser plate. However, this calibration proved problematical as discussed by Herman et al. (1991).

6.6. The current literature suggests a real dearth of two–stream methods available to the unsuspecting investigator. In the atmospheric sciences literature, no less than three review articles have been published over the last dozen years [such as Meador and Weaver (1980) and King and Harshvardhan (1986)]. Despite the apparent

numerous forms of two–stream "models," there is really only one underlying "conceptual" model and all others are deviations from this
single concept.

The actual discussion of the two–stream method described in
this section is approached from the phenomenological viewpoint and
thus potentially offers the greatest insight into the relationship between the scattered radiation and the medium. A second approach is
that found more typically in the modern literature and demonstrates
more clearly the mathematical simplification that the two–stream
method introduces to the transfer equation is explored in Problem
6.10. Stackhouse and Stephens (1991) describe a two–stream model
that has both the solar and thermal emission sources included as
part of their solutions. A similar model is also described in Toon et
al. (1989).

The analysis of the effects of droplet concentration on the albedo
of clouds was taken from the work of Twomey (1991). The susceptibility factor $(d\mathcal{R}/dN_o)_w$ shown in Fig. 6.20 approaches 1% (per
cm^{-3}); that value would mean a reflectance change of 0.01 for a
concentration change of just 1 cm^{-3}. To produce such a change up
to a height of 1 km requires about 50 tons of material for the whole
global atmosphere (taking the mass of a nucleus as 10^{-16} g); to maintain it, even with a residence time of only 2 days, an injection rate
of about 1–10 kilotons annually would suffice (i.e., very much less
than the hundreds of megatons of condensation nuclei material currently being injected either naturally or anthropogenically. Twomey
estimates that increasing N_o in the more susceptible clean locations
by a lot less than the current spread of measured values N_o could
change the global albedo of the planet by as much as 1 or 2%.

Sensing the microphysical properties of clouds using reflected
sunlight is a topic that has interested a number of investigators over
many years. Early pioneers of this topic for terrestrial clouds are
Hansen and Pollack (1970), and Twomey (1971). The general concept, that of measuring the reflection at a visible nonabsorbing wavelength to estimate the optical depth, followed by measurement at another slightly absorbing wavelength to extract particle size effects,
is unfortunately plagued by our inability to reconcile measurements
of the reflection spectra with our theoretical understanding of these
spectra. This general problem is reviewed in the paper of Stephens
and Tsay (1990). Examples of other methods that use reflected sunlight, largely in the window located at 3.7 μm, are those of Arking
and Childs (1985) and Stone et al. (1990), and more recently by

Schmidt et al. (1992). Rossow (1989) provides a review that largely deals with the topic of detectibility of clouds from space. This is a topic that we will return to later in Chapter 7.

The idea of sensing cloud–top pressure seems to have originated in the early 1960s when Hanel (1961) first proposed that cloud–top pressure may be estimated from measurements of absorption of reflected sunlight by well mixed gases. Hanel's original idea was to use CO_2, but Yamamoto and Wark (1961) proposed that the A band was better suited and free of contamination by water vapor. A more extensive review on the topic, including more recent developments, is provided by O'Brien and Mitchell (1991).

6.8 Problems

6.1. Explain or interpret the following in just a few sentences:

    a. The exponent α characterizing the wavelength variation of aerosol optical depth has a typical value of 1.3 prior to a volcanic eruption and a value much smaller than this after a volcanic eruption.

    b. Values of NDVI corrected for molecular scattering are larger than values derived from uncorrected AVHRR intensity data (refer to Plate 2 at the front of the book).

    c. The wavelength $\lambda = 0.3125\ \mu m$ is a better choice for sensing total column ozone from satellite measurements of reflected sunlight than is the wavelength $\lambda = 0.3025\ \mu m$.

    d. Light deep in crevasses of ice appears bluish–green.

    e. The sky color is blue because of the preferential scattering of blue light by Rayleigh scatterers (the molecules of N_2 and O_2). However, the color of horizon skylight is much whiter than blue skylight observed, say, in the zenith. [*Hint: Consider the quantity $d\mathcal{R}/d\lambda$ in (6.59a) as a function of $d\tau^*/d\lambda$.*]

    f. Wet sand is darker than drier sand.

6.2. Aerosol turbidity:

    a. The Ångström turbidity factor (as opposed to coefficient) is the optical depth of aerosol. Assuming that the scattering cross section of the aerosol is twice their geometric cross–section, what density of aerosol of diameter 1 μm is required in the lowest kilometer of atmosphere to produce a turbidity factor of unity?

    b. It is commonly assumed that the Ångström turbidity factor

varies with wavelength with an exponent of 1.3, as noted
in reference to (6.9). For an atmosphere with optical depth
of aerosol of 0.3 at a wavelength of 0.5 μm, what is the turbidity coefficient at 0.3 and 0.7 μm respectively? From the
results of problem 5.2, estimate the total transmission due
to Rayleigh plus aerosol optical depth at the three wavelengths, assuming $\theta_o = 30°$.

c. Derive the Linke turbidity parameter relevant to (b) above
for a wavelength of 0.5 μm.

6.3 Solid angle:

a. Given the following data, the diameter of the sun, moon,
and Earth: 1.39×10^6 km, 3.48×10^3 km, and 12.74×10^3
km, respectively, and the distance from Earth; sun, $1.49 \times$
10^8 km and moon, 3.8×10^5 km, calculate the solid angles
subtended by both the sun and moon as seen from the
center of the Earth. What are the differences when seen
from the surface of the Earth?

b. If the incident direction is $\vec{\xi} = (\theta, \phi)$ and the scattered
direction is $\vec{\xi'} = (\theta', \phi')$, show that the scattering angle Θ
is

$$\cos \Theta = \cos \theta \cos \theta' + \sin \theta \sin \theta' \cos(\phi' - \phi)$$

c. Using your answers from (b) (i) derive an expression for the
scattering angle associated with vertically incident sunlight,
and (ii) derive an expression for the scattering angle for
sunlight scattered horizontally as a function of the zenith
and azimuth angles of the sun.

6.4. Repeat the example given in the text to illustrate (6.39) and
calculate j assuming $\tilde{\omega}_o = 0.999$, $\tilde{\omega}_o = 0.99$, $\tilde{\omega}_o = 0.9$, $\tilde{\omega}_o = 0.6$, and
$\tilde{\omega}_o = 0.3$, the 1% criterion used in the worked example, and the
two values of optical depth used in the text. Contemplate these
answers with reference to Figs. 6.12a, b and Fig. 6.13.

6.5. Consider a zenith viewing instrument at some level z. Show
that the primary scattered sunlight sensed by this instrument is

$$I^1(z, \mu = 1) = \frac{\tilde{\omega}_o F_o}{4\pi m} P(\vec{\xi}_o, \vec{\xi}) [e^{-\sigma_{ext}(z_t - z)/\mu_o} - e^{-\sigma_{ext}(z_t - z)}]$$

where $m = 1/\mu_o - 1$.

6.6. Using the ozone distribution specified in Problem 3.10, derive
the vertical profile (in pressure coordinates) of the normalized
weighting function relevant to measurements of reflected UV
sunlight based on (6.49) for the following parameters:

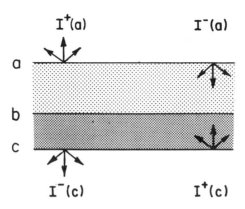

Figure 6.24 Two scattering layers.

(i) $\lambda = 0.3125\mu m, \theta = 0^\circ, \theta_o = 60^\circ, \tilde{\omega}_o = 1, F_o = 1, k_\lambda = 0.0016DU^{-1}$

(ii) $\lambda = 0.3025\mu m, \theta = 0^\circ, \theta_o = 60^\circ, \tilde{\omega}_o = 1, F_o = 1, k_\lambda = 0.0066DU^{-1}$

selecting your value of r_p in Problem (3.10) such that $X_o = 341$ DU. You may assume that $\tau_{Ray} = 0.0088\lambda^{-4.15}$ applies at sea level where λ is in μm. Compare your results to those presented in Fig. 6.20.

6.7. Using the results from Problem 6.5 with $z = 0$, derive an expression for the profile of the weighting function associated with measurements from an instrument located at the ground observing the zenith UV skylight. Calculate the profiles for the parameters given in Problem 6.6, normalize them, and compare your results.

6.8. Consider two scattering layers overlying each other as shown in Fig 6.24. You may assume that each layer is identical such that the reflection and transmission properties defined by $\mathcal{R}_1$ and $\mathcal{T}_1$ for the upper layer are the same as those of $\mathcal{R}_2$ and $\mathcal{T}_2$ of the lower layer (Fig. 6.28).

   a. Deduce the emergent intensities $I^+(a)$ and $I^-(c)$ as a function of the incoming intensities $I^-(a)$ (assuming no light is incident from below) and the reflection and transmission properties of the layers.

   b. Deduce the intensities $I^\pm(b)$ at the interface between the two layers in terms of incoming intensities $I^-(a)$ (again assuming no light is incident from below) and the reflection and transmission properties of the layers.

c. Suppose that the lower layer is opaque such that $T_2 = 0$ as for a land surface, compare your result for $I^+(a)$ with (6.53).

6.9. The intensity in the atmosphere is often considered to be a weak function of azimuth and so the azimuthally averaged intensity

$$\bar{I} = \frac{1}{2\pi} \int_{2\pi} I(\theta, \phi) d\phi$$

is often used. A convenient representation of this quantity is

$$\bar{I}(\mu) = I_o + \sum_n I_n \mu^n$$

where $\mu = \cos\theta$. Assuming the following specific forms of this expansion,
  a. $\bar{I}(\mu) = I_o + I_1 \mu$
  b. $\bar{I}(\mu) = I_o$
calculate the following hemispheric integral quantities
$h = 2\pi \int_0^1 \bar{I} d\mu$, $F = 2\pi \int_0^1 \bar{I} \mu d\mu$,
$K = 2\pi \int_0^1 \bar{I} \mu^2 d\mu$ and the diffusivity $D = F/h$.

6.10. The aim of this problem is to derive the two–stream equations in a form similar to (6.57). A helpful reference for this problem is that of Meador and Weaver (1980). We begin with (6.55) written in an appropriate plane parallel form (you may want to show this)

$$\mu \frac{dI}{d\tau} = -I + \frac{\tilde{\omega}_o}{4\pi} \int_0^{2\pi} \int_{-1}^1 PI d\mu d\phi \qquad (6.63)$$

where $d\Omega = -d\mu d\phi$, $ds = dz/\mu$, $d\tau = -\sigma_{ext} dz$ and where the position and angular dependences on quantities are omitted (but understood).

  a. Write down the azimuthally averaged form of this equation and replace the azimuthally averaged phase function with

$$\bar{P} = \frac{1}{2}(1 + 3g\mu\mu').$$

This follows from (5.43) with $N = 1$ and from the addition theorem for spherical harmonics (e.g., Liou, 1980, p. 179).

b. Assume the form of $\bar{I}$ given in Problem 6.9a and carry out the following integral transforms on each term of (6.63): $2\pi \int_0^1 ...\mu d\mu$ and $2\pi \int_{-1}^0 ...\mu d\mu$ noting the definition of flux

$$F^+ = 2\pi \int_0^1 \bar{I}\mu d\mu \quad \text{and} \quad F^- = 2\pi \int_{-1}^0 \bar{I}\mu d\mu$$

c. Derive the equations for F^+ and F^- and compare to (6.58). The equation you have derived is known as *Eddington's two-stream model* and you will note how it resembles (6.58) except that the coefficients of $F^\pm$ differ.

6.11.

a. Estimate the value of $\mathcal{R}_\infty$ appropriate to a wavelength of 0.7 μm, assuming $\tilde{\omega}_o = 0.9997$ and $g = 0.85$, which is somewhat typical of stratus clouds with 10μm–sized cloud droplets.

b. For this stratus cloud, what is the albedo, absorption, and transmission when $\tau^*=20$?

c. What optical depth is required of a cloud composed of isotropic scatterers to produce the same properties as (b)?

6.12. At 10.6 μm, the wavelength of the CO_2 laser, $\tilde{\omega}_o = 0.36$, b=0.1 for fog droplets. Find $\mathcal{R}_\infty$ and the transmissivity of diffuse radiation at this wavelength in a 100m thick fog having the same properties as the stratus cloud considered in Problem 5.8b? (*Note that although the cloud emits radiation at these wavelengths and calculation of this effect requires an emission source term, this contribution to the total radiation field simply adds to the contributions associated with diffuse scattering.*)

6.13. In the visible portion of the spectrum, the albedo of Venus is 0.9. Suppose the clouds of Venus are very deep and that the scattering is isotropic, what is the minimum value of $\tilde{\omega}_o$ for the cloud particles?

7

Passive Sensing — Emission

This chapter describes a theory of radiative transfer that includes the process of emission and introduces a number of remote sensing topics that are based on the measurement of this emission. Emission by the atmosphere is the source of atmospheric microwave and infrared radiation. We are able to deduce a great number of properties of the atmosphere from its measurement, including the amount and distribution of water vapor, the temperature profile, the amount of cloud liquid water, rainfall, and sea surface temperature.

Figure 7.1 is an example of a measured intensity spectrum obtained from the IRIS instrument flown on Nimbus 4. It provides a framework for thinking about how radiative transfer takes place in an absorbing–emitting atmosphere. Superimposed on the measured spectra are the blackbody curves for selected temperatures. Also highlighted are spectral positions of the absorption bands of the predominant absorbing gases. This diagram more clearly shows how emissions from different levels in the atmosphere (and therefore at different temperatures) combine to produce the observed spectra. For instance, emission in the central portions of the 9.6μm ozone band occurs at temperatures below about 250 K, and emission in the 15μm CO_2 band varies throughout the atmosphere according to the spectral position relative to the band center. For both O_3 and CO_2, the increase in emitted radiation in the center of the absorption band occurs higher up in the atmosphere than in the neighboring spectral regions where the absorption is weaker is indicative of the increase in temperature with increasing altitude in the stratosphere. Also noteworthy is the water vapor emission that is confined to the lower atmosphere (emission by the vibration and rotation bands is broadly characterized by the 275 K blackbody curve). An important spectral region is the atmospheric window between about 800 and $1200\,\mathrm{cm}^{-1}$ in which the atmosphere is almost transparent (except for the ozone band) and the emission originates from levels close to the surface. This spectral region is extensively used in remote sensing of surface and cloud properties as described shortly.

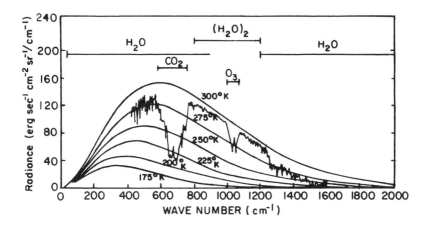

Figure 7.1 The infrared spectrum of radiation emitted from the atmosphere measured by the IRIS instrument. Shown are the blackbody curves derived for the different temperatures indicated as well as the principal absorbing gases that contribute to the spectrum (from Hanel, 1983).

7.1   Radiative Transfer with Emission

The increase of radiation along a path ds due to emission is

$$dI_\lambda = k_{\lambda,v} \mathcal{J}_\lambda ds \qquad (7.1)$$

where $\mathcal{J}_\lambda$ is the *source function*. When this emission takes place in the lower atmosphere where thermodynamic equilibrium occurs, $\mathcal{J}_\lambda = \mathcal{B}_\lambda$. The net change in radiation along ds due to the combination of emission and extinction is

$$dI_\lambda = dI_\lambda(\text{emission}) + dI_\lambda(\text{extinction}) \qquad (7.2)$$

It is reasonable to neglect scattering in many problems of infrared and microwave radiative transfer especially for clear sky conditions and, for most circumstances, for cloudy skies as well. Then substituting (6.1) and (7.1) in (7.2), we obtain the following transfer equation

$$\frac{dI_\lambda}{ds} = -k_{\lambda,v}[I_\lambda - \mathcal{B}_\lambda] \qquad (7.3)$$

which is the mathematical relationship that describes how radiation is transferred from one layer to another layer as a result of absorption

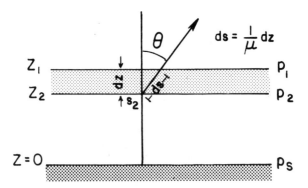

Figure 7.2 The geometric setting for the integral transfer equation in a plane parallel vertically stratified atmosphere.

and emission. The amount of radiation leaving the end of the path is a function of the distribution of absorber along the path (through the presence of $k_{\lambda,v}$) and the distribution of temperature (largely through the presence of $\mathcal{B}_\lambda$). We therefore expect that the measurement of intensity at the endpoint of the path contains information about both of these distributions. This is the principle upon which emission–based sounding of constituent concentration and temperature is based and this principle is a topic described in more detail later in this chapter.

The interactions between radiation and the gases of the atmosphere are generally weak enough that the photon mean free path is larger than the mean free path of molecules. Hence, the radiative transfer in the atmosphere tends to be nonlocal with significant contributions arising from different points along the path. This path–integrated emission is represented by the integral form of the radiative transfer equation. We obtain this equation by first making use of the definition, $d\tau_\lambda(s) = -k_{\lambda,v}(s)ds$, for an element of the optical thickness (the reason for the negative sign in this definition of optical thickness becomes evident later) and then multiply each side of (7.3) by the factor $\exp[-\tau_\lambda(s)]$. Combining terms gives

$$\frac{dI_\lambda e^{-\tau_\lambda(s)}}{d\tau_\lambda} = -\mathcal{B}_\lambda e^{-\tau_\lambda(s)} \tag{7.4}$$

Integration of this equation along a path extending from some point $s = s'$ to an end point $s = s''$ yields

$$I_\lambda(s'')e^{-\tau_\lambda(s'')} - I_\lambda(s')e^{-\tau_\lambda(s')} = \int_{\tau(s'')}^{\tau(s')} \mathcal{B}_\lambda(s)e^{-\tau_\lambda(s)}d\tau(s)$$

which, on rearrangement, gives

$$I(s'') = I(s')e^{-[\tau(s')-\tau(s'')]} + \int_{s'}^{s''} B(s)e^{-[\tau(s)-\tau(s'')]}d\tau(s) \quad (7.5)$$

where the wavelength dependence of all factors in (7.5) is taken to be understood. The first term on the right–hand side of this equation represents the radiation, originally incident at s', that is transmitted to s''. We will refer to this as the surface term and describe one application where this term is used to obtain surface temperature. The integral term represents the emitted radiation from all points along the path and transmitted to s''.

When (7.5) is applied to the atmosphere, it is customary, but not necessarily realistic, to assume that the atmosphere is plane parallel and horizontally homogeneous. For such a stratified atmosphere, the integral equation (7.5) can be expressed in terms of optical depth $\tau(z)$ [rather than optical thickness $\tau(s)$]. It is conventional to define the optical depth such that $\tau = 0$ at the top of the atmosphere and $\tau = \tau^*$ at the surface.[1] For slant paths, the expression relating optical depth to optical path is

$$\tau(s) = \tau(z)/\cos\theta$$

producing

$$I(\tau, +\mu) = I(\tau^*, \mu)e^{-(\tau^* - \tau)/\mu} + \int_{\tau}^{\tau^*} B(t)e^{-(t-\tau)/\mu}\frac{dt}{\mu} \quad (7.6a)$$

for $0 < \mu < 1$ which defines radiation that upwells from the atmosphere, and

$$I(\tau, -\mu) = I(0, -\mu)e^{-\tau/|\mu|} + \int_{0}^{\tau^*} B(t)e^{-(\tau-t)/|\mu|)}\frac{dt}{|\mu|} \quad (7.6b)$$

for $0 > \mu > -1$ for downwelling radiation.

---

[1] The convention that τ increases downward from the top of the atmosphere has roots in the traditional astrophysics literature on radiative transfer where τ is taken to increase along the direction of sunlight entering the atmosphere of distant planets. Optical depth increases in the opposite sense to z and hence the negative sign in its definition.

Excursus: The Angular Properties of Emitted Radiation

Consider an isothermal atmosphere with $B = B_o$ where B_o is a constant. Now imagine that a radiometer on the ground scans across the sky. The relevant equation that describes the radiation sensed by this instrument is (7.6b) which simplifies to

$$I(\tau^*; -\mu) = B_o[1 - e^{-\tau^*/|\mu|}] \qquad (7.7)$$

assuming that the cosmic term $I(0, -\mu) = 0$. This expression predicts that, for a given value of τ^*, the downwelling intensity is a minimum at $\mu = 1$ (directly downward), and increases to a value of B_o as the instrument scans out to the horizon. This increase is referred to as *limb brightening* and arises from the fact that the effective level of the emission decreases in the atmosphere as the zenith angle increases.

Consider the same instrument on a satellite and suppose it scans as it looks down at Earth. Again for an isothermal atmosphere, direct integration of (7.6a) gives

$$I(0; \mu) = B_o e^{-\tau^*/\mu} + B_o[1 - e^{-\tau^*/\mu}] = B_o \qquad (7.8)$$

Thus the intensity is uniform, regardless of the direction of observation (in this case it is said to be *isotropic*). This occurs because the portion of radiation emitted by the surface and absorbed by the atmosphere is precisely compensated by the emission from the atmosphere. Figure 7.3 presents a plot of each term of the right-hand side of (7.8) as a function of τ^* assuming a value of $B_o = 1$. The second term represents the atmospheric emission and increases as τ^* increases. This characteristic increase is referred to as a curve of growth of I with respect to τ^* and is important to a number of emission-based remote sensing methods that are described later.

Figure 7.3 serves to illustrate two important elements of emission based sensing of the atmosphere. The first of these is that it is necessary to be able to distinguish between the atmospheric and surface contributions. Circumstances that provide a large contrast between the surface and atmosphere, such as occurs at microwave frequencies over water surfaces, allow us to isolate the atmospheric signature more or less unambiguously. The second aspect is that the key parameter in the emission sensing is the optical depth and

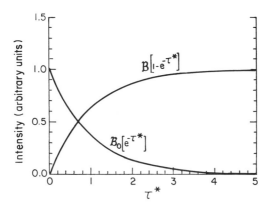

Figure 7.3 Contributions to the upwelling intensity at $\mu = 1$ at the top on an atmosphere of optical depth τ^* from the top of an isothermal atmosphere due to emission from the surface and emission from the atmosphere. The intensity is expressed in normalized units.

all methods that derive atmospheric properties from emission measurements establish a relationship between τ^* and the property of interest. Several examples of these relationships are discussed later.

For a nonisothermal atmosphere, solutions of either (7.6a) or (7.6b) are usually carried out using numerical integration methods. For illustrative purposes, however, the example of a Planck function that is a linear function of optical depth yields both simple and useful solutions to these equations[2]. Let $\mathcal{B}_o$ and $\mathcal{B}^*$ be the values of the Planck function at the top and bottom of the atmosphere, respectively. It follows from direct integration of (7.6a) and (7.6b) with $\mathcal{B}_o + (\mathcal{B}^* - \mathcal{B}_o)\tau/\tau^*$ substituted for $\mathcal{B}(\tau)$ that the upward looking instrument sees radiation that varies with μ according to

$$I(\tau^*; -\mu) = \mathcal{B}_o \left(1 - e^{-\tau^*/\mu}\right) + (\mathcal{B}^* - \mathcal{B}_o)\left[1 - \frac{\mu}{\tau^*}\left(1 - e^{-\tau^*/\mu}\right)\right]$$

(7.9)

which again predicts limb brightening. The downward looking instrument detects an angular variation in intensity of the form

---

[2] There is quite a subtle, but nevertheless important, point to recognize when the Planck function is approximated in this way. The radiative transfer equations described throughout this chapter apply to a single wavelength. A linear-in-τ Planck function implies that the variation of temperature with z varies unrealistically from one wavelength to another.

$$I(0; \mu) = B^* e^{-\tau^*/\mu} + B^* \left(1 - e^{-\tau^*/\mu}\right) -$$

$$(B^* - B_o) \left[1 - \frac{\mu}{\tau^*} \left(1 - e^{\tau^*/\mu}\right)\right] \qquad (7.10)$$

which is no longer isotropic but decreases as the instrument scans toward the horizon (*limb darkening*).

In general, we learn from these simple examples that the amount of limb darkening or brightening depends upon the vertical distribution of temperature (that is on both B^* and B_o) and on τ^*. There is also generally less angular variation in the upwelling intensity than for the downwelling field, particularly at wavelengths where τ^* is small, because the atmosphere transmits some of the radiation incident upon it from the underlying surface. As the optical depth increases, the angular variation of the downward intensity decreases because most of the radiation originates nearer the detector. The upward intensity generally changes relatively slowly with angle, except in the limb.

7.2   The Remote Sensing of Sea Surface Temperature

Except for a few selected examples, most topics in this book deal with the remote sensing of atmospheric properties. There are many surface properties that influence the atmosphere in significant ways. It is also important for studies of the atmosphere to monitor these properties. Sea and land surface temperatures, for example, fundamentally impact upon both the weather and the climate of Earth. In fact, global monitoring of sea surface temperature (SST), in particular, looms as one of the most important exercises in the study of the Earth's climate and the potential for climate change. Some appreciation of the need for remote measurements of SST is provided in Fig. 7.4 which exemplifies the current distribution of in situ SST observations from ships and buoys. The density of observations is greatest in the oceans of the northern hemisphere, and data coverage is sparse over the southern hemisphere oceans. This situation is obviously inadequate for climate studies of SSTs and is clearly inadequate for monitoring global temperature.

The remote sensing of surface temperature is commonly done using measurements of the Earth–atmosphere emission at various wavelengths. The general idea is to measure the emission at wavelengths where the atmosphere is most transparent. Unfortunately,

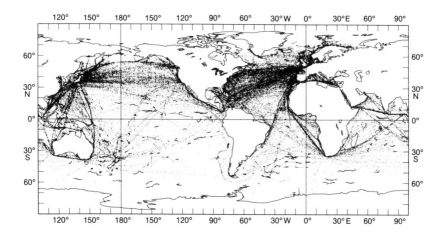

Figure 7.4 Distribution of surface marine in situ (ship and buoy) observations for October 1986. Drifting buoys are distinguished as nearly continuous wiggly lines (after Reynolds, 1988).

the atmosphere is never completely transparent in the main "windows" even for clear skies; therefore, some method is needed to correct for the atmospheric emission. A technique that attempts to provide such a correction is the "split–window" approach which uses measurements at multiple wavelengths. This multichannel approach (referred to as MCSST) originates from the spectral transfer equation, (7.6a), written in the form

$$I(\tau \approx 0, +\mu) = \mathcal{B}(T_s)\mathcal{T}(\tau^*, \mu) + \mathcal{B}(T_a)[1 - \mathcal{T}(\tau^*, \mu)] \qquad (7.11)$$

where the emission from the surface is approximated by the blackbody function $\mathcal{B}(T_s)$ at the surface temperature T_s. In (7.11), $\mathcal{T}(\tau^*, \mu) = e^{-\tau^*/\mu}$ is the transmittance of the atmosphere along the slant path defined by μ and extending from the surface to the satellite. We also express the emission by the atmosphere in terms of an effective blackbody temperature T_a defined from the relationship

$$\mathcal{B}(T_a) = [1 - \mathcal{T}(\tau^*, \mu)]^{-1} \int_0^{\tau^*} \mathcal{B}(t)e^{-t/\mu} \frac{dt}{\mu} \qquad (7.12)$$

The split–window technique uses observations at two channels in an attempt to eliminate the term containing T_a from (7.11) and solve for T_s. To explore this approach, suppose we measure the two intensities at adjacent wavelengths λ_1 and λ_2. We represent these measurements by I_1 and I_2 and it follows from (7.11) that

$$I_1 = \mathcal{B}_1(T_s)\mathcal{T}_1(\tau_1^*, \mu) + \mathcal{B}_1(T_a)[1 - \mathcal{T}_1(\tau_1^*, \mu)] \qquad (7.13a)$$

$$I_2 = \mathcal{B}_2(T_s)\mathcal{T}_2(\tau_2^*, \mu) + \mathcal{B}_2(T_a)[1 - \mathcal{T}_2(\tau_2^*, \mu)] \qquad (7.13b)$$

Prabhakara et al. (1974) claim that the value of T_a varies by less than 1 K across the 10.5–12.5 μm window so it is also reasonable to assume a common value for T_a. It is also reasonable to expect that the surface emissivities are the same at adjacent wavelengths (here we conveniently assume this emissivity to be unity).

To solve for T_s, we make use of Taylor's theorem to arrive at the following

$$\mathcal{B}_\lambda(T) \approx \mathcal{B}_\lambda(T_a) + \frac{\partial \mathcal{B}_\lambda}{\partial T}(T - T_a)$$

where the partial derivative is evaluated at $T = T_a$. Application of this equation to both wavelengths and further elimination of the $T - T_a$ factor yields

$$\mathcal{B}_2(T) \approx \mathcal{B}_2(T_a) + \frac{\partial \mathcal{B}_2/\partial T}{\partial \mathcal{B}_1/\partial T}[\mathcal{B}_1(T) - \mathcal{B}_1(T_a)] \qquad (7.15)$$

We use this expression twice, once to approximate our observation I_2, which we write in terms of a brightness temperature $T_{b,2}$ according to $I_2 = \mathcal{B}_2(T_{b,2})$, and a second time to approximate $\mathcal{B}_2(T_s)$ to obtain

$$\mathcal{B}_1(T_{b,2}) = \mathcal{B}_1(T_s)\mathcal{T}_2 + \mathcal{B}_1(T_a)[1 - \mathcal{T}_2] \qquad (7.16)$$

Eliminating $\mathcal{B}_1(T_a)$ from (7.13a) and (7.16) yields the split–window equation

$$\mathcal{B}_1(T_s) = I_1 + \eta[I_1 - \mathcal{B}_1(T_{b,2})] \qquad (7.17)$$

where

$$\eta = \frac{1 - \mathcal{T}_1}{\mathcal{T}_1 - \mathcal{T}_2} \qquad (7.18)$$

This relationship is further approximated and linearized in terms of the brightness temperatures,

$$T_s \approx T_{b,1} + \eta[T_{b,1} - T_{b,2}] \qquad (7.19)$$

In practice, the actual split–window technique is rarely used in the form given by either (7.17) or (7.19) but these serve merely as a justification for regressing the surface temperature as a linear function of the measured brightness temperatures. The more typical form of this regression is

$$SST = aT_{11} + b(T_{11} - T_{12}) - c \qquad (7.20)$$

where the coefficients a, b, and c are empirically derived from in situ observations such as obtained from drifting buoys. In this expression, $T_{11,12}$ are the brightness temperatures of the $11\mu m$ and $12\mu m$ channels of the AVHRR instrument. When comparing the satellite observations to conventional observations, it is important to note that the satellite temperatures correspond to the temperature of a surface layer just a few millimeters thick (this is referred to as the "skin" temperature). In situ measurements, on the other hand, are bulk measurements of the temperature of a layer of water perhaps a few meters deep. The regression approach of the MCSST is an attempt, in part, to tune the satellite skin temperatures to the in situ bulk temperature. Part of this tuning also accounts for the fact that the atmosphere is not completely transparent at the window wavelengths, especially over the moist tropics (discussion of Table 3.1 and Problem 7.6 offer more quantitative perspectives of these atmospheric effects).

The multichannel approach described here is presently used operationally by NOAA; the precise details of the operational algorithm are described by McClain et al. (1985). The approach actually uses both the $11\mu m$ and $12\mu m$ channels of the AVHRR and the $3.7\mu m$ window channel and has a form similar to that of (7.20). An important part of the algorithm, and one of relevance to many retrieval problems involving infrared emission measurements, is the need to establish whether or not clouds appear in the field of view. Undetected clouds are a source of bias when temperatures associated with the emission from clouds are mistakenly mixed with the SSTs.

Another problem associated with the MCSST approach is that the retrieval of SST is affected by other changes in the atmospheric infrared opacity that occur, for example, with increased concentrations of stratospheric aerosol from volcanic eruptions. Figure 7.5 shows a monthly time series of in situ and satellite observations for the 1982–1986 period for two eastern Pacific regions: one in the tropics (labeled Nino–3) and one in the northern midlatitudes (Namais). The satellite–derived SSTs are lower overall than are the in situ values by about 0.5 degrees; however, the bias is approximately 2 degrees during the period of the El Chichon eruption.

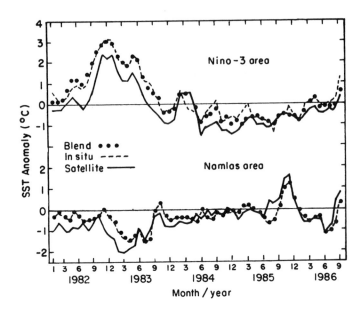

Figure 7.5 Time series of in situ (dashed) and satellite (solid line) SST anomalies for the period shown. The blended data are given as dotted curves. The tropical Pacific region is referred to as Nino–3 and the midlatitude region is referred to as Namais (after Reynolds, 1988).

Reynolds (1988) describes a method which attempts to blend the SST observations from ships and buoys with the satellite observations in an attempt to remove biases of this type. The philosophy is to use the in situ observations as a kind of "benchmark" and the satellite observations are used to define the shape of fields in regions where there is little in situ data. This approach attempts to overcome these biases and is used to develop a climatological SST data set. An example of the comparison of the blended product to the in situ and satellite data is also shown in Fig. 7.5 for the two east Pacific regions.

7.3 Examples of Path–Integrated Quantities

The microwave radiation emitted from the surface of the Earth and sensed by an instrument on a satellite is modulated by processes of emission, absorption, and scattering in the atmosphere; thus, it depends upon the properties of the Earth's surface, atmospheric constituents, and hydrometeors (water droplets, rainfall and ice crystals). The microwave emission from the surface of the Earth depends both on the physical temperature and the emissivity of the surface. Over oceans, values of emissivity typically vary between 0.4 to 0.5, and the emissivity depends on the surface wind speed (refer to Section 4.5). Thus the emission from the ocean is relatively constant and provides a low–level background signal for observing atmospheric emission. Over land, however, both temperature and emissivity are highly variable. The temperature of the land varies with solar insolation, both diurnally and in response to clouds. Land temperature is also a function of the surface albedo, evaporation and evapotranspiration, wind speed, and other factors. The land–surface emissivity is dependent on the thickness, type, and water content of the vegetation canopy and the moisture content and type of the soil. This makes the microwave background signal over land highly variable and measurements subsequently more difficult to interpret. Because of this difficulty, the methods now described are applied only over oceans or over large water bodies on land.

A method that provides vertically integrated water vapor (we refer to this as the *precipitable water*) and the vertically integrated cloud liquid water amount is now described. This method makes use of measurements of the microwave emission by the atmosphere and Fig. 7.6 offers a physical perspective on the approach. Shown are theoretically derived brightness temperature spectra from 6.6 to 37 GHz calculated by a radiative transfer model assuming conditions relevant to a tropical atmosphere. The solid, almost straight, line is the brightness temperature that would be measured over the ocean in the absence of any atmospheric absorption. A calm ocean surface therefore appears cold and the overlying atmosphere appears warm against this cold background due to the increased emission by water vapor and liquid water. The amount of warming is related to the amount of increased emission by the atmosphere associated with the amount of water vapor and liquid water distributed along the path. The solid curve indicates the absorption line centered at 22.235 GHz and is derived for a water vapor overburden of 34.2 kgm^{-2}. The dashed curve is the spectrum produced when a cloud layer containing 0.5 kgm^{-2} of liquid water droplets (a value perhaps typical of low–level stratiform clouds) is added to the atmosphere. Cloud droplets at these wavelengths are assumed to be Rayleigh particles (Section 5.3) and produce a systematic increase in emission at all frequencies over and above that due to water vapor. The dif-

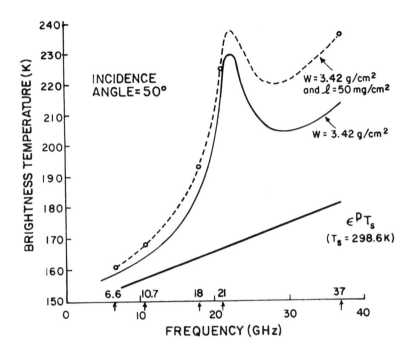

Figure 7.6 Brightness temperature spectra between 6 and 37 GHz for no atmosphere (solid line), a tropical model atmosphere with 34.2 kgm^{-2} of precipitable water (solid curve), and the addition of 0.5 kgm^{-2} of cloud liquid water (from Prabhakara et al., 1982).

ference in brightness temperature associated with the emission by cloud droplets and water vapor increases as the spectral frequency increases. Also shown for reference are the spectral positions of the channels of two satellite microwave radiometers, the Scanning Multi-channel Microwave Radiometer (SMMR) flown on the experimental Nimbus 7 satellite (upward pointing arrows) and the Special Sensor Microwave Imager (SSM/I – downward pointing arrows) which has been used operationally since July 1987 as part of the Defense Military Satellite Program.

7.3.1   Microwave Radiative Transfer

It is important to realize that a significant portion of the microwave intensity emitted from the atmosphere toward the ocean surface is reflected back to the atmosphere. Furthermore, the emissivity is polarized by an amount dependent on the viewing direction (Section

4.5). Equation (7.6a) can be readily modified to account for these factors in the following way,

$$T_B = \epsilon^p T_s e^{\tau^*/\mu} + \int_0^{\tau^*} T(t) e^{-(t-\tau)/\mu} d(\frac{t}{\mu})$$

$$+ \mathcal{R}^p e^{-\tau^*/\mu} \int_{\tau^*}^0 T(t) e^{-(\tau^*-t)/\mu} d(\frac{t}{\mu}) \qquad (7.21)$$

where T_B is the brightness temperature measured at the satellite altitude,[3] T_s is the sea surface temperature, ϵ^p is the emissivity of the ocean surface with the given polarization state p, and $\mathcal{R}^p = (1 - \epsilon^p)$ is the surface reflectivity. The first term of the right–hand side of (7.21) is the surface emission term, the second defines the integrated atmosphere emission, and the third corresponds to the downwelling radiation emitted by the atmosphere, reflected at the surface and then transmitted to the satellite sensor. A simplification to (7.21) can be made if it is assumed that the absorption by water vapor is confined to the boundary layer. Thus

$$\int_0^{\tau^*} T e^{-(t-\tau)/\mu} d(\frac{t}{\mu}) \approx T_s \int_0^{\tau^*} e^{-(t-\tau)/\mu} d(\frac{t}{\mu}) \approx T_s \left(1 - e^{-\tau^*/\mu}\right)$$

$$(7.22)$$

where we have specifically assumed that the emitting temperature of the water vapor is the same as the sea surface temperature. Refinements to this approximation to include vertical variations of temperature are relatively simple to make but lead only to relatively small corrections to what is presented below so these are omitted here. On substitution of (7.22) into (7.21) we obtain

$$T_B \approx T_s \left[1 - \mathcal{T}^2(\mu)(1 - \epsilon^p)\right] \qquad (7.23)$$

where $\mathcal{T}(\mu) = e^{-\tau^*/\mu}$ is the transmissivity along the direction defined by μ.

---

[3] We can invoke the Rayleigh–Jeans distribution for $\mathcal{B}$ for many of the microwave frequencies of interest in this book and thus replace this function simply by the temperature T (refer to the discussion in Section 2.5 for more details).

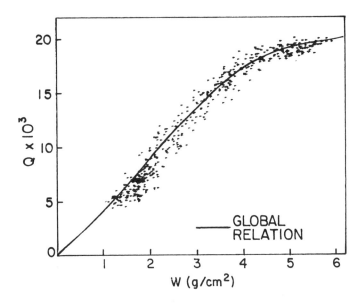

Figure 7.7 Scatter plot of sea–level water vapor mixing ratio and precipitable water for July data in the northern hemisphere and for January data in the southern hemisphere (Liu, 1986).

Excursus: Surface–Level Humidity and Heat Fluxes Over the Ocean

Atmospheric scientists have long been interested in the relationship between vertically integrated path quantities, like precipitable water, and surface observations of properties such as the dew–point temperature, mixing ratio, and vapor pressure. Over the past few years, the association between precipitable water derived from microwave emission measurements and surface level mixing ratio has been studied. A relationship between these quantities is particularly important over the ocean regions where observations are lacking and where this information is crucial for studies of air–sea interaction.

Using 17 years of sounding data from 49 ocean stations, Liu (1986) proposes an empirical relationship between mixing ratio near the surface and precipitable water, an example of which is shown in Fig. 7.7. This relationship provides us with a way of estimating the flux of water vapor E via the bulk parameterization

$$E = \rho_{air} C_E v (q_s - q)$$

where ρ_{air} is the density of surface air, C_E is a specified transfer coefficient, q_s is the saturation humidity at the surface, v is the near–surface wind speed and q is the humidity of air at some level just above the surface (usually within the surface layer such that v and q are observed at the same height). This value of q is determined from satellite observations of precipitable water using the a priori relationship between q and w. The latent heat flux then follows as the product of E and the latent heat of vaporization L. In principle, the wind speed at the surface can also be estimated from spaceborne sensors using methods like those described earlier. It is therefore possible to estimate E and LE from satellite observations provided we know C_E and ρ_{air}. Liu and Niiler (1984) propose this approach for estimating the monthly mean latent heat fluxes and claim an accuracy of approximately 20 Wm^{-2} for these fluxes.

7.3.2 A Microwave Method for Integrated Water Vapor and Cloud Water

Both the vertically integrated water vapor and cloud liquid water contents are important quantities in the study of the Earth's climate system and especially relevant to topical problems of global change. Figure 7.6 also illustrates how the emission of microwave radiation arises from emission by vapor and by cloud liquid water droplets. This emission is unpolarized whereas the microwave radiation from the ocean surface is polarized to some degree (as described in Section 4.5). The stronger the absorption and thus the stronger the emission by the atmosphere, the more obscured is this polarization. A scheme that exploits this property of emission to derive vertically integrated water vapor and cloud liquid water contents will now be described. The method begins with the approximation (7.23) to analyze the 19.35 GHz and 37 GHz brightness temperatures for the two measured polarization states. This equation is differenced relative to these two polarizations to produce

$$\Delta T_B = T_s \left(\mathcal{R}^V - \mathcal{R}^H \right) \mathcal{T}^2 \qquad (7.24)$$

at each frequency where $\mathcal{R}^{V,H} = 1 - \epsilon^{V,H}$ represents the surface reflectivity in the horizontal (H) and vertical (V) polarization states. These surface reflectivities are functions of wind speed as we have previously described. The square of the atmospheric transmission is

$$\mathcal{T}^2 = \mathcal{T}_w^2 \mathcal{T}_W^2 \mathcal{T}_{ox}^2$$

which includes the transmission factor for O_2 and factors for cloud liquid water $\mathcal{T}_W = \exp(-k_\ell W/\mu)$ and water vapor $\mathcal{T}_w =$

$\exp(-k_w w/\mu)$ where k_ℓ is the absorption coefficient of liquid water (the form of this coefficient is described in Section 5.3) and k_w is the absorption coefficient for vapor. With some rearrangement of (7.24), it follows that

$$k_\ell W + k_w w = -\frac{\mu}{2} \ln \frac{\Delta T_B}{T_s(\mathcal{R}^V - \mathcal{R}^H)\mathcal{T}_{ox}^2} \qquad (7.25)$$

This equation, applied to both the 19 GHz and 37 GHz measurements, constitutes a set of linear simultaneous equations which can be solved for w and W given values of $\mathcal{T}_{ox,19}$, $\mathcal{T}_{ox,37}$, Δk_w, and Δk_W and modeled values of $\mathcal{R}^V$ and $\mathcal{R}^H$. These surface reflectivities are functions of wind speed as previously noted in Chapter 4.

Knowing just what values to use for the absorption coefficients is perhaps the greatest source of uncertainty not only to the retrieval described here but also to most retrievals of water vapor based on measurements of atmospheric emission. Values of k_ℓ can be taken from a number of sources and the values of k_W in principle follow from the particle scattering theories discussed in Chapter 5. The temperature dependence of k_W arising from the temperature dependence of the refractive index of water at these microwave frequencies is yet another source of uncertainty.

Examples of the precipitable water w derived from this retrieval method are given in Figs. 7.8a and b as four–year seasonal averages of w for December–January–February and June–July–August, respectively. The uncertainty in w is estimated to be about 3 kgm^{-2} based on comparisons with near coincident radiosonde data. The distribution of monthly mean w broadly follows the distribution of SST (Stephens, 1990) with the largest amount over the warmest waters of the equatorial western Pacific Ocean.

One significant factor that has limited the wide use of microwave liquid water data is the general lack of independent data to verify retrievals. Greenwald et al. (1992) attempt such a verification using a limited amount of independent ground–based microwave measurements of W as well as an estimate derived from AVHRR measurements of reflected sunlight. Comparisons between near coincident SSM/I satellite values of W with those of both AVHRR and the surface microwave measurements are presented in the form of a scatter plot in Figs. 7.9a and b, respectively, for stratocumulus clouds off the west coast of California. The error bars are meant to signify the extent of spatial variability associated with the measurements and represent one standard deviation above and below the mean value.

DJF 1987–1991 PWC

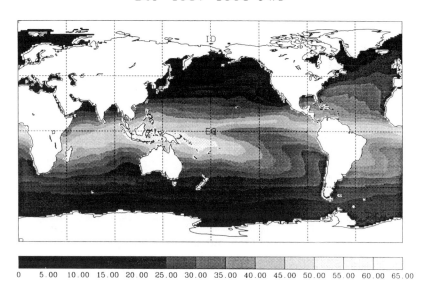

0   5.00   10.00   15.00   20.00   25.00   30.00   35.00   40.00   45.00   50.00   55.00   60.00   65.00

JJA 1987–1991 PWC

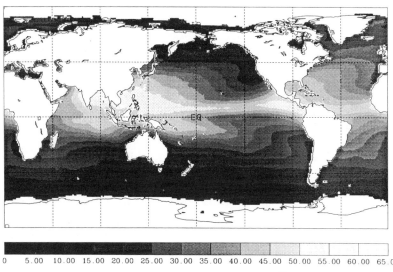

0   5.00   10.00   15.00   20.00   25.00   30.00   35.00   40.00   45.00   50.00   55.00   60.00   65.00

Precipitable Water ($kg\,m^{-2}$)

Figure 7.8 (a) The DJF averaged precipitable water derived as an average of DJFs from four years of data. (b) Same as (a) but for the JJA season (Jackson and Stephens, 1993).

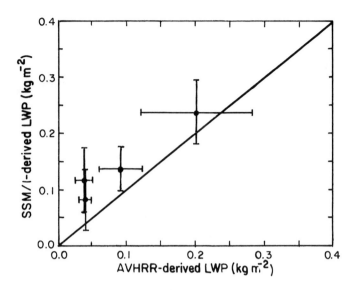

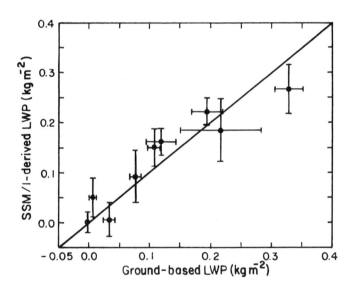

Figure 7.9 Scatter plot of W derived from the microwave retrieval method based on SSM/I measurements with those derived from AVHRR (a) and surface microwave measurements (b). The horizontal and vertical lines about each point are meant to convey the extent of variability in the measurements (Greenwald et al., 1993).

It is yet to be shown if this level of agreement is reasonable and these comparisons do not constitute a rigorous test of the approach, especially when applied to provide global distributions of W. With this in mind, the distribution of W derived from emission measurements over oceans are shown in Fig. 7.10 for August 1987 and February 1988. The main characteristics of global cloudiness are reproduced in these maps including the predominant regions of cloudiness associated with the Inter–Tropical Convergence Zone (ITCZ), the South Pacific Convergence Zone (SPCZ) and the cloudiness associated with midlatitude storminess.

7.4   Passive Sensing of Rainfall

Precipitation, in the form of rainfall, is a key component of the Earth's hydrological cycle. The need to obtain better observations of precipitation globally as well as the need to incorporate more realistic cloud processes in models of the large scale atmospheric motion has prompted more interest in the possibility of space–borne observations of precipitation.

A number of methods exist for the remote sensing of rainfall. Two of these, based on radiation emitted by clouds and rainfall, will now be described and Fig. 7.11 provides a convenient framework for contrasting these two techniques. The diagram presents measurements of radiation at a number of frequencies as a function of elapsed time along a flight path of an instrumented aircraft. The emissions at 18, 37, 92, and 183 GHz are presented in terms of brightness temperature as is the emission at 11 μm. The bottom shaded portion of each panel is the rainfall rate deduced by radar (methods for estimating rainfall from radar measurements are described in Chapter 8). As the aircraft overflies the cloud system, the 11μm brightness temperature decreases from about 245 to 205 K. This decrease occurs at the edge of the deepening outflow anvil from the thunderstorm complex. The microwave temperatures do not show a corresponding drop indicating that the cloud is transparent at these frequencies and presumably composed of ice crystals much smaller than the wavelength of the radiation in question. Passage of the aircraft across the coastline from ocean to land is indicated by an increase in the microwave brightness temperatures due to the sharp rise in surface emissivity at these frequencies. Over the area of deep convection and heavy rain, small decreases (10 to 15 K) in the 11μm brightness temperatures associated with overshooting cells are ob-

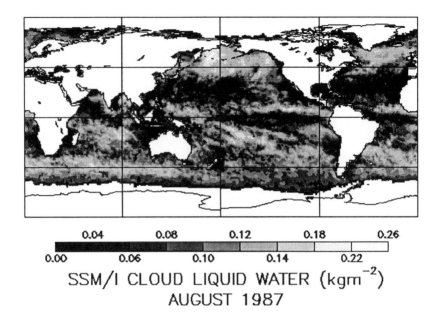

SSM/I CLOUD LIQUID WATER (kgm^{-2})
AUGUST 1987

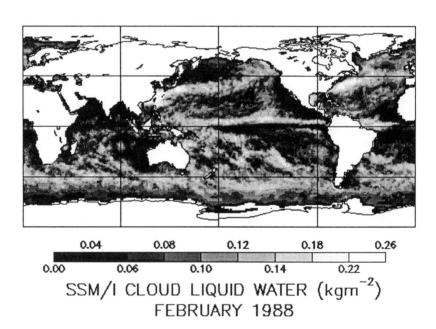

SSM/I CLOUD LIQUID WATER (kgm^{-2})
FEBRUARY 1988

Figure 7.10 Global distributions of cloud liquid water derived from SSM/I observations for August 1987 (top) and February 1988 (bottom) (Greenwald et al., 1993).

served, but the correlation between the decreased emission in these regions and precipitation as indicated by radar measurements is obviously weak. A deep layer of cloud composed of relatively large ice particles leads to significant depressions in the measured microwave temperatures over these deep cores. Values of this depression are as large as 70 to 100 K for the higher frequency 92 and 183 GHz channels due to scattering of radiation at these frequencies by large ice particles. Although this scattering increases with increasing frequency, its effect is even apparent in the 18 and 37 GHz channels. The minima in the 18 GHz time series broadly correspond to the peaks in convective rainfall (these are denoted by the letters A, D, E, and G). It is this general relationship between microwave brightness temperature and rainfall that serves as the basis for microwave sensing of precipitation. These results, however, demonstrate how ice particle scattering complicates microwave emission estimates of rainfall especially for frequencies greater than about 37 GHz.

7.4.1 Rainfall Estimates from Outgoing Longwave Radiation (OLR)

Satellite infrared image data have been used extensively over the past two decades in an attempt to derive rainfall. The basic idea of this approach assumes that the rainfall rate is related in some way to the depth of the cloud and thus to cloud top temperature. Since measurements of the IR emission at window wavelengths provide a way of estimating this temperature, the OLR is assumed to be related to rainfall. Estimates of the outgoing longwave radiation have been available since 1974 from the window channel measurements available from operational NOAA polar–orbiting satellites. It was quickly realized that OLR provides a qualitative index of tropical convection and over large, fixed areas in excess of 10^4 km^2 the OLR is found to be significantly correlated to rainfall. Arkin (1979) and Richards and Arkin (1981) estimate that 50 to 70% of the variance of areally averaged rainfall accumulations measured during the GARP Atlantic Tropical Experiment (GATE) is explained by a linear function of the mean fraction of the area covered by cloud with equivalent black body temperatures colder than thresholds ranging between 220 and 250 K. Largest values of explained variance were found for a brightness temperature threshold of 235 K and for areal averages defined by a grid 2.5 degrees latitude/longitude on a side. These gross statistical correlations suggest that OLR precipitation

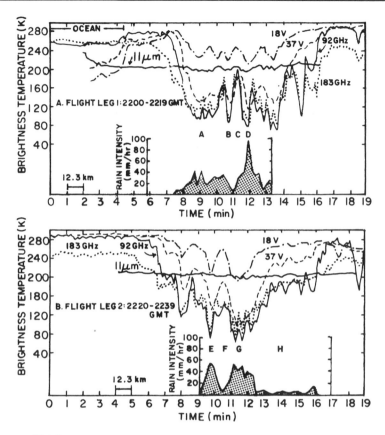

Figure 7.11 Observed brightness temperatures at 11 μm, 18 and 37 GHz, and at 92 and 181 GHz as a function of NASA's ER-2 flight path for two different time segments. Also shown is the low–level radar–based rain rate for the portion of the flight path where radar data are available (Adler et al., 1990).

estimates may play some useful role in climatological studies of precipitation

Arkin and Meisner (1987) applied this threshold method to IR imagery from the GOES satellite to arrive at a rainfall estimate according to

$$GPI = 3A_c t$$

where GPI is the rainfall estimate referred to as the GOES Precipitation Index and is expressed in millimeters, A_c is the fractional area (dimensionless between 0 and 1) of cloud colder than 235 K in a $2.5° \times 2.5°$ box, and t is the length of the period (hours) for which A_c was the mean fractional cloudiness. The product of A_c with duration t is referred to as the area–time–integral (ATI) and we will return to this product quantity in Chapter 8.

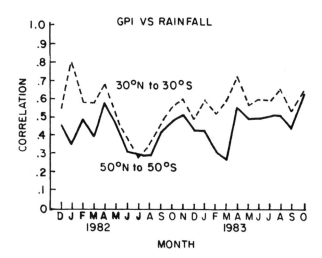

Figure 7.12 Time series of the spatial correlation between monthly rainfall observed at 465 stations and the GPI for the entire 30°N to 30°S region and for the entire 50°N to 50°S region (Arkin and Meisner, 1987).

Figure 7.12 presents the correlation of the GPI with independent station measurements of precipitation for the region extending from 30°N to 30°S and from 50°N to 50°S. The station data used in this comparison were limited to only 465 stations which is only a very small percentage of the study area. The secular variation in the correlation for the two regions demonstrates that OLR and rainfall are more highly correlated in the tropics throughout the period of study possibly because of the higher frequency of convective rainfall in that region.

This analysis serves to illustrate one of the main problems with present satellite rainfall retrieval methods. There is a general lack of extensive ground truth observations for verification. This is also an inherent problem for the remote sensing of SST over southern oceans and for the retrieval of cloud liquid water, although it is perhaps even more severe for precipitation due to the inherent small–scale spatial variability of rainfall. Despite this problem, efforts are underway to produce more global distributions of rainfall deduced from satellite measurements. Figure 7.13 is an example of such an attempt and shows maps of the GPI over the tropics. The regions not covered by the GOES were filled in using NOAA polar orbiting satellite data to produce these maps. The anticipated characteristics of tropical precipitation seem to be well reproduced in this analysis,

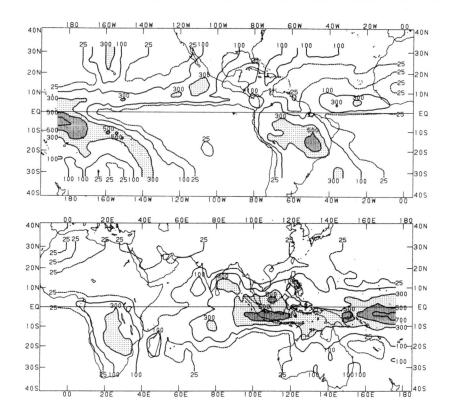

Figure 7.13 Estimated rainfall (millimeters), on a 2.5° × 2.5° grid for the period December 2–31. Areas of missing geostationary satellite data have been filled in using OLR data from the NOAA polar–orbiting satellite (Arkin and Ardanuy, 1989).

including heavy precipitation in areas of deep convection located over the Indonesian maritime continent, the Amazon and Congo basins as well as the precipitation associated with the ITCZ and the SPCZ.

7.4.2   Microwave Emission Methods for Sensing Rainfall

Although retrieval of rainfall from microwave emission measurements suffers from its own ambiguities, as highlighted in Fig. 7.11, these measurements in principle offer a more direct way of estimating rainfall than do OLR measurements. The microwave technique exploits the direct consequences of the interaction between microwave radiation and precipitation–sized hydrometeors. This interaction is characterized by the optical depth τ^* associated with the emitting rain drops and the basis for estimating rainfall lies in an assumed relationship between this optical depth and the rainfall rate and a suitable contrast between the emission from the surface and the emission from the raindrops. Emission from raindrops, when viewed against a cold ocean background, increases with increasing optical depth in a manner illustrated by the curve of growth highlighted in Fig. 7.3. This increased emission, measured as an increased brightness temperature, is then associated with the rain rate $\Re$. We can explore a specific example of the relationship between $\Re$ and τ^* by assuming a Marshall–Palmer size distribution

$$n(D) = N_o e^{-\Lambda D} \tag{7.26}$$

where $N_o = 0.16 \text{ cm}^{-4}$ and

$$\Lambda = 81.56\Re^{-0.21} \tag{7.27}$$

where the rain rate $\Re$ is expressed in units of millimeters per hour. It therefore follows by definition that the volume extinction coefficient is

$$\sigma_{ext} = \frac{N_o}{4} \int_0^\infty n(D)\pi D^2 Q_{ext} dD. \tag{7.28}$$

Figure 7.14 presents calculations of σ_{ext} as a function of rain rate assuming the dielectric constant for water at $\lambda = 1.55 \text{ cm}$ (19.3 GHz) and $T = 273 \text{ K}$. The calculations are based on the Lorenz–Mie theory and present both the extinction and scattering coefficients. These results show how the extinction is defined by the rain rate (under the assumption of the Marshall–Palmer distribution) and furthermore how the extinction at this frequency is largely governed by absorption processes since the scattering coefficient is almost an order of magnitude smaller than the absorption coefficient.

It is straightforward in principle to derive the brightness temperature associated with upwelling radiation as a function of τ^* and thus as a function of $\Re$. Figures 7.15a and b show examples of these relationships from synthetic data using output from a numerical cloud model as input into a numerical radiative transfer model. Each point is the microwave brightness temperature representative of a single square grid point of the model which is approximately 1.5 km on its side. The calculated brightness temperature is presented as a function of the modeled rain rate. The scatter diagram of Fig.

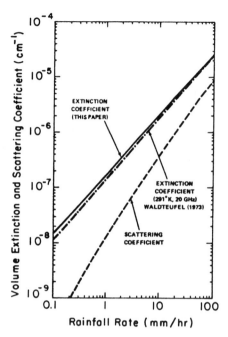

Figure 7.14 The volume scattering coefficient and volume extinction coefficient at $\lambda = 1.55$ cm for a Marshall–Palmer distribution of raindrops as a function of rain rate. The dielectric constant used in the calculations applies to water at T=273K (after Wilheit et al., 1977).

7.15a corresponding to 10 GHz exhibits the type of curve of growth expected from our simple model results graphed in Fig. 7.3. However, the equivalent relationship for 19 GHz is more complex for a number of reasons. In this case, scattering by the large ice crystals over regions of heaviest rainfall confuses matters. Radiation emitted from the underlying rain is scattered downward by the overlying layer of ice particles and away from any instrument looking downward above the cloud. This increase in scattering as $\Re$ increases leads to a decrease in brightness temperature.

Real data exhibit even more scatter than shown in Fig. 7.15b due to the highly variable spatial distribution of rainfall within an instrument field of view (this is referred to as the problem of *beam filling*). This is not treated in the results of Fig. 7.15 since the brightness temperatures are based on radiative transfer calculations that assume the rainfall to be horizontally homogeneous. Estimating rainfall from microwave emission measurements and understanding how the factors mentioned here, and others, affect these estimates remain topics of active research.

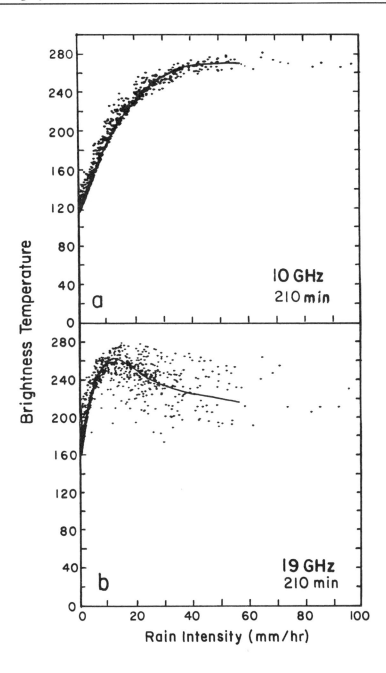

Figure 7.15 Scatterplots of microwave brightness temperature at 10 GHz (upper) and 19 GHz (lower) as a function of rain rate. The time indicated represents time into a cloud model simulation (from Adler et al., 1991).

7.5   Principles of Sounding by Emission

An important application of the particular type of radiative transfer discussed in this chapter centers on the inversion of the radiative transfer equation to retrieve vertical distributions of temperature and trace gas concentrations. The general basis of this inversion was outlined in the introductory chapter and much more can be found in the literature cited at the back of this chapter. This section intends only to provide the reader with a broad understanding of the physical basis for temperature sounding.

The basic approach of emission sounding of temperature is to detect the radiation emitted by gases of known distribution, like that of the uniformly mixed gases and specifically by the carbon dioxide molecule at wavelengths centered at 15 μm and by the oxygen molecule at frequencies around 60 GHz.

In discussing how temperature is retrieved from spectral intensity measurements, consider the following two experiments. The first setup has a radiometer at the ground and the second has a radiometer on a satellite that orbits the Earth. It is convenient to introduce

$$T(t, \tau, \mu) = e^{-(t-\tau)/\mu} \tag{7.29}$$

as the transmittance along the path from the optical depth t to the optical depth τ along the direction defined by μ which we take to be the cosine of the zenith angle of observation. Therefore,

$$\frac{dT}{dt}(t, \tau, \mu) = -\frac{1}{\mu} e^{-(t-\tau)/\mu} \tag{7.30}$$

and, on substitution into (7.6a), we obtain

$$I(\tau, \mu) = I(\tau^*, \mu) T(\tau^*, \tau, \mu) + \int_{\tau^*}^{\tau} B(t) \frac{dT}{dt}(t, \tau, \mu) dt \tag{7.31}$$

The following notation

$$W(\tilde{z}_1, \tilde{z}_2) = \frac{dT}{d\tilde{z}}(\tilde{z}_1, \tilde{z}_2)$$

is used to refer to the quantity known as the *weighting function* for the reasons that become apparent later. In this definition, $\tilde{z}$ is taken to be an arbitrary vertical coordinate system such as altitude z, optical depth τ as in the case of (7.31), pressure p or $\ln p$. Therefore, the weighting function is defined at some level $\tilde{z}_1$ relative to the measurement level $\tilde{z}_2$. In terms of this general vertical coordinate, (7.31) becomes

$$I(\tilde{z}, \mu) = I(\tilde{z}^*, \mu)T(\tilde{z}^*, \tilde{z}, \mu) + \int_{\tilde{z}^*}^{\tilde{z}} B(\tilde{z}')W(\tilde{z}', \tilde{z}, \mu)d\tilde{z}' \quad (7.32)$$

from which the meaning of the weighting function W emerges. Within the context of (7.32), the contribution to the intensity measured by a radiometer located at $\tilde{z}$ due to the emission from the layer $\Delta\tilde{z}'$ centered at $\tilde{z}'$ is determined from the local layer black body emission $B(\tilde{z}')$ *weighted* by the factor $W(\tilde{z}', \tilde{z})\Delta\tilde{z}'$. The functional form of W is of fundamental importance to vertical sounding problems and simple models are now described in an attempt to build an understanding of the general properties of W and how it depends on the strength and distribution of absorber.

7.5.1 Weighting Functions for Nadir Sounding

Absorption by molecules in the atmosphere below 50 km is strongly influenced by the collision broadening processes expressed in terms of the Lorentz line shape (refer to Chapter 3),

$$k_\nu = \frac{S\tilde{p}\alpha_{Lo}/\pi}{(\nu - \nu_o)^2 + \alpha_{Lo}^2\tilde{p}^2} \quad (7.33)$$

where α_{Lo} is the line width defined at a pressure $p_o = 1$ atmosphere, $\tilde{p} = p/p_o$ where p is the pressure in the same unit, S is the line strength and ν_o is the frequency of the line center. For a gas uniformly mixed in the atmosphere with a mass mixing ratio r, it follows that the optical depth of a layer between defined pressure levels p' and p'' is (e.g., Section 3.3c)

$$\tau_\nu = \frac{r}{2p_og} \int_{p'}^{p''} \frac{S\alpha_{Lo}/\pi}{(\nu - \nu_o)^2 + \alpha_{Lo}^2 p^2}dp \quad (7.34)$$

and for absorption in the line wings where $\nu - \nu_o > \alpha_{Lo}p$,

$$\tau_\nu \approx \frac{k_o r}{g}[p''^2 - p'^2] \quad (7.35)$$

where $k_o = S\alpha_{Lo}/\pi(\nu - \nu_o)^2$. For an atmosphere in hydrostatic equilibrium, $p = p_o e^{-z/H}$ where H is the atmospheric scale height, and it follows from (7.35) that

$$t(z) = \tau^* e^{-2z/H} \quad (7.36)$$

where $p' = 0$, and $\tau^* = k_o r p_o / 2g$. Let us define the weighting function relative to the satellite altitude $(z = \infty)$ as

$$W(z, \infty) = \frac{dT}{dz}(z, \infty) \qquad (7.37)$$

then substituting $\exp(-t)$ for $T(z, \infty)$ and with (7.36) into (7.37), we obtain

$$W(z, \infty) = \frac{dT(z, \infty)}{dt} \frac{dt}{dz} = \frac{2\tau^*}{H} \exp\left[-\frac{2z}{H} - \tau^* e^{-2z/H} \right] \qquad (7.38)$$

The characteristic relationship between W and z given by this simple formula is shown in Fig. 7.16a for three different values of τ^*. The shape of the weighting function is governed by two factors: the factor $\exp(-2z/H)$ which decreases with increasing z and the factor $\exp[-\tau^* \exp(-2z/H)]$ which increases as z increases. The first of these factors represents the decrease in absorber gas with changing z and the second factor characterizes the increase in transmission as the path decreases to 0 as z approaches the satellite altitude. These combine to produce a familiar bell–shaped weighting function curve. The width of this curve ultimately characterizes the vertical resolution of the retrieval (refer to Problem 7.5 as an example of this) and the peak of the curve occurs at

$$z_{max} = \frac{H}{2} \ln \tau^*$$

which follows by setting $dW/dz = 0$. Therefore, measurements at different frequencies characterized by different values of τ^* allow us to sample the emission from different layers in the atmosphere. These layers broadly correspond to the layers that surround the peaks of the corresponding weighting functions. The physical interpretation of the general bell shape of the weighting function curves was discussed in Chapter 3. Near the line center, τ^* is large and the weighting function is more sharply peaked with a maximum higher in the atmosphere. As we move away from the line center toward the wings, τ^* decreases by virtue of the decreasing absorption coefficient. If far enough out from the line center, the maximum of the weighting function actually occurs at the surface (Fig. 7.16a).

The width of lines in the 15 μm CO_2 band varies from about 0.1 to 0.001 cm^{-1} over the range of atmospheric pressures of interest. Most radiometers measure the spectral intensity with a spectral

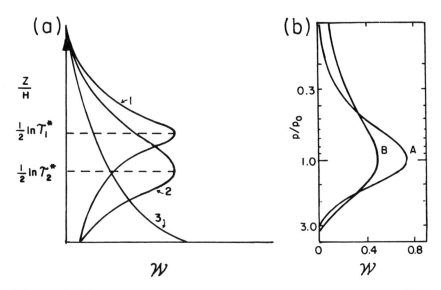

Figure 7.16 (a) The behavior of a monochromatic weighting function W in the wing of a collision broadened line corresponding to a hydrostatic atmosphere for different value ranges of τ^*, the total optical depth of the atmosphere. Curves labeled as 1: $\tau^* > 1$, 2: $\tau \geq 1$ and 3: $\tau^* \leq 1$. (b) The weighting functions for a monochromatic frequency in the wing of a collision broadened line (A), and for a nonmonochromatic band of frequencies containing strong absorption lines (B) (Houghton et al., 1984).

bandwidth that is significantly broader than the width of individual lines so the radiation actually detected by such an instrument results from the collective emissions by a band of hundreds to thousands of lines. The weighting function for a broad interval containing many overlapping absorption lines can be thought of as a superposition of weighting functions of individual lines each characterized by different values of τ^*. The result is a weighting function that is smeared out over several layers and one that is broader than a weighting function derived for single lines. Figure 7.16b provides an actual example of weighting functions calculated using a special type of transmission model to represent the absorption of a spectral interval that contains many spectral lines.

7.5.2 Weighting Functions for Zenith Sounding

The weighting functions characterizing the emission from the atmosphere as measured by an instrument on the ground looking up are very different from those for nadir viewing instruments. To examine the characteristic shape of these functions, consider the zenith intensity measured at the ground

$$I(\tau^*) = \int_0^{\tau^*} B(t)e^{-(\tau^*-t)}dt \tag{7.39}$$

which follows from (7.6b) with $I(0, -\mu) = 0$. The transmittance $T(\tau^*, t) = \exp[-(\tau^* - t)]$ assuming $\mu = 1$ which becomes

$$T(z,0) = \exp[-\tau^* + \tau^* e^{-2z/H}] \tag{7.40}$$

by virtue of (7.36). It simply follows that the weighting function

$$W(z,0) = \frac{dT}{dz} = \frac{2z}{H}e^{-\tau^*} \exp\left[-\frac{2z}{H} + \tau^* e^{-2z/H}\right] \tag{7.41}$$

where the major difference between (7.41) and (7.38) is the change in sign inside the exponential, which leads to $dW/dz < 0$ and predicts that the shape of the weighting function is always of the type labeled 3 in Fig. 7.16a regardless of the value of τ^*. Two factors also dictate the general shape of these weighting functions. The first of these factors, $\exp(-2z/H)$, characterizes the change in absorber with z and the second factor, $\exp[\tau^* \exp -2z/H]$, is the decrease of transmission as z increases upward away from the instrument. For the distribution of absorbing gas considered, an instrument looking up always receives most of its radiation from layers adjacent to the ground. In the line centers, this radiation originates almost exclusively from layers near the surface whereas radiation is received from higher up in the atmosphere for wavelengths located in the line wings although the maximum emission still occurs near the surface where pressure is greatest and the line is broadest.

7.5.3   Weighting Functions for Limb Sounding

An example of the geometric configuration for limb sounding is given in Fig. 7.17. The intensity measured at the satellite can be expressed as the integral of the emission along a line–of–sight that is defined at the tangent altitude h. We write this equation in the form

$$I(h) = \int_\infty^0 B(s)\frac{dT}{ds}(s,0)ds \tag{7.42}$$

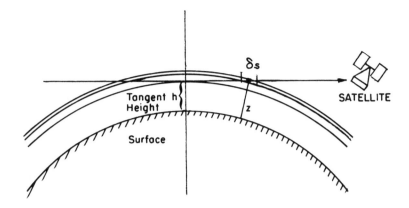

Figure 7.17 Limb viewing geometry. The satellite instrument scans through the atmosphere along the limb at a tangent height h. Intensity is received from segments δs of the path at height z.

where $\mathcal{T}(s,0)$ is the transmission along the path of length s from the outer levels of the atmosphere at $s = \infty$ through the tangent height to the instrument at $s = 0$. As earlier, our analysis considers only the hypothetical case of monochromatic radiation whereas real applications deal with a frequency average over the band pass of the instrument. In reality, an instrument also has a finite field of view and measures the radiation over some finite vertical layer. In the case of vertical sounding, it is usually assumed that the intensity is uniform across the instrument field of view. Although this generally does not apply for limb sounding because the intensity typically varies rapidly across the field of view, it is nevertheless assumed here merely to simplify matters.

It is also possible to carry out a simple analysis of the properties of the weighting functions for limb sounders assuming a small field of view of the instrument and a single frequency response. We use the approach developed earlier for nadir sounding and rewrite (7.42) in the form

$$I(h) = \int_h^\infty B(z') \frac{d\mathcal{T}}{ds}(z',\infty) \frac{ds}{dz'} dz' \qquad (7.43)$$

where the integration is now over altitude z rather than over the tangent path s. We can relate s to z using simple geometric arguments to obtain the approximation

$$s^2 \approx 2R(z - h)$$

With a vertical weighting function defined as

$$W(h; z, \infty) = \frac{dT}{ds}(z, \infty)\sqrt{R/2(z - h)} \quad z > h$$

$$W(h; z, \infty) = 0 \quad z < h, \tag{7.44}$$

then (7.43) becomes

$$I(h) = \int_h^\infty B(z')W(h; z', \infty)dz' \tag{7.45}$$

The square root factor in the definition of W represents the enhancement of the tangent path relative to a vertical path. Because of this factor, the tangent path defines an absorber amount that is several times larger than that of the vertical path. This enhancement of the absorber path increases the sensitivity of the emission from trace gases such as CO, NO, N_2O, and ClO providing the sensitivity necessary for sounding these gases.

To calculate the weighting functions relevant to limb sounders, the contribution to the measured intensity due to the emission by a layer Δz thick located at z is required. Assuming the geometry of Fig. 7.17, the emission from the layer located about height z follows from (7.43) as

$$\Delta I(z) = \frac{d\bar{T}}{ds}(z, \infty)\sqrt{R/2(z - h)}[e^{-\tau_1} + e^{-\tau_2}]B(z)\Delta z \tag{7.46}$$

where

$$\tau_1 = r \int_z^\infty k_m(z')\rho_{air}(z')\sqrt{R/2(z' - h)}dz'$$

and

$$\tau_2 = \tau_1 + 2r \int_h^z k_m(z')\rho_{air}(z')\sqrt{R/2(z' - h)}dz'$$

Here the atmosphere has been divided into two regions; one is the region above the reference level z and nearer to the satellite than the tangent point and the other is that portion of the atmosphere behind the reference level including the path through the layer defined by h and z. τ_1 is the optical path for the first of these paths and τ_2 corresponds to the second and longer path. The factor of two results from the path symmetry about h through the layer defined by h

and z. Using the hydrostatic assumption, together with an assumed Lozentz line profile,[4]

$$\tau_1 = \frac{\tau^*}{2}\sqrt{\pi HR}[1 - \text{erf}(\sqrt{2(z-h)/H})]\exp(-2h/H), \quad (7.47a)$$

and

$$\tau_2 = \tau_1 + \tau^*\sqrt{\pi HR} \ \text{erf}(\sqrt{2(z-h)/H})\exp(-2h/H), \quad (7.47b)$$

where erf is the error function. The weighting function then follows from (7.45) and (7.46) as

$$W(h; z, \infty) = \tau^* \sqrt{R/2(z-h)}\exp(-2z/H)[e^{-\tau_1} + e^{-\tau_2}]. \quad (7.48)$$

Figure 7.18a provides examples of this weighting function for different values of τ^*, with the tangent height $h = 12$ km and other parameters as given. For the larger optical depths, the emission arises from broad layers above the tangent height and closer to the sensor and the contribution from the layers near the tangent height is reduced. For $\tau^* = 0.1$, the main contribution comes from the layer immediately above the tangent height. In reality, the observations are obtained by operating in the optically thin region of the stratosphere and by scanning the sensor field of view. This offers a way of obtaining a high vertical resolution in the sounding. Figure 7.17b presents examples of actual weighting functions derived for a relatively wide spectral interval covering much of the 15μm band of carbon dioxide for various values of the tangent height h.

Some distinct advantages of limb sounding by emission are:
• Relatively high vertical resolution. None of the emitted radiation originates below the tangent point. From simple geometric arguments, the atmospheric shell immediately above the tangent height h contains the longest ray path of any layer and since pressure decreases exponentially with height, a large fraction of the outgoing radiation originates from a layer which is typically a few kilometers above the tangent height. As a result of these factors, the weighting functions are spiked.

---

[4] Note the assumption of a Lorentz line shape, which is used for convenience here, is not strictly appropriate since Doppler broadening mechanisms will also be important at typical stratospheric pressures and at the wavelengths of relevance to most limb sounders.

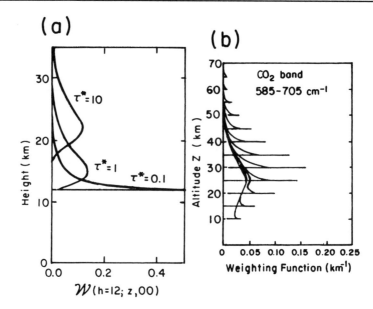

Figure 7.18 (a) The weighting function for limb sounding derived from (7.48) with $H=7$ km, $R=6370$ km, and $h=12$ km for three values of τ^*. (b) A set of weighting functions for a limb sounder based on the ideal case of an instrument with an infinitesimal vertical field of view. The weighting functions are computed for the spectral band 585–705 cm^{-1} which covers most of the 15 μm absorption band of carbon dioxide (after Gille and House, 1971).

- No surface influence on the measured intensities.
- Based on the geometry, considerably more emitting gas (up to about 60 times) exists along grazing paths than along vertical paths. This means that there can be significant contributions by the emission of gases of low concentration than occurs for along vertical paths.

The technique does have certain disadvantages. It is sensitive to the presence of aerosol in the lower stratosphere and generally cannot be reliably used to probe below the tropopause. Limb sounding also requires relatively precise information about the field of view and spacecraft attitudes so that instrument pointing can be accurately determined.

7.5.4 Weighting Functions of an Operational Sounder System

The NOAA TIROS–N Operational Vertical Sounder (TOVS) contains three sounders: the Microwave Sounding Unit (MSU), the

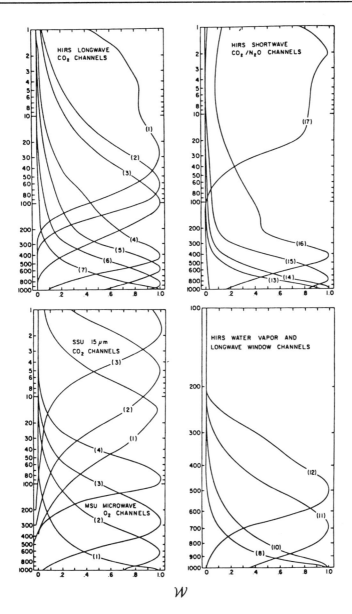

Figure 7.19 Weighting functions for the TOVS sounders (Smith et al., 1979).

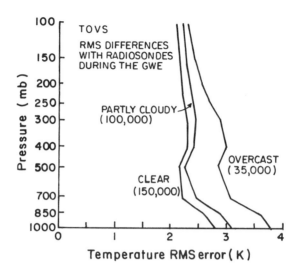

Figure 7.20 Root–mean–square differences between TOVS and radiosonde soundings for an entire year. The collocation and time differences are generally within 200 km and ± 3 hours (Smith, 1991).

High–resolution Infrared Radiation Sounder (HIRS), and the Stratospheric Sounding Unit (SSU). Figure 7.19 presents characteristic weighting functions for each sounder and illustrates the general features predicted by our simple analysis described above.

The actual numerical methods used to retrieve temperature from spectral measurements are not described any further here as these are summarized in some detail by Houghton et al. (1984) and other references noted later. These details are also the subject of the inversion project introduced in Appendix 2. A general assessment of the accuracy of retrieved temperatures is also provided in a number of studies and an example of such an assessment is presented in Fig. 7.20 in the form of the root–mean–square difference between the TIROS–N and radiosonde layer–mean temperatures. This rms difference is around 2–3 K for clear skies but is poorer for partially cloudy conditions and worst for overcast conditions. The reduced accuracy of the overcast sky retrievals is perhaps due to the limited number of tropospheric sounding microwave channels and their poor vertical resolution in the lower troposphere compared to the infrared channels. Although the rms differences of a few degrees shown in Fig. 7.20 seems small, individual differences from radiosondes may be as high as 10 K. These differences typically occur as a result of the coarse vertical resolution of TOVS soundings and the inability to re-

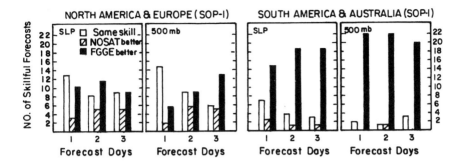

Figure 7.21 Number of cases of positive and negative forecast impacts of the FGGE observing system with and without satellite data during the period January 5 – March 5, 1979 (Kalnay et al., 1985).

solve sharp temperature gradients. Because of this problem, the rms difference between TOVS and radiosonde measured temperatures tend to be largest near the surface and near the tropopause where vertical temperature gradients are capable of varying significantly.

Many of the major operational forecast centers of the world, as well as a number of numerical modeling research laboratories, have investigated the impact of the TOVS data on numerical forecasts. The general findings from this research are conveniently summarized in Fig. 7.21 which is taken from the study of Kalnay et al. (1985). This diagram shows the number of times one of two observing systems, the FGGE, which includes satellite data, and the NOSAT, which excludes these data, led to a better forecast when compared to each other[5]. Whereas the results for the northern hemisphere are mixed, satellite data proved to be indispensable for analyses in low latitudes and in the southern hemisphere. Even in the northern hemisphere, satellite data improved the analyses over remote areas like the northern Pacific region and helped improve forecasts downstream of that region.

---

[5] FGGE is the acronym for the First GARP (Global Atmospheric Research Program) Global Experiment. One of the aims of this experiment was to develop an international observing system for studying the atmosphere and to test these systems during special observing periods throughout an operational year that began in 1978.

7.6   Sensing Clouds by Emission: Windows and Arches

The general topic of remote sensing of clouds has recently loomed as an important aspect of global climate change research. The field is not as highly developed as other topics of remote sensing discussed in this book although a number of new satellite initiatives through the 1990s is expected to accelerate research on this topic and produce new quantitative techniques for sensing clouds. Most of the existing methods unfortunately provide only gross and highly ambiguous properties of clouds (such as how much cloud exists, how often it occurs and some measure of an "effective" particle size among others). A method providing this type of information about cirrus clouds using infrared emission measurements will now be described.

Emission spectra between 9.1 and 16.7 μm measured by an interferometer flown on a high flying aircraft over a cirrus cloud of varying thickness are shown in Fig. 7.22. The feature to note from these spectra is how the emission, particularly in the window region, changes as conditions vary from clear to cloudy skies. The most obvious change is the decrease in emission as the optical thickness of the cloud increases, especially in the atmospheric window region between 10 and 13 μm. The radiation sensed by the instrument over thin cloud, for example, is a mixture of the radiation that is emitted below the cloud by the surface and underlying atmosphere and the radiation emitted by the cloud itself (both by the cloud particles and the emitting gases between these particles). A simple model of this mixture, given the assumption that scattering by cloud particles is negligible for these wavelengths, follows from (7.8) which can be written as

$$I_{obs}(0, \mu) = I_s e^{-\tau^*/\mu} + B(T_c)[1 - e^{-\tau^*/\mu}] \qquad (7.49)$$

where $I_{obs}(0, \mu)$ is the intensity observed at cloud top, I_s is the upward intensity at cloud base, τ^* is the optical thickness of the cloud, and $e^{-\tau^*/\mu} = T(\tau^*, 0, \mu)$ is the transmission through the cloud. This simple model considers the cloud to be isothermal at a temperature T_c. Since scattering is assumed to be negligible, it follows that

$$[1 - e^{-\tau^*/\mu}] = A(\tau^*, 0, \mu) = \varepsilon(\tau^*, \mu)$$

where $A(\tau^*, 0, \mu)$ is the absorptivity of the cloud (also known as the cloud emissivity). The stronger the absorption, the larger is τ^*, the

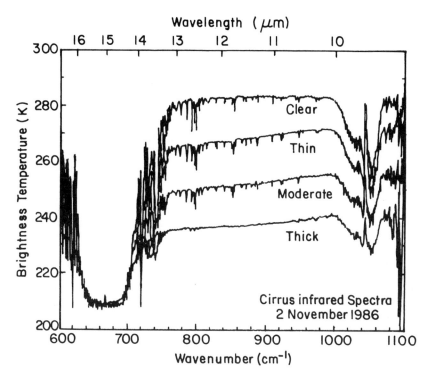

Figure 7.22 The emission spectra between 9.1 and 16.7 μm expressed in terms of brightness temperature measured over a clear scene and cloudy scenes of thin, moderate, and optically thick cirrus clouds observed by an interferometer on a research aircraft (Smith, personal communication).

closer ε is to unity and the smaller is the surface contribution to the observed intensity.

A second but less notable feature of the emission spectra shown in Fig. 7.22 is the change in the difference between the emission at 10 and 12μm and the relative change in this difference from clear to cloudy skies. A convenient way to study this change is to consider the slope in the emission spectrum from one wavelength, such as at 10.8μm to another longer wavelength such as at 12μm. The brightness temperature difference $\Delta T = T_{10.8} - T_{12}$ is a measure of this slope and is near zero both for clear skies and for overcast skies filled with optically thick clouds.[6] A plot of this temperature difference as a function of one of the measured brightness temperatures, say $T_{10.8}$, resembles an arch shape like that shown in Fig. 7.23. One

_____

[6] This brightness temperature difference is 0, by definition, for blackbodies.

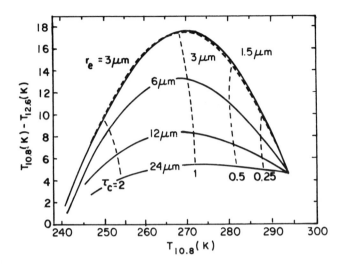

Fig. 7.23 Theoretical brightness temperature differences between 10.8 and 12.6μm as a function of the corresponding brightness temperature at 10.8μm for high level spherical ice crystal cirrus clouds as a function of effective radius and optical thickness (from King et al., 1992).

foot of the arch defines the temperature of clear skies and the other foot establishes the temperature associated with the emission by the densest portion of the cloud. The variation of ΔT between each foot depends on a number of factors including the size and shapes of the emitting ice particles and the optical thickness of the cloud. Figure 7.23 shows theoretically based calculations of the brightness temperature differences as a function of $T_{10.8}$ for calculations assuming spherical ice particles and a distribution characterized by different values of effective radius r_e and different values of optical thickness.

A convenient way to highlight the sensitivity of the emission spectrum to particle size is to invert (7.49) for optical depth

$$\tau^* = -\mu \ln[\frac{I_{obs} - \mathcal{B}(T_c)}{I_{clr} - \mathcal{B}(T_c)}] \tag{7.50}$$

where the clear sky intensity I_{clr} is substituted for I_s. The right–hand side of (7.50) can be evaluated using the measured intensities although estimating T_c is not easy for satellite applications. One approach to do this is to assume that this temperature corresponds to the intercept of the ΔT curve in Fig. 7.23 with the $T_{10.8}$ axis.

The clear sky intensity may also be estimated from observations of clear sky areas adjacent to the cloud (these areas correspond to the other foot of the arch in Fig. 7.23). Another way of estimating the clear sky brightness temperature is discussed later.

Having derived τ^* according to (7.50), the wavelength ratio

$$\gamma = \frac{\tau^*_{12.6}}{\tau^*_{10.8}} \tag{7.51}$$

is then established and we can see how this ratio depends on particle size by considering the following. In the absence of scattering (and in the absence of any molecular absorption both of which are reasonable assumptions for the wavelengths under consideration), the optical depth is

$$\tau^* = \pi \int Q_{abs} n(r) r^2 dr \Delta z$$

for spherical particles of radius r uniformly distributed throughout the cloud layer of depth Δz. For this specific case, and for a cloud composed of N_o particles of radius $r = a$,

$$\tau^* = \pi N_o a^2 Q_{abs} \Delta z$$

such that the optical depth ratio becomes

$$\gamma = \frac{Q_{abs,1}}{Q_{abs,2}}$$

The sensitivity of this ratio to particle radius a can be explored with the model for Q_{abs} introduced in section 5.5. Application of (5.58) leads to

$$\gamma = \frac{K(n_1^2(1 - c_1^3)v_1)}{K(n_2^2(1 - c_2^3)v_2)} \tag{7.52}$$

where

$$c_{1,2} = \frac{(n_{1,2}^2 - 1)^{1/2}}{n_{1,2}}$$

and where $n_{1,2}$ is the real part of the refractive index with the subscripts referring to the two selected wavelengths. For the specific example given later, the subscript 1 refers to $\lambda = 12\mu m$ and subscript 2 to $\lambda = 10.8\mu m$. Figure 7.24a presents a plot of the relationship defined by (7.52) as a function of the particle radius a

Figure 7.24b presents γ derived from Lorenz–Mie theory and from scattering solutions for cylindrical particles as a function of particle radius. In Fig. 7.24a γ is expressed as a function of the radius of a single particle and of the effective radius r_e of an assumed size distribution in Fig. 7.24b. In each diagram, γ decreases rapidly to values near unity as a increases to 15 μm or as r_e increases to 20 μm at which point γ remains relatively insensitive to further increases in particle radius. The general variation of γ as a function of particle size can be understood in terms of the general dependence of Q_{abs} on the similarity parameter v as described previously in Section 5.5. Values of v_1 and v_2 corresponding to selected values of particle radius are given in Fig. 7.23a for reference. The lack of sensitivity of γ for large particles occurs when values of both v_1 and v_2 are large enough (in this case both exceed unity) such that Q_{abs} approaches the large particle asymptotic limit where $Q_{abs,1} \approx Q_{abs,2}$.

The problem with the approach described here for sensing ice particle size, and one common to most passive methods for sensing ice crystals, is that the measured brightness temperature differences are sensitive to the size of the crystals as well as to their shape and other factors. Unfortunately, it is generally not possible to separate these effects unambiguously. Methods to derive ice crystal size in terms of spheres of some equivalent radius and any subsequent interpretation of this size should be viewed cautiously. Another factor of relevance to the approach described earlier is that scattering of infrared radiation is not entirely negligible and the quantity γ is itself ambiguous and not necessarily a measure of the ratio of absorption optical depths as assumed earlier.

Excursus: Golden Arches

One of the advantages of satellites as observational platforms of clouds is their ability to record patterns and structures of clouds over wide ranges of space and time. A method that exploits this particular advantage, as well as using the properties of cloud emission, is the spatial coherence technique introduced by Coakley and Bretherton (1983). The idea behind the approach is portrayed in the upper panel of Fig. 7.25 which schematically presents a group of 2×2 neighboring pixels of 11μm radiances expressed in this specific example as brightness temperature. These pixels are processed to provide the average 11μm brightness temperature of the group and the standard deviation about this average. The latter is a measure

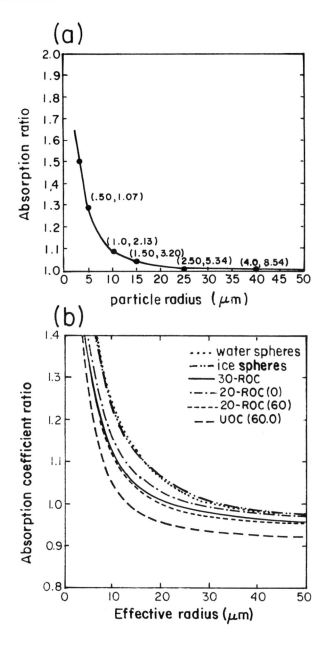

Figure 7.24 (a) The ratio of the absorption coefficient γ as a function of particle radius a. The values of v_1 and v_2 correspond to a single particle of specified radius with the refractive index 1: $\lambda=10.8$, m= 1.0905–i0.1710, 2: $\lambda=12$, m=1.2457–i0.4023. (b) The results of Parol et al. (1991) for γ as a function of effective radius r_e for water and ice spheres and different orientation averages of ice cylinders.

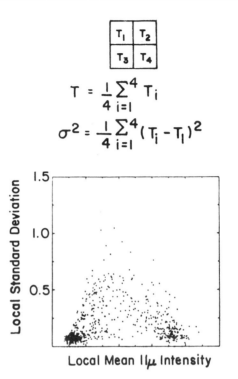

$$T = \frac{1}{4} \sum_{i=1}^{4} T_i$$

$$\sigma^2 = \frac{1}{4} \sum_{i=1}^{4} (T_i - T_1)^2$$

Figure 7.25 A schematic demonstration of the method of spatial coherence (upper panel) as it might be applied to brightness temperatures. Spatial coherence analysis of 11μm intensities for a $(250 \text{ km})^2$ region over stratus clouds off the west coast of California. Each point represents values for a 4×4 array of $(1 \text{ km})^2$ AVHRR pixels (Coakley, 1991).

of the texture of the image on the scale of the pixel array chosen. These two pixel group quantities are then plotted on a scatter diagram in the fashion given by the example of Fig. 7.24b. The satellite data used to construct this scatter plot are the 11μm radiances obtained with the NOAA–9 overpass at 2242 GMT on July 7 obtained from the AVHRR viewing marine stratus clouds off the west coast of California (Coakley, 1991).

The scatter of points on the diagram resembles an arch. The feet of the arch contain important information about those regions of the image that are relatively homogeneous across the group of neighboring pixels. One foot is associated with the relatively clear sky portion of the scene and the other to the pixel groups that are completely filled by a cloud with the same temperature. This provides a way of discriminating clear sky brightness temperatures (T_{clr}) from partially cloudy skies (T_{broken}) and from the brightness temperature (T_{cld}) of a homogeneous layered cloud. For the case shown, only two effectively homogeneous surfaces exist, one is the clear sky background and the other is that of the solid cloud portions of the image. The points in the arch correspond to a partially filled pixel group of cloud cover N and it is assumed that these points can be expressed as

$$T_{broken} = (1 - N)T_{clr} + NT_{cld} \qquad (7.53)$$

from which the cloud cover follows

$$N = \frac{T_{broken} - T_{clr}}{T_{cld} - T_{clr}}$$

Application of this approach alone provides limited quantitative information about clouds and applies only to single layer clouds. The approach relies on the empirical relationship (7.53) and on the statistical nature of the observations that can be used to identify both T_{clr} and T_{cld}.

7.7 More on Clouds: Slicing up the Atmosphere

Another approach for sensing cloud amount and the height of clouds (or specifically cloud top pressure) is the method referred to as CO_2 slicing. This method has been used by Wylie and Menzel (1989) to study cirrus clouds although the idea was described much earlier by Smith et al. (1974), McCleese and Wilson (1976) and others. The slicing method seeks to take advantage of differences of the absorption of the atmosphere at several spectral regions within one of the CO_2 molecular absorption bands. Whether clouds appear in images of these channels depends on the particular channel and the altitude of the cloud. Figure 7.26 gives examples of three weighting functions corresponding to three CO_2 channels of the VAS. Only clouds that emit radiation above the 350mb level contribute significantly to the radiance measured by the satellite in channel 3, whereas channel 4 is capable of differentiating the emission by clouds from that of the molecular atmosphere down to 700mb.

The method of CO_2 slicing begins with the radiative transfer equation

$$I(p, \mu) = I(p_s, \mu)\mathcal{T}(p_s, p, \mu) + \int_p^{p_s} \mathcal{B}(p')\mathcal{W}(p, p')dp' \qquad (7.54)$$

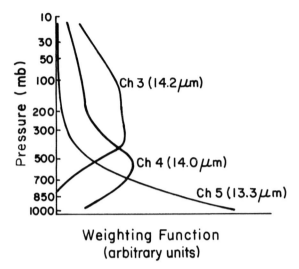

Figure 7.26 The weighting functions for three VAS CO$_2$ spectral bands centered at 14.2, 14.0, and 13.3 μm used in CO$_2$ slicing (Wylie and Menzel, 1989).

for the upwelling intensity at pressure level p. The radiation received at $p = 0$ then follows as[7]

$$I(0, \mu) = I(p, \mu)T(p, 0, \mu) + \int_0^p B(p')W(0, p')dp' \qquad (7.55)$$

or in terms of the surface emission $I(p_s, \mu)$ as

$$I(0, \mu) = I(p_s, \mu)T(p_s, p, \mu)T(p, 0, \mu)+$$

$$T(p, 0, \mu)\int_p^{p_s} B(p')W(p, p')dp' + \int_0^p B(p')W(0, p')dp' \qquad (7.56)$$

With this form of transfer equation, it follows that the intensity difference between clear sky and a cloudy sky is

$$\Delta I = I_{clear} - I_{cldy} = \varepsilon \int_{p_s}^{p_c} T(0, p')\frac{dB(p')}{dp'}dp' \qquad (7.57)$$

---

[7] There are a number of subtle assumptions in going from (7.54) to (7.56). The form of transfer equation given by (7.56) is referred to as the stacked layer equation. This is only valid when transmission functions multiply

$$T(p_s, 0, \mu) = T(p_s, p, \mu)T(p, 0, \mu)$$

such as for transmission functions that are pure exponential functions of path.

where ε is the cloud emissivity and the wavelength ratio of this intensity difference defines a function

$$G(p_c) = \frac{\Delta I_1}{\Delta I_2} = \frac{\varepsilon_1 \int_{p_s}^{p_c} T \frac{dB_1}{dp'} dp'}{\varepsilon_2 \int_{p_s}^{p_c} T \frac{dB_2}{dp'} dp'}$$

that depends on cloud top pressure. This function is referred to as the cloud top pressure function and its general properties can be examined as follows. Suppose the optical depth of the atmosphere has the form

$$t(\tilde{p}) = \tau^* \tilde{p}^2 \tag{7.58}$$

which follows directly from (7.36) with $\tilde{p} = p/p_s$ and suppose also that the Planck function varies with pressure in the following simple way

$$B = B_o + B^* \tilde{p}.$$

Substitution of (7.58) into the definition $T(0, p) = \exp(-t(\tilde{p})/\mu)$ together with differentiation of the Planck function yields

$$G(p_c) = \frac{B_1^* \int_{\tilde{p}_s}^{\tilde{p}_c} \tilde{p}' \exp[-\tau_1^* \tilde{p}'^2] d\tilde{p}'}{B_2^* \int_{\tilde{p}_s}^{\tilde{p}_c} \tilde{p}' \exp[-\tau_2^* \tilde{p}'^2] d\tilde{p}'} \tag{7.59}$$

for $\mu = 1$. It is also convenient to assume that $B_1^*/B_2^* = 1$, that $\tilde{p}_s = p_s/p_o = 1$ and that the two wavelengths are close enough that $\varepsilon_1 = \varepsilon_2$. It follows from integration of (7.59) that

$$G(p_c) = \left(\frac{\tau_2^*}{\tau_1^*}\right) \frac{\exp[-\tau_1^* \tilde{p}_c^2] - \exp[-\tau_1^*]}{\exp[-\tau_2^* \tilde{p}_c^2] - \exp[-\tau_2^*]} \tag{7.60}$$

and examples of $G(p_c)$ derived from this formula are presented in Fig. 7.27a for different values of atmospheric optical depth τ_1^* and τ_2^*. The curves indicate how the cloud–top function depends on the ratio of optical depths as well as on the magnitude of these optical depths. In an optically thin atmosphere (curve 1), the function $G(p_c)$ only weakly depends on cloud–top pressure since emission by the clouds affects both channels more or less equally throughout most of the atmosphere. The other extreme is an optically thick atmosphere (curve 4) in which case the cloud–top pressure function is extremely sensitive to clouds with low cloud–top pressures. In this case, the peak of the weighting function of the optically thicker atmosphere occurs higher up and only emission by higher level clouds produces any significant effect on the intensity relative to clear sky intensities.

Figure 7.27b illustrates the cloud–top function calculated numerically from (7.57) assuming more realistic atmospheric profiles of temperature and humidity and two channels of the HIRS (channels 6 and 7). This diagram helps emphasize some of the problems in using the slicing method in polar regions where low–level temperature inversions occur. The variation of this function with pressure under these conditions is complicated. The existence of a temperature inversion below 700 mb in this profile causes cloudy sky intensities to exceed the clear sky intensities so that G exceeds unity. Lower in this atmosphere, a point is reached where the temperature of the cloud and ground approach one another such that $\Delta I_2 \to 0$ and $G \to \pm\infty$. This occurs between 800–850 mb in the example given in Fig. 7.27b.

The approach of Wylie and Menzel (1989) is to calculate $G(p_c)$ using temperature and moisture data representative of the region of interest and to estimate the cloud pressure from the ratio of observed intensities of two VAS channels. A comparison of their approach with cloud–top pressures deduced from lidar and from stereo parallax analyses is given in Fig. 7.28. The VAS cloud–top pressures typically exceed the lidar estimates by 70 mb. This comparison serves to emphasis the ambiguity of cloud–top pressure itself and how this quantity in fact may have a different interpretation depending on how it is derived. In the case of intensity slicing, it is assumed that the cloud occupies a thin layer defined by a unique value of p_c. In reality, the radiation emerging from cloud–top is the accumulation of radiation emitted throughout the depth of the cloud as well as transmitted from below the cloud. A simple way of viewing this radiation is to consider that it originates from some effective level below cloud–top. This level only approximates the cloud top surface for the densest portions of the cloud.

7.8   Intensity Classification of Clouds

Clouds provide a first–order effect on the radiative budgets and water exchanges in the atmosphere. They also play a fundamental role in studies of climate and climatic change. Several attempts have been made to classify the global distribution of clouds based on measurements obtained from radiometers flown on satellites. Two examples of these radiometric classifications of clouds will now be discussed.

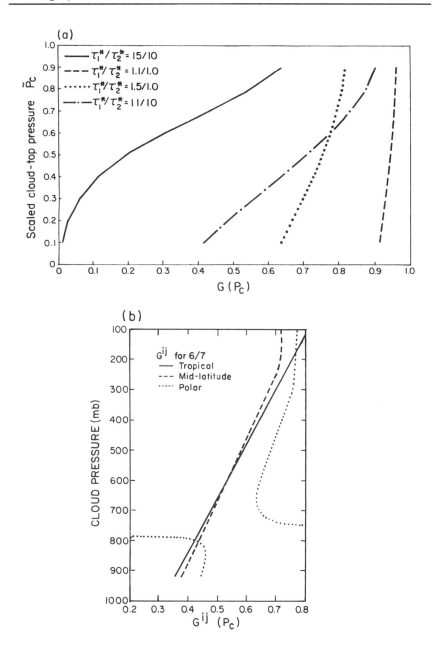

Figure 7.27 (a) The cloud–top pressure function as derived from (7.56) for the following ratios of $\tau_1^* : \tau_2^*$ curve 1 1.1:1.0, curve 2 1.5:1.0, curve 3 11:10, and curve 4 15:10. (b) The cloud–top pressure function calculated for ratios of channels 6 and 7 of the HIRS (the channels centered on 13.64 and 13.25 μm) using tropical, midlatitude, and polar climatological temperature and moisture profiles (Wielicki and Coakley, 1981).

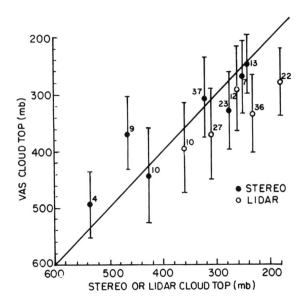

Fig. 7.28 The mean and standard deviation of all VAS cloud–top pressure data compared to lidar and satellite stereo cloud–top pressures. The error bars are 1 SD from the mean for the VAS derived pressures (Wylie and Menzel, 1989).

7.8.1   Emission Classification in the Split Window

Inoue (1989) developed a simple way to classify clouds according to the difference in their emission properties at 11 and 12 μm. As mentioned earlier, $\Delta T = T_{10.8} - T_{12}$ is a good indicator of the opacity of clouds. Thick clouds, radiating approximately like a blackbody, possess small values of ΔT whereas thin clouds exhibit more variable values of ΔT as described earlier in a way that depends on particle size and other factors.

Inoue's classification scheme is based on threshold analyses of the $T - \Delta T$ diagram like that shown in Fig. 7.23 and an example of this scheme is shown in Fig. 7.29. Two threshold values of ΔT can be identified, one at $\Delta T = 1$ K corresponding to optically thick clouds and another is set to a slightly larger value corresponding to the clear sky value of ΔT. Two thresholds values of the brightness temperature $T_{10.8}$ are also introduced in the Inoue scheme; one is the high cloud threshold which is set at -20° C and the other corresponds to clear sky temperatures. Data representing different cloud types fall in the different classification boxes. For example, cumulonimbus

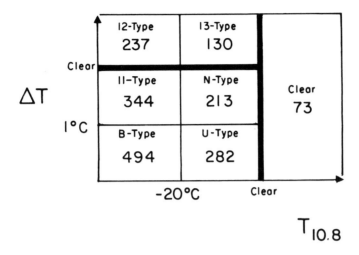

Figure 7.29 An example of a cloud type classification diagram introduced by Inoue (1989).

clouds are thick, possess ΔTs less than 1 K, and are cold. These fall in the type B category. Low level cumulus and stratocumulus clouds fall into Inoue's category U. Thin cirrus clouds are characterized by values of ΔT that exceed the clear sky threshold value and fall in categories I2 and I3 for thick and thin clouds respectively.

7.8.2 The International Satellite Cloud Climatology Project (ISCCP)

ISCCP formerly began in 1983 with the collection of the first internationally coordinated satellite intensity data. The original plan called for this collection for only a five–year period but the ISCCP has since been extended to 1995. This program was the first of its kind involving routine collection of operational satellite data. Many key problems needed attention, and these are will now be highlighted since they offer important lessons to the remote sensing of the atmosphere on the global scale.

Data Coverage and Data Management

Obtaining global coverage of cloudiness using satellite observations, while resolving diurnal variations of cloudiness, is difficult. The IS-CCP notionally planned to make use of data from five geostationary satellites as well as data from a single polar orbiter in an effort to fill in the data voids over polar regions. Actual coverage during the first

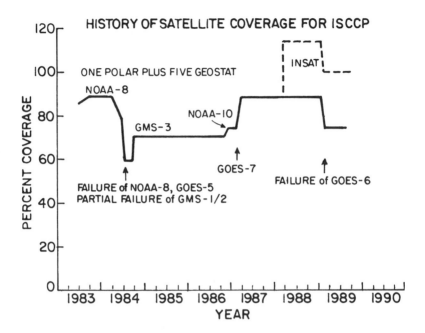

Figure 7.30 History of satellite coverage for the ISCCP. The coverage is defined to be at 100% for five geostationary satellites and one polar orbiter satellite, representing eight observations per day (although the actual observation frequency is smaller for polar orbiters). The initial complement of satellites included the NOAA–7, METEOSAT–2, GMS–2, GOES–5, and GOES–6. Failures and replacements of satellites are indicated. Time is given in quarter years (from Rossow and Schiffer, 1991).

six years of the ISCCP, measured against this hypothetical ideal of six satellites, is presented in Fig. 7.30. Because of the loss of satellites throughout this period, coverage was limited to about 90% for about three of the six years although with the availability of IN-SAT data, the data coverage could exceed that originally sought for ISCCP. Figure 7.30 graphically highlights the difficulty in providing observations of the global atmosphere and provides a commentary on those programs that propose to do so using single satellite platforms.

Satellite Radiance Calibration

The NOAA–7, 8, and 9 polar orbiting satellites provide ISCCP with AVHRR imagery data. These data play a crucial role in the ISCCP. They not only provide a completion of the data coverage over the poles and other regions that are not covered by geostationary satellites but the multispectral observations of the AVHRR also offer the potential for better analysis of clouds, such as cirrus and polar stratus clouds, that are difficult to identify with visible and far–infrared radiances. Perhaps the most important application of AVHRR data lies in its use as an intersatellite calibrator of radiance data. The procedure is to compare data routinely from each geostationary satellite and the afternoon NOAA polar orbiter, and normalize all radiance measurements to a single satellite. This normalization is then adjusted to that of the NOAA–7 in July 1983 as a way of monitoring long–term drifts during the lifetimes of the different satellites. Absolute calibration of this NOAA–7 standard is provided by aircraft measurements over a well–defined reflecting desert surface (Whitlock et al., 1989).

Figure 7.31 portrays the history of VIS and IR calibrations over the first five years of the project. The corrections that were required for the NOAA–9 visible radiances are shown in Fig. 7.31a. According to Fig. 7.31b, output from the IR channel varies with time and the operational calibration procedure generally corrects for these changes with only a small adjustment necessary from late 1987 through 1988.

Cloud Detection and Analyses

The cloud detection scheme is different from that described above and uses both visible reflection information and emitted radiation. The detection approach examines all of the data for one month to collect statistics on the space/time variations of the VIS and IR intensities. The key assumptions used in the analysis are that the intensities in clear scenes are less variable than those in cloudy scenes and that it is the clear scenes that compose the darker and warmer parts of the VIS and IR intensity distributions, respectively. Estimates of the clear sky values of VIS and IR intensities for each location and time are made and composited into maps (these are referred to as the "clear sky composites"). This approach is novel in two respects. First, all of the complicated tests usually used to detect cloudiness directly, many of which were first proposed by other investigators, are used here to identify clear scenes. The use of time variations at one location to identify clear scenes also differs from many other methods .

The differences between the intensities measured and the estimated clear sky intensities are compared to the uncertainties in estimating the clear intensities. If the differences are larger than

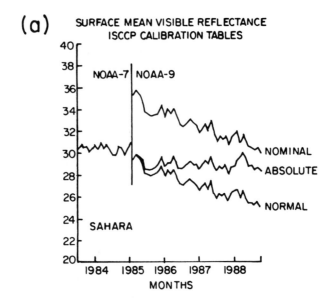

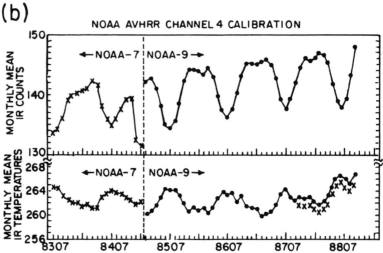

Figure 7.31 History of the AVHRR (a) visible channel and (b) the IR channel 4 for the NOAA–7 and NOAA–9 satellites. The nominal calibration is the calibration originally supplied by the satellite operator; the normalized calibration is that used to match NOAA–9 to NOAA–7 and the absolute is the final adjustment. The IR calibration is illustrated by showing the global, monthly mean IR brightness temperatures (from Rossow and Schiffer, 1991).

this uncertainty and in the "cloudy direction" at either wavelength (colder IR or brighter VIS), then the pixel is labeled cloudy. Once each pixel is classified as clear or cloudy, the measured intensities are compared to radiative transfer model calculations that include the effects of the atmosphere, surface and clouds. The intensity data are then converted into two cloud properties— the "visible" optical thickness (defined at 0.6 μm) and a cloud–top pressure. The optical thickness parameter determines the amount and angular distribution of sunlight reflected by the cloud layer (the full effects of multiple scattering are included in the model)— the cloud–top pressure is supposed to account for cloud emissivities less than 1. At night, when only IR intensities are measured, no cloud optical thickness is reported and IR variations are associated with the cloud–top brightness temperature.

Thus, the ISCCP clouds are categorized in terms of cloud–top pressure and optical depth properties as schematically shown in Fig. 7.32a. Two examples of this two–dimensional distribution for July 1983 are presented in Figs 7.32b and c for two different latitude zones. In the subtropics during winter (Fig. 7.32 b), the predominant cloud type has low tops and relatively low optical depths (probably associated with highly broken cloud). The tropical distribution is more complicated showing a prevalence of high, optically thick clouds and low, relatively thin clouds associated with highly broken low–level cloud.

7.9   Notes and Comments

7.1. The radiative transfer described in this chapter and the remote sensing methods based on this transfer ignore scattering processes. This is clearly a valid assumption when we consider only the transfer of infrared and microwave radiation in a molecular atmosphere but it is not always valid when dealing with transfer in clouds. In the infrared, scattering by clouds tends to be less dominant that the absorption and emission be cloud particles but it is not always negligible especially when a very cold cloud overlies a warm underlying surface (Stephens, 1980). Scattering by ice particles in the microwave becomes an increasingly important consideration as the frequencies increase beyond approximately 80 GHz.

7.2. A basic issue for almost all radiometric studies of the troposphere and the underlying surface is the need to establish clear sky intensities and thus determine whether a scene contains cloud. This

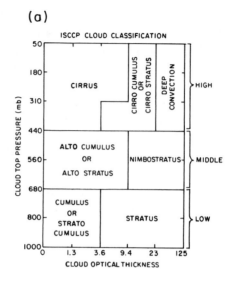

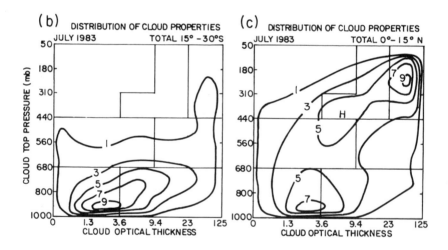

Fig. 7.32 (a) Radiometric classification of cloudy pixels in terms of optical thickness and cloud–top pressure. (b) The frequency distribution of cloud optical thickness and cloud–top pressure for July 1983 for the southern subtropics and (c) the northern tropics (from Rossow and Schiffer, 1991).

is the first step required in sensing the surface (this section), sounding the atmosphere (Section 7.5), and studying and classifying clouds (Sections 7.6–7.8). Different ways of establishing the clear sky intensity threshold exist, and some of these are described in Sections 7.7 and 7.8. One method that is used operationally as part of TOVS sounding is the cloud clearing scheme of McMillan and Dean (1982) which masks clear sky pixels so that data from these pixels can be subsequently used in soundings and SST retrievals.

An interesting but more difficult problem is the remote sensing of land surface temperature. The problem is that the emissivities of these surfaces (both their spectral and angular variation) is not well known [Prata (1993) and references therein].

7.3. A number of different methods exist for the retrieval of PWC from the SSM/I. Some are Alishouse et al. (1990), Schluessel and Emery (1990), and Petty and Katsaros (1990). Early microwave measurements from the Cosmos 243 satellite (Basharinov et al., 1969) determined total water content over the oceans from observations at 1.35cm wavelength. Measurements on and near the water vapor absorption line at 22 GHz were also obtained from the NEMS on the Nimbus 5 satellite (Staelin et al., 1976), SCAMS on Nimbus 6 (Grody et al., 1980), the SMMR on Nimbus 7 (Prabhakhara et al., 1982) before the SSM/I. The PWC has also been inferred from infrared sensors on satellites (such as from the IRIS instruments on the Nimbus 3 and 4; Conrath, 1969; Prabhakara et al., 1979).

7.4. A general overview of the methods for deriving rainfall from space sensors is given by Browning (1990). The OLR–precipitation approach described in this section forms the basis of the Global Precipitation Climatology Project (GPCP) under the sponsorship of the World Climate Research program (WCRP). The GPCP is also examining ways of merging the OLR methods and the passive microwave methods of precipitation retrieval.

The general concept for the remote sensing of hydrometeors from microwave emission measurements was suggested by Buettner (1963) and explored with aircraft observations by Singer and Williams (1968) and Kriess (1969) among others and quantified by Wilheit and collaborators during the late 1970s and 1980s (Wilheit et al, 1977; Wilheit, 1986) and Adler and Rodgers (1977). This is now a vigorous topic of research (e.g., Adler et al., 1991, Mugnai et al., 1990, and Kummerow et al., 1989) spurred on by the Tropical

Rainfall Measurement Mission (TRMM) to be launched in the late 1990s (Simpson et al., 1988).

The microwave methods described in this chapter are based on measuring the emission from the rainfall. These measurements are typically made around 19 GHz or so. Other passive microwave methods exist based on scattering of microwave radiation at higher frequencies. These methods, like that of OLR, are more indirect in that the radiation sensed by the radiometer is modulated by scattering from the ice particles at cloud–top. This is an interesting approach and requires a combined understanding of both emission and multiple scattering which is a topic that is not discussed in this book. An example of the method is contained in the work of Spencer et al. (1989). A principal advantage of the approach is that it applies equally well over land and ocean unlike the microwave emission approaches. Spencer (1993) has also introduced a largely empirical scheme for retrieving global rainfall from the emission sensed in the 60 GHz oxygen band by the MSU instrument.

7.5. Atmospheric sounding was one of the main motivations of the meteorological satellite program. Even before SPUTNIK, it was recognized that air temperature observations might best be made on a global scale using radiometric measurements from a satellite (King 1958; Kaplan 1959; Houghton 1961). In the United States, two experiments were developed in the 1960s for flight on the NIMBUS 3 experimental weather satellite: (1) a Michelson interferometer, called IRIS (refer to Chapter 3 for further discussion of this instrument), (2) a grating spectrometer called SIRS which measured radiation in eight distinct spectral bands selected specifically for temperature sounding. The SIRS absolute radiometric accuracy of 1 % and signal–to–noise ratio of 4000–1 was a phenomenal achievement for its time.

A major problem with the SIRS data was created by its large geographical field of view (225 km square). It is estimated that clouds interfered with the measurements more than 90% of the time. In order to account for the influence of cloud, a radiative correction to the measurement was calculated using a "guess" temperature profile and the tropospheric radiance observations (Smith et al., 1970). The guess profile was generated from statistical relationships with uncontaminated stratospheric–channel radiances. Although the procedure gave reasonable results for a single layer of cloud when the guess profile was close to the true profile, the cloudy–condition retrieval were in fact unreliable, and to a degree dependent on the cloud situation.

The TIROS–N satellite, the first of the current series of operational polar–orbiting satellites, was launched into orbit on October 13, 1978, just prior to the initiation of the FGGE. The second spacecraft in the series, NOAA–6, was launched into orbit on June 27, 1979, midway through the FGGE year.

An excellent overview of the topic of satellite sounding, including its historical perspective, limitations, and impacts is given in the review article of Smith (1991). A text largely devoted to sounding is that of Houghton et al. (1984). An excellent review of the topic of sounding inversion, including definition of resolution and other issues, is found in Rodgers (1976).

7.8. A plan to study clouds within the perspective of global climate was laid down at a meeting in Oxford in 1978. This plan identified a program to address a number of key problems associated with the role of clouds in climate and became known as the ISCCP. The basic idea of this project is a simple one, namely to make use of the international network of geostationary satellites to provide a satellite–based climatology of cloudiness. The goals of the project are:

- To produce a global, reduced resolution, calibrated, and normalized infrared and visible intensity data set, along with basic information on the radiative properties of the atmosphere, from which cloud parameters can be derived.
- To coordinate basic research on techniques for inferring the physical properties of clouds from satellite intensity data.
- To derive and validate a global cloud climatology
- To promote research using ISCCP data to improve parameterizations of clouds in climate models.
- To improve our understanding of the Earth's radiation budget (defined at the top of the atmosphere and at the surface) and hydrological cycle.

Important references describing ISCCP through its formative stages and its many aspects are conveniently summarized in Rossow and Schiffer (1991). Rossow and Garder (1992) describe the ISCCP cloud detection scheme in more detail than described in this book and Brest and Rossow (1989) discuss the topic of calibration of the AVHRR which is used as an intercalibration of radiometers on different geostationary satellites.

7.10   Problems

7.1 Explain or interpret the following:

    a. In Fig. 3.26, emission spectra measured by an interferometer on the ground and by an interferometer on an aircraft are shown. What gross inferences about the vertical distribution of temperature can you make by studying these spectra?

    b. The AVHRR instrument has a channel centered at $3.9\mu m$ which is also located in an atmospheric window. A satellite image of a cloud generally appears colder than its immediate surroundings at night but warmer than its surroundings by day.

    c. Measurements of emission from the atmosphere at $15\mu m$ at a spectral resolution typical of that provided by an interferometer, in principle, yields higher vertical resolution sounding relative to an instrument that measures this emission over relatively broad channels.

    d. Low clouds emit more infrared radiation than high clouds of comparable optical thickness.

    e. The relative change in emission by thin clouds across the $10\text{–}13\mu m$ window region compared to a clear sky background emission varies according to the size of the cloud particles.

    f. For the same total column optical thickness, the weighting functions of water vapor are more peaked than corresponding weighting functions for CO_2.

    g. Qualitatively, what is the nature of the weighting function corresponding to emission in the $9.6\mu m$ band for measurements at the ground?

    h. The "effective" level of emission determined from the brightness temperature of an optically thin cloud layer may actually not reside within the cloud layer itself.

7.2 Derive (7.9) and (7.10) from (7.6b) and (7.6a), respectively, assuming that

$$B(\tau) = B_o + (B^* - B_o)\tau/\tau^*$$

7.3 Assume a Lorentz line shape for the 22.235 GHz water vapor line. On the line center,

a. Show that the optical depth due to water vapor is

$$\tau^* = \frac{k_o r_s p_s}{3g}$$

where $k_o = S/\pi\alpha_s$, α_s is the line half–width defined at the surface pressure p_s and r_s is the surface value of mixing ratio. Assume the form $r = r_s\tilde{p}^3$ in your derivation where $\tilde{p} = p/p_s$ (compare this with Problem 3.7).

b. Using the results obtained from Problem 3.7 (a), express this optical depth as a function of precipitable water content w.

c. Using (7.23), estimate T_b for five values of w assuming 1-ϵ=0.6, T_{OX}=0.98, and $k_o = 1 \times 10^{-2}$ m^{2}kg^{-1}. Use the 5 values of T_s and corresponding values r_s listed in Table 3.2, the latter based on a surface relative humidity of 70 %.

7.4 Repeat the analysis but assume the measurements are at a frequency in the wing of the absorption line. Assume this frequency to be 19 GHz and express your answer in terms of $k_o = S\alpha_s/\pi(\nu - \nu_o)^2$ and assume a value of 4.15×10^{-3} m^{2}kg^{-1} for this quantity.

7.5 The volume absorption coefficient of the water vapor line at 22 GHz is approximated by

$$k \approx 6 \times 10^{-3} \rho_v \left(\frac{300}{T}\right)^2 \text{km}^{-1}$$

where ρ_v is the water vapor density in gm^{-3}, and T is the temperature in Kelvin. A commonly used standard atmosphere has the properties

$$\rho_v = \rho_o e^{-z/2}$$

$$T = T_o - 6.5z$$

where z is in km and ρ_o and T_o are sea–level values (say 1.19 kgm^{-3} and 292 K). Assuming the atmosphere lies in the lowermost 10 km, calculate the brightness temperatures at 22 GHz observed by an instrument at the top of the atmosphere looking down and at the base of the atmosphere looking up.

7.6 Consider a downward–looking, nadir– pointing radiometer observing the ocean surface from an airborne platform above a 2 km thick cloud with a water content $w = 1.5\text{gm}^{-3}$. The volume absorption coefficient of water is given approximately by

$$k_v = 2.4 \times 10^{-4} \nu^{1.95} w \qquad \text{km}^{-1}$$

where ν is in GHz and w gm^{-3}. For an ocean brightness temperature of 150 K, calculate and plot the observed brightness temperature as a function of frequency from 1 to 30 GHz assuming the physical temperature of the cloud is 275 K. For this exercise, ignore the effects of water vapor absorption and compare your answer at 22 GHz with your answer for Problem 7.5.

7.7 A simple measure of the vertical resolution of a single sounding channel may be defined as

$$\Delta z = z_{max} - z_{1/2}$$

where z_{max} is the height of the maximum of the weighting function and $z_{1/2}$ is one of the heights corresponding to the altitude where the weighting function is half its maximum value. Following the analysis presented in Section 7.5 (a), show that $z_{1/2}$ is obtained from the solution of

$$\frac{H}{2} \tau^* e^{-2z/H} + z = \frac{H}{2} \ln[2e\tau^*]$$

for z. Solve this equation and obtain an estimate of Δz assuming $H = 8$ km and values of $\tau^* = 0.1$, 1, and 10. *[Hint: The solution to the equation*

$$ae^{-bx} + x = c$$

has the form

$$x = \frac{W(-abe^{-bc}) + bc}{b}$$

where $W(y)$ satisfies

$$W(y)e^{W(y)} = y.$$

Use the graphical representation of $W(y)$ presented in Fig. 7.33 to obtain your answer.]

7.8 Express (7.6a) in the form

$$I = \int_1^0 B(T)\frac{dT}{d\tilde{p}}d\tilde{p} + B(T_s)T(\tilde{p}_s)$$

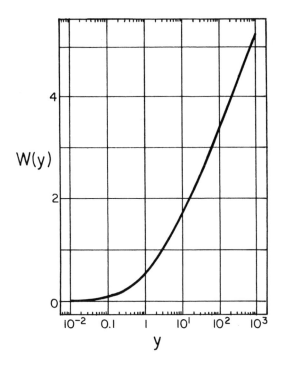

Fig. 7.33 The function $W(y)$.

assuming $\mu=1$. Using the transmission function derived in Problem 3.7, and assuming that the Planck function varies as

$$B(\tilde{p}) = B^*\tilde{p}$$

where B^* is the blackbody function at the surface where $\tilde{p} = \tilde{p}_s = 1$, obtain the approximate solution

$$I \approx B(T_s)[1 - \frac{\beta}{8} + \frac{\beta^2}{30}]$$

for $\beta < 1$ neglecting all terms $O(\beta^3)$ and above. Check the temperature corrections listed in Table 3.2 for the five values of r_s and T_s assuming $\Delta T = T_s - T^*$ where T^* is the $11\mu m$ brightness temperature $I = B(T^*)$.

7.9 For an atmosphere represented by

$$p = p_o e^{-z/H}$$

estimate the ratio of the limb to vertical absorption paths for a uniformly mixed gas. Assume a tangent height h and derive an expression for the vertical path from the top of the atmosphere to this tangent height. Parameters you will need are $H = 7$ km, the radius of Earth $R = 6.37 \times 10^3$ km. Plot the ratio of the slant to vertical paths as a function of h ranging from 20 to 80 km.

7.10 Derive (7.57). You will need to consider the cloud as a thin layer located at pressure p_c and approximate the radiative transfer through this layer by (7.49).

7.11 Show that $G(p_c)$ is independent of p_c for an isothermal atmosphere.

7.12 Consider a two–layer atmosphere, the upper layer is 2km thick and contains a cloud with a mean temperature of 220 K. The lower blackbody surface has a temperature of 285 K and the intervening layer is transparent to the radiation under consideration. Using (7.55), estimate the 11μm brightness temperature observed by a radiometer on a satellite assuming $\varepsilon = 0.2$, 0.6, 0.8 and 1.0. If you assume that the lapse rate of the atmosphere is 6.5 K km^{-1}, what is the effective height of this emission?

8
Active Sensing

Active sensing is a powerful way of observing the atmosphere. Measurements from active systems are now used extensively in research as well as routinely as part of the operational observations provided by national weather services worldwide. Figure 8.1 portrays the essential elements of a generic active system. In the operation of such a system, electromagnetic energy is transmitted into the atmosphere, absorbed and scattered by the intervening aerosols and gases, and then scattered by a target volume at some determined range. The energy scattered by this volume and returned to a receiver is referred to as the echo. This echo is then processed to provide the information of interest. Two basically different modes of detection distinguish the capabilities of the system. For example, an active system might detect only the intensity of the backscattered radiation in which case the system is said to be *incoherent*. Other active systems may have detection capabilities that provide a measure of both the phase and amplitude of the backscattered electromagnetic wave. These systems, sometimes referred to as *coherent* systems, are much more complex in both their design and in their operation. Coherent systems are used, for example, to measure atmospheric motions based on measurements of the Doppler shift of the returned signal.

This chapter reviews the principles of remote sensing of two active systems: radar and lidar. Radars transmit a pulse of microwave energy to the atmosphere whereas lidars transmit shorter wavelength UV, visible, or infrared radiation. Although the design and operation of each is very different, the broad principles of remote sensing by these systems are ostensibly the same.

Figure 8.2 contrasts various operating wavelengths of lidars and radars against a typical atmospheric molecular absorption–transmission spectrum. Most weather radars operate at longer microwave wavelengths (operational weather radars operate in the S band, for example). Clear air radars, on the other hand, operate at VHF and UHF frequencies (50 MHz to 1000 MHz) with wavelengths ranging from about 30 cm to 6 m. Both systems are largely unaffected by gaseous absorption but this is not the case for radars

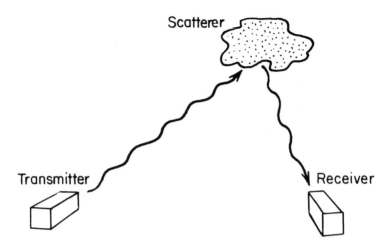

Figure 8.1 The basic concept of an active system. A pulse of radiation travels from a transmitter to the target and back to a receiver. The amount of radiation received depends on properties of the particles in the illuminated volume and any attenuation between the receiver, transmitter and scatering volume.

operating at shorter wavelengths. The utility of different lidar systems as a remote detector of atmospheric properties also depends on the availability of lasers with sufficient output power to provide an adequate lidar backscatter signal. The choice of a desirable laser thus depends on the compromise reached between the power needed to detect a signal from a suitable range and the tunability or spectral coverage of the laser. It is because of their output power that ruby, dye, neodymium, yttrium–aluminum–garnet (YAG), and CO_2 lasers figure predominantly in present day lidar systems. These lasers offer sufficient energy for the detection of most atmospheric species at moderate range and generally at wavelengths that are not appreciably attenuated. Some of the lasers are line tunable only (i.e., they can only operate at a limited number of discrete wavelengths) while other are continuously tunable within a given spectral range.

Topics described in this chapter involve the following properties of the scattering of microwave and laser radiation by the atmosphere:
• The strength of backscattering which is related to the density, size distribution, and shape of the particles in a scattering volume.

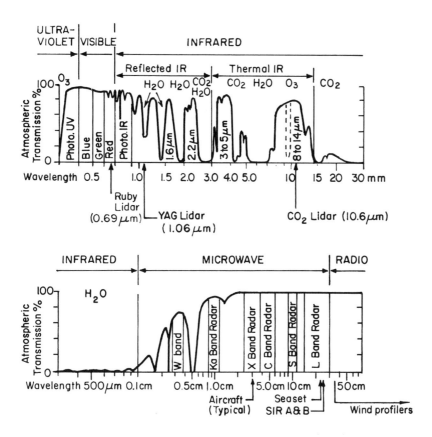

Figure 8.2 Expanded diagrams of the visible and infrared regions (upper) and the microwave region (lower) showing various atmospheric windows. Wavelengths associated with some radars and lidars are highlighted on this spectrum. Gases responsible for the atmospheric absorption are also shown.

● The polarization state of the backscattered radiation which is affected by attenuation along the path and the properties of the backscattering volume. Scattering by nonspherical particles, for example, changes the polarization properties of the incident beam in a way that is related to the shape of the scatterer.

● The attenuation of the pulse along its path to the target volume and back to the receiver. At a given wavelength, the attenuation is related to the concentration of attenuator and its distribution along the path.

• The phase change suffered by a pulse of EM radiation after being scattered off a moving particle.

• The different spectral broadening of pulses after scattering from particles of different masses.

8.1   Basic Operational Considerations

8.1.1   Range Resolving Systems

Radar and lidar systems operate either in a constant wave (CW) mode, in which a continuous beam of electromagnetic energy is transmitted and received, or in a pulsed mode, whereby pulses of extremely short duration (typically 10^{-6} - 10^{-8} s in length) are transmitted. The distance between the backscattering element and the system, called the range and hereafter denoted as R, is simply determined by measuring the time of return of the pulse. Systems with these capabilities are known as *range resolved systems*. By contrast, the echoes measured by CW systems provide information about the path integrated backscattering whereas range resolved systems provide information about the scattering and attenuation as a function of range. The latter systems are therefore useful for studying the three–dimensional structure of the atmosphere.

Pulsed systems send a short pulse of energy to the atmosphere along a path and echoes are received from scattering at all points along this path. The range capabilities of the system depends on the transmitted power of the pulse and hence on the pulse duration. Since the pulse has a finite duration, the received signal is spread out in time. By sampling these echoes at equally spaced periods of time, information about the atmosphere is obtained at equally spaced positions along the path (Fig. 8.3a). This sampling approach is referred to as *range gating*.

If the transmitted pulse has a finite length h, then the leading edge of the pulse scattered from the point at $R + h/2$ returns to the receiver at precisely the same time that the trailing edge arrives from the point R. Therefore, the received signal is a result of scattering from a small volume of atmosphere spread out in range. This spread establishes the volume which is called the *resolution* and the depth of this volume is equal to $h/2$. The sampling period must be set so $t \geq ch/2$ for the sampled points to be independent of each other. The resolution of most weather radars and many lidars varies between

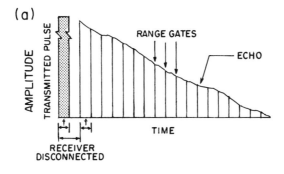

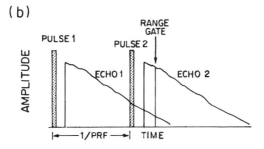

Figure 8.3 (a) After each pulse is transmitted, the echo is sampled at selected times (referred to as range gates). The echo received at any range gate for a vertically pointing system is spread over a range of heights, and the amount of spread is called the resolution, which is one half the pulse length. (b) Examples of a pulse repetition frequency (PRF) that is too large. Echoes are received from the first pulse after those from the next pulse start to arrive. In the overlap range gate, information about backscatter from two different heights is received simultaneously. This is referred to as range aliasing.

about 50 m and 500 m although the actual volume sampled by each system is much different.

Another factor important to pulsed systems is the rate at which pulses are transmitted into the atmosphere. This rate, specified in terms of the *pulse repetition frequency* (PRF), is one factor that determines the range of the instrument. If the period between pulses is too short, echoes from successive pulses overlap (Fig. 8.3b) producing *range aliasing* or *range folding*. Under these circumstances, it is impossible to interpret the echoes. Thus the time between pulses must be at least as long as the delay between transmission of a pulse and the reception of its echo from the volume located at the predetermined maximum range of the system.

8.1.2   Scanning Modes

A single active system is able to sample the three–dimensional atmosphere by scanning across the volume of interest. A number of scan concepts are used and four of these are illustrated in Fig. 8.4. Perhaps the most popular scan configuration is that obtained by scanning the instrument in azimuth around the local vertical at a fixed elevation. More about this particular scan configuration is described shortly.

There are several ways to display scan data. These displays are called *indicators*. The most extensively used indicator of weather radar data is the plan–position indicator (PPI) illustrated in Fig. 8.5a. This display presents a plan view of the received signals on a polar coordinate system. From this display, one immediately obtains the range and bearing of the target. The PPI is also intensity modulated so that the intensity of the reflectivity is related to the strength of the returned signal. Another common indicator of radar and lidar data is the range–height indicator (RHI). The RHI presents a view of the vertical cross–section of backscattered data on a co–ordinate system with range as the abscissa and altitude as the ordinate (Fig. 8.5b). A third type of indicator is the constant–altitude plan–position indicator (CAPPI) which is a presentation of backscattering echo patterns on a constant altitude plane. This is achieved by synthesizing PPI scans at progressively higher elevation angles. Still other more specialized types of indicators can be used to present information depending on the particular application of the data.

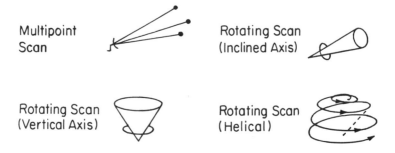

Figure 8.4 Various scanning methods to sample a three–dimensional volume of atmosphere.

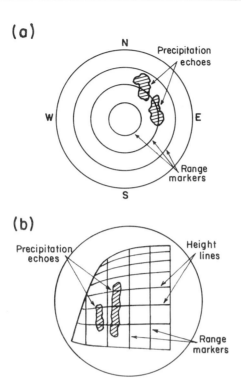

(a)

(b)

Figure 8.5 Examples of two indicators commonly used to display backscattering data: (a) plan–position indicator (PPI), (b) range–height indicator (RHI), (adapted from Battan, 1973).

8.2 Conventional Weather Radar

Radars were originally developed during the second world war to locate enemy aircraft invading British shores. The systems employed at that time, however, were subject to one serious environmental problem — when heavy rain fell between the radar and the target, the radar was no longer able to detect enemy aircraft. The potential of radar for studying atmospheric precipitation was then realized. Since then radars have become more powerful, their beams have become more directional, and their receivers have become more sensitive. Radars are now extensively used to examine the structure of thunderstorms, to map the Earth's surface and to explore the properties of the other planets of the solar system.

Before we attempt to understand how to apply radars to study the lower atmosphere, it is first necessary to consider some basic features of microwave transmitters. The antenna emits an electromagnetic wave which we characterize in terms of the power $\mathcal{P}_t$ transmitted by the antenna. In analyses of radar backscattering, it is convenient to introduce the following parameter

$$G = \frac{I_p}{\mathcal{P}_t/4\pi R^2} \tag{8.1}$$

at a specific range R. This parameter is referred to as *the antenna gain* and is defined as the ratio of the intensity at the peak of the transmission pattern (I_p) to an isotropic intensity that is derived assuming that the total power is distributed equally in all directions. Since the difference between the power transmitted by an antenna to the power received from backscattering is typically several orders of magnitude, it is convenient to express the received signal defined in decibels (dB) which refers to the difference between two power levels $\mathcal{P}$ and $\mathcal{P}_t$ according to

$$\mathcal{P}(dB) = 10\log\frac{\mathcal{P}}{\mathcal{P}_t} \tag{8.2}$$

For example, the peak power transmitted by a radar may be of the order of 100 kW and the returned signal may for instance be of the order of a few milliwatts, or $\approx$ -100 dB. Weather radars, for example, are sensitive enough to detect a single raindrop from 20 km away or a bumblebee in flight (Problem 8.4).

Suppose for the moment that the radar transmits with a total power $\mathcal{P}_t$ equally in all directions. At a range R, the isotropic intensity is $\mathcal{P}_t/4\pi R^2$. In reality, however, the microwave pulse emitted by the antenna is not isotropic but is focused to some prescribed direction. The directionality of the antenna is characterized by its radiation pattern;[1] an example of such a pattern for a hypothetical antenna is shown in Fig. 8.6a. The beamwidth $\Delta\psi$ is a quantity used to characterize the width of the primary lobe (i.e., the largest lobe) and is defined as the angle at which the microwave radiation is one–half its peak intensity. It is obviously desirable to make the primary lobe as narrow as possible so that it can be focused onto a small volume of atmosphere. It is also desirable to minimize the

---

[1] The radiation pattern of the antenna is directly analogous to the radiation pattern of particles which we describe by the scattering phase function (Section 5.4). This formal analogy was already noted in discussion of a dipole antenna introduced in Section 5.2

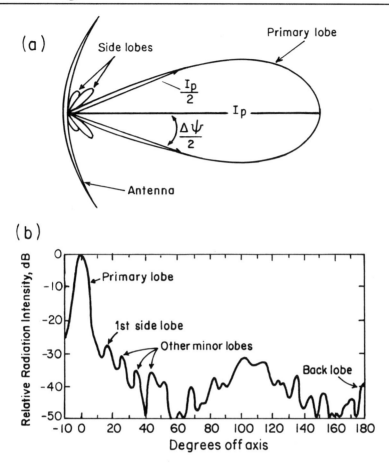

Figure 8.6 (a) An idealization of the cross–section of the intensity pattern of a radar beam emitted from a parabolic reflector. The angles are exaggerated to show the lobe structure. (b) Typical one–half radiation pattern from a radar with a paraboloid detector (from Cutler et al., 1947).

power in the side lobes to avoid contamination by reflections from other parts of the scene illuminated by these side lobes (Fig. 8.6b).

8.2.1 The Radar Equation

Analyses of both radar and lidar echoes are based on mathematical expressions that relate the received power to the transmitted power. This expression includes the effects of backscattering and attenuation processes along the path to and from the target volume. The mathematical relationship for radar is referred to as the *radar equation*; the mathematical expression for lidar is known as the *lidar*

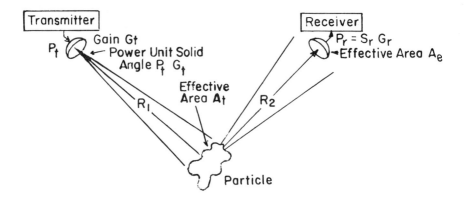

Figure 8.7 The geometric setting for the general derivation of the bistatic radar equation.

equation. Both equations are a form of radiative transfer equation with the specific assumption that the atmosphere is sufficiently tenuous within the volume to ignore multiple scattering.

A relatively simple radar equation follows from reference to Fig. 8.7. The power intercepted by a target volume at range R_1 with a cross–sectional area A_t intercepts is $\mathcal{P}_{inc} = I_p A_t$. Using the definition of gain, the incident power illuminating the volume at R_1 is

$$\mathcal{P}_{inc} = \frac{P_t G A_t}{4\pi R_1^2} \tag{8.3}$$

According to (5.30), the intensity of radiation scattered by a single particle in the direction Θ to a point R_2 from the particle is

$$I_{sca} = \frac{|S(\Theta)|^2 I_{inc}}{k^2 R_2^2} \tag{8.4}$$

where $I_{inc} = \mathcal{P}_{inc}/A_t$ is the incident intensity. From the definition of backscatter cross–section given by (5.62c), it follows that

$$I_{sca} = \frac{I_{inc}}{k^2 R_2^2} |S(\Theta)|^2 = \frac{I_{inc}}{4\pi R_2^2} C_b(\Theta) \tag{8.5}$$

Suppose that the illuminated volume is occupied by several particles with a size distribution that is represented in the usual way by $n(r)$ (these are sometimes referred to as distributed targets). Suppose also that these particles in the volume scatter incoherently. The total radiation scattered by such a volume and received at the detector is then

$$I_r = \frac{I_{inc}}{4\pi R_2^2} dV \int n(r) C_b(r) dr \tag{8.6}$$

where the integration applies to all particle sizes contained in the volume element dV, and where it is also assumed that the particle size distribution is uniform within dV. Expressing (8.6) in terms of power leads to

$$P_r = \frac{P_t G}{4\pi R_1^2} \frac{A_e}{4\pi R_2^2} dV \int n(r) C_b(r) dr \tag{8.7}$$

where (8.3) is used together with the introduction of A_e as the cross–sectional area of the receiving antenna.[2] For monostatic sytems, $R_1 = R_2 = R$ and the radar equation becomes

$$\frac{P_r}{P_t} = \frac{G^2 \lambda^2}{(4\pi)^3 R^4} dV \int n(r) C_b(r) dr. \tag{8.8}$$

We shall refer to the integral quantity $\int n(r) C_b(r) dr$ as the backscatter coefficient, β, which is typically expressed in units of inverse kilometer.

From beam geometry considerations and with some approximation, it follows that the illuminated radar volume is (Ishamari, 1978)

$$dV \approx R^2 \Delta\phi\Delta\psi \ h/2 \tag{8.9}$$

where $\Delta\phi$ and $\Delta\psi$ are half–power beam widths (the beam is not taken to be symmetrical) and h is the length of the emitted pulse so that $h/2$ is the radar resolution. Thus we can write the radar equation in terms of these radar factors by substituting (8.9) into (8.8) to obtain

$$\frac{P_r}{P_t} = \frac{G^2 \lambda^2}{(4\pi)^3 R^2} \frac{h\Delta\phi\Delta\psi}{2} \int n(r) C_b(r) dr \tag{8.10}$$

If the radar wavelengths are large enough that the scatterers behave as Rayleigh particles, then the backscattering by these particles

---

[2] The definition of the effective cross sectional area is $A_e = G\lambda^2/4\pi$.

is proportional to the sixth power of particle diameter D according to (5.65),

$$C_b(r) = \frac{\pi^5}{\lambda^4} \mid K \mid^2 D^6 \tag{8.11}$$

Substitution of (8.11) into (8.10) then yields

$$\frac{P_r}{P_t} = \pi^2 G^2 \frac{\Delta\phi\Delta\psi\ h}{128R^2} \mid K \mid^2 \int n(D)D^6 dD \tag{8.12}$$

which we write as

$$P_r = C \frac{\mid K \mid^2}{R^2} Z \tag{8.13}$$

and call this the radar equation. In this equation, the factor C is appropriately known as the radar constant and incorporates the various antenna factors $(P_t, \Delta\phi, \Delta\psi, h$ and $G)$ specific to the radar system in question. The factor

$$Z = \int n(D)D^6 dD$$

is referred to as the *radar reflectivity* and we see from a comparison of (8.13) and (8.10) that this reflectivity is related to β according to

$$\beta = \int n(r)C_b(r)dr = \left(\frac{\pi^5 \times 10^{-7}}{\lambda^4 \mid K \mid^2}\right) Z$$

where β is in km^{-1}, λ is in cm, and Z has units of mm^{6}m^{-3}. Since the magnitudes of Z commonly encountered for cloud systems span several orders, it is also convenient to express reflectivity relative to a value of 1 mm^{6}m^{-3} in a logarithmic scale. This is referred to as dBZ$(= 10\log_{10} Z)$.

Radar echoes produced by rain drops vary from pulse to pulse merely by the movement of these drops in the sampling volume during the time between pulses. Since the phase of the electromagnetic wave backscattered to the receiver depends on the distance between the drop and receiver, tiny movements of the scatterer result in a change in phase relative to the previously scattered pulse. These small random movements of the particles between pulses produce random phase shifts and we refer to this as the incoherent part of the phase shift[3]. This produces a variability in the backscattered intensity between pulses that is removed by averaging a great number of echoes (typically this average is done over a few seconds for thousands of pulses to obtain reflectivities to within ±1dB). The radar equation may then be viewed as an expression for the average of many pulses.

Another correction to (8.13) is applied when the Rayleigh scattering assumption is violated because the particles are too large relative to the wavelength and/or because the particles are nonspherical such as in the case of ice crystals. Under these circumstances we express the returned power in terms of an *effective radar reflectivity factor* Z_e which we can think of as the reflectivity factor of a population of liquid, spherical particles satisfying the Rayleigh approximation and producing a signal of the same power. For ice, this is

$$Z_e = \frac{|K|^2}{|K_i|^2} Z$$

where we assign a value of 0.93 for $|K|^2$ which is the value for liquid water for wavelengths of conventional weather radars and where $|K_i|^2 = 0.176$.

Figure 8.8 presents examples of RHI scans of radar reflectivities measured in tropical convection during the wet season of northern Australia. This wet season is characterized by monsoonal–type convection associated with the synoptic–scale monsoon trough and continental convection of the monsoon "break" period (Williams et al., 1991). Convection during the break period is primarily forced by local surface heating whereas the forcing of the monsoonal convection is synoptic in scale. The vertical development of convection deduced from these two radar cross sections suggest fundamentally different precipitation microphysics and related differences in lightning activity (Williams et al., 1991) between the monsoon and monsoon break periods. Deep convection to 15–20 km occurs during both regimes with the monsoonal convection typically imbedded in extensive stratiform cloud layers. Reflectivities of 10–30 dBZ are typical in the mixed phase region of monsoonal cloud systems (i.e., in the region between 0° and −40°C), whereas reflectivities in the

---

[3] If a component of the movement of the particles is such that they all move in some direction due to advection by wind, for example (with small random movements superimposed), then this movement produces a coherent phase shift that is measured by Doppler systems.

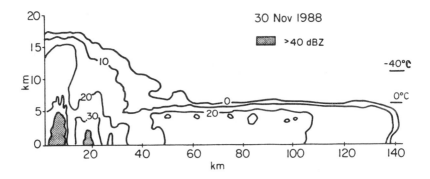

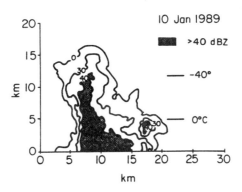

Figure 8.8 Vertical cross section of radar reflectivity in dBZ for the monsoon case of November 30, 1988, obtained from the TOGA radar (upper) and cross section during a break period obtained from the MIT radar (Rutledge et al., 1992).

mixed phase region of continental towers are often as large as 30–50 dBZ. These radar observations suggest that the bulk of the precipitation in monsoon systems lies beneath the melting layer in regions of warm rain processes and over a large fraction of the precipitating area. This is in direct contrast to the break convection, in which case mixed phase precipitation processes are the primary mechanism producing rainfall.

Excursus: Reflectivities and Precipitation

Measurement of precipitation was one of the first applications of weather radar to receive close scrutiny by radar meteorologists. The popular approach uses a relationship between the intensity of the backscattered radiation, expressed in terms of either Z or Z_e, and the rain rate $\Re$ in the form

$$Z = A\Re^b \qquad (8.14)$$

where $\Re$ is typically measured in millimeters of rain per hour (mmhr^{-1}). A practical problem in deriving the $Z - \Re$ relationships lies in the difficulty of calibrating the radar precisely; the maintenance of a stable radar system is extremely important for this particular application.

It is convenient to discuss the $Z - \Re$ relationships within the context of a Marshall–Palmer distribution

$$n(D) = N_o e^{-\Lambda D} \qquad (8.15)$$

which has two parameters (namely, N_o and $\Lambda = 1/D_o$) that define it. The reflectivity is thus

$$Z = N_o \int_0^\infty e^{-\Lambda D} D^6 dD$$

which becomes[4]

$$Z = N_o(6!)\Lambda^{-7} \qquad (8.16)$$

The rain rate is defined as

$$\Re = \frac{1}{\rho_L} \int_0^\infty m(D)n(D)v(D)dD \qquad (8.17)$$

where $m(D)$ is the mass of a drop of diameter in the range D to $D + dD$ and $v(D)$ is the fall speed of drops within the same size range. Again assuming a Marshall–Palmer distribution, then

$$\Re = \frac{\pi}{6}\rho_L N_o \int_0^\infty D^3 e^{-\Lambda D} a D^b dD$$

---

[4] Here we use the definite integral

$$\int_0^\infty x^{\nu-1} e^{-\mu x} dx = \frac{\Gamma(\nu)}{\mu^\nu}$$

where for an integer n, the gamma function is $\Gamma(n+1) = n\Gamma(n); \ \Gamma(1) = 1$.

where we use the relation $v(D) = aD^b$. Integration leads to

$$\Re = \frac{\pi}{6} N_o a \frac{\Gamma(4+b)}{\Lambda^{4+b}} \qquad (8.18)$$

and it is a simple matter to combine (8.16) with (8.18) to obtain a relationship of the form given by (8.14).

Obviously, the $Z - \Re$ relationship varies with the size distribution. We do not therefore expect a universal relationship connecting Z to $\Re$ although it is common that larger reflectivities are associated with the more intense rainfall. Attempts to classify $Z - \Re$ relationships in terms of the type of rain (stratiform rain versus convective rain, for example) have generally been unsuccessful as several $Z - \Re$ relationships are typically found for rain conditions that are supposedly the same. This variability is largely a consequence of the natural variability of the precipitation microphysics.

An example of the differences between the $Z - \Re$ relationships for stratiform and convective rainfall is given in Fig. 8.9. Figure 8.9a is a time series of one–minute rainfall recorded by a tipping bucket raingage during the passage of a squall line through Darwin, Australia. This time series shows the characteristics of the precipitation of such a system with intense rainfall over a short period of time associated with the convective part of the weather system and the lighter but steadier rainfall from the trailing stratiform portion of the system. A scatter diagram of Z versus $\Re$ is presented in Fig. 8.9b. Stratiform reflectivities are 3–5 dBZ higher than the convective reflectivities for the same rain rate. This shift in the $Z - \Re$ relationship is related to differences in the microphysics of rain in the convective and stratiform part of the cloud. Figure 8.9c supports this proposition and shows a five–minute average of the raindrop size spectrum for the rain rate of 5.3 mmhr^{-1} measured at the tail end of the convective rain and the beginning of the stratiform rain for the event shown in Fig. 8.9a. These distributions confirm that the stratiform rain is composed of larger drops and thus larger reflectivities than convective rain.

8.2.2   Radar Attenuation

The radar equation, (8.13), assumes that empty space exists between the radar and the illuminated backscattering volume. This is generally unrealistic because of the possibility of attenuation of the beam along its path. At any particular wavelength, this attenuation may arise from absorption by atmospheric gases, absorption by

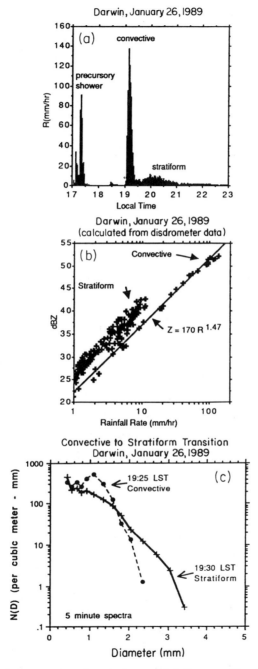

Figure 8.9 (a) Time series of one–minute rainfall rates, (b) rainfall rate versus radar reflectivity, and (c) raindrop distributions at the end of convective rain and the beginning of stratiform rain ($\Re=5.2$ mmhr^{-1}) during the squall line event that produced the rain depicted in (a) (from Short et al., 1992) .

cloud droplets, or scattering and absorption by precipitation. The attenuation may also arise from a combination of all processes. Thus (8.13) requires modification to include these attenuation processes. We do this by appealing to the basic law of extinction defined by (6.1) and written as:

$$\Delta \bar{P} = -2\sigma_{ext}\bar{P}\Delta r \tag{8.19}$$

In this equation Δr is the range resolution of the instrument ($= h/2$) and the factor of 2 appears by virtue of the folding of the path. Thus integration of (8.19) yields

$$\bar{P}(R) = \bar{P}_t \exp[-2 \int_0^R \sigma_{ext}(r')dr']. \tag{8.20}$$

and it follows from a combination of (8.13) and (8.20) that

$$\bar{P}_r = C\frac{|K|^2}{R^2}\bar{Z}\,\exp[-2 \int_0^R \sigma_{ext}(r')dr'] \tag{8.21}$$

Radar attenuation is usually expressed in terms of the decadic extinction coefficient (units of decibels per unit length) defined as $\sigma_{ext}^D = 10\,(\log e)\,\sigma_{ext}$. Shorter wavelength radars suffer from attenuation by absorption by molecular oxygen and water vapor, as well as cloud droplet absorption. According to (5.26), the absorption coefficient for cloud droplets in the Rayleigh scattering limit is

$$\sigma_{ext}^D = -0.4343\frac{6\pi}{\rho_L\lambda}\Im m\,[K]\,w \tag{8.22}$$

where w is the cloud liquid water content. If we write this coefficient in the form

$$\sigma_{ext}^D = k_c w \tag{8.23}$$

then k_c may be interpreted as an attenuation coefficient in units of dB/km/g/m^3. Values of k_c corresponding to water and ice particles for three wavelengths and for different temperatures are presented in Table 8.1a. These values indicate that the attenuation decreases as the wavelength increases with values for 3 cm being considerably smaller than values for 1 cm. Attenuations at even longer wavelengths, such as at 5 or 10 cm (results for these wavelengths are not included), are negligible. Cloud droplet attenuation is also strongly

Table 8.1a Attenuation coefficients for cloud k_c.

Temperature	Wavelength (cm)		
(°C)	0.9	1.8	3.2
Water Cloud			
20	0.647	0.128	0.0483
0	0.99	0.267	0.0858
-8	1.25	0.34	0.122
Ice Cloud	$\times\ 10^3$		
0	8.74	4.36	2.46
-20	2	1	0.563

Table 8.1b Attenuation coefficients for rain k_p.

Wavelength cm	M–P (°C)	Modified M–P (°C)	Mueller– Jones (°C)
0.86	0.27	0.31	0.39
3.21	$0.011\mathfrak{R}^{0.15}$	$0.013\mathfrak{R}^{0.15}$	0.018
5.5	0.003–0.004	0.0031	0.0033
10	0.0009–0.0007	0.00082	0.00092

k_c is expressed in units of dB/km/g/m^3 and k_p is expressed in units of dB/km/mm/hr (modified from Battan, 1973).

temperature dependent because of the temperature dependence of the dielectric constant of water at radar wavelengths. The differences of the dielectric properties between water and ice gives rise to the different attenuations of ice and water hydrometeors listed in the table. We conclude that attenuation of microwave radiation by ice particles is negligible.

Attenuation by precipitation–sized particles is also important at some wavelengths. Theoretical and observational studies establish an empirical relationship of the type

$$\sigma_{ext}^D = k_p \mathfrak{R}^\gamma \tag{8.24}$$

for this extinction, where both k_p and γ depend on the radar wavelength, the temperature of raindrops, and the particle size distri-

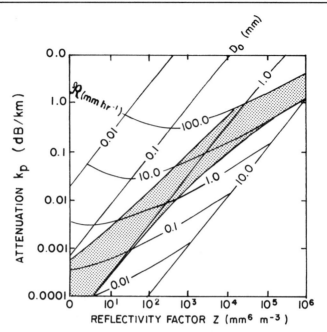

Figure 8.10 The rain parameter diagram for exponential drop size distributions. The shaded area contains 69 $Z - \Re$ relationships summarized by Battan
(1973). Rain rate isopleths are indicated as are the $k_p - Z$ relationships for
given slopes of the size distribution (i.e., for given values of D_o) (from Atlas and
Ulbrich, 1974).

bution. Table 8.1b presents values of k_p (expressed in units of
dB/km/mm/hr) for four radar wavelengths. This attenuation decreases as the wavelength of interest increases. Attenuation by rain
is negligibly small for the longer 10cm wavelength weather radar,
but it is significant at the shorter wavelengths where it is dependent
on the intensity of the rainfall.

Figure 8.10 is a diagram that summarizes the interrelationships
among $k_p, Z, \Re$, and D_o [as defined in relation to (8.15)]. This
diagram is known as the rain parameter diagram and applies for
$\lambda = 3.22$ cm and $T = 10°C$. A measurement of $k_p = 0.1$ dBkm^{-1} and
$Z = 2 \times 10^3$ mm^{6}m^{-3}, for example, corresponds to a rainfall rate
in excess of 10 mmhr^{-1} and D_o less than 1 mm. Commonly used
empirical $Z - \Re$ relationships are also represented on this diagram
by the shaded region. These relationships spread over an almost ten-
fold range in $\Re$ for a given Z (indicated by the extent of the shading
relative to the $\Re$ isopleths). Note how the changes in k_p are small
compared with changes in Z for a given value of $\Re$. Stated differ-

ently, for wavelengths as large as 3 cm, k_p is a measure of $\Re$ that is relatively insensitive to changes in droplet size spectra in contrast to $Z - \Re$ relationships which are much more dependent on precipitation microphysics.

8.2.3 Area–Time–Integral Methods for Convective Rain

Common experience tells us that there is a certain correlation between the area occupied by convective clouds, their height, life duration, and the total rainfall from these clouds. For instance, it was described in Chapter 7 how the relationship between cloud–top as defined by the OLR and areal rainfall may be used as an estimator of rainfall over a sufficiently large area. Similarly, the convective area of a storm defined as the area of radar reflectivities exceeding some threshold also correlates to the total rainfall of the storms (Lopez et al., 1983). However, for instantaneous measurements on a small number of cells, the correlation between rainfall and reflectivity is unconvincing because of the natural variance of rainfall. This variance is eventually reduced when a sufficiently large sample is integrated in time and over space.

To examine how this arises, we follow the arguments of Atlas et al. (1990) who attempted to provide a theoretical foundation for the observed relation between area and time averages of radar reflectivity and the total rainfall. Suppose that a single instantaneous rainfall distribution occurring at time t, covering an area A in the x, y plane, is $\Re(x, y, t)$. We can talk about this rainfall in terms of the probability density function (pdf) of rain denoted as $p(\Re)$. Empirical evidence suggests that this pdf is well behaved and can be constructed either from a single snap shot of many storms or from observations of a single storm considered during its entire time of activity.

The pdf $p(\Re)$ represents the percentage of the area where the rain rate is between $\Re$ and $\Re + d\Re$. The average rain rate above some threshold $\Re_0$ follows from the pdf as

$$\bar{\Re}_0 = \frac{\int_{\Re_0}^{\infty} \Re\, p(\Re) d\Re}{\int_{\Re_0}^{\infty} p(\Re) d\Re}$$

If the area of rain that exceeds the threshold rate $\Re_0$ is A_0, then

$$V_0 = A_0 \Re_0 \tag{8.25}$$

is the rate at which a volume of rain above the threshold $\mathfrak{R}_0$ falls. However, this is only a fraction f of the total rain volume where

$$f = \frac{\int_{\mathfrak{R}_0}^{\infty} \mathfrak{R} \, p(\mathfrak{R}) d\mathfrak{R}}{\int_0^{\infty} \mathfrak{R} \, p(\mathfrak{R}) d\mathfrak{R}}$$

Thus, the total volumetric rainfall V is

$$V = \frac{V_0}{f} = A_0 \frac{\bar{\mathfrak{R}}_0}{f} = A_0 S_0 \qquad (8.26)$$

where, for a given threshold $\mathfrak{R}_0$, $S_0 = \bar{\mathfrak{R}}_0/f$. If F_0 is the fractional area covered by rain exceeding the threshold $\mathfrak{R}_0$ (this is simply A_0/A), then the average rain rate is

$$< \mathfrak{R} >= F_0 \bar{\mathfrak{R}}_0/f$$

or alternatively,

$$< \mathfrak{R} >= F_0 S_0 \qquad (8.27a)$$

which again applies to a specific value of the threshold $\mathfrak{R}_0$.

The procedure is then simple and direct. The first step is to estimate the fractional area F_0 covered by reflectivities above a threshold Z_0, or from a prescribed $Z - \mathfrak{R}$ relationship, above a rainfall threshold $\mathfrak{R}_0$. The proposal of Atlas et al. (1990) is to employ a $Z - \mathfrak{R}$ relationship adapted from climatological data for the given region under study. The rainfall pdf can subsequently be derived from these radar reflectivity data either from instantaneous data or from data accumulated from a climatology. Once the rainfall pdf is known, S_0 follows from

$$S_0 = \frac{\int_0^{\infty} \mathfrak{R} \, p(\mathfrak{R}) d\mathfrak{R}}{\int_{\mathfrak{R}_0}^{\infty} p(\mathfrak{R}) d\mathfrak{R}} \qquad (8.27b)$$

Having then estimated S_0 and having deduced F_0, $< \mathfrak{R} >$ follows from (8.27a).

The validity of the method requires that a sensible climatological $Z - \mathfrak{R}$ relationship exists and that any one snapshot of the rain distribution contains a representative number of storm cells to accumulate meaningful pdf statistics. Rosenfeld et al. (1990) suggest that a 100×100 km domain provides a sufficient sample with an error

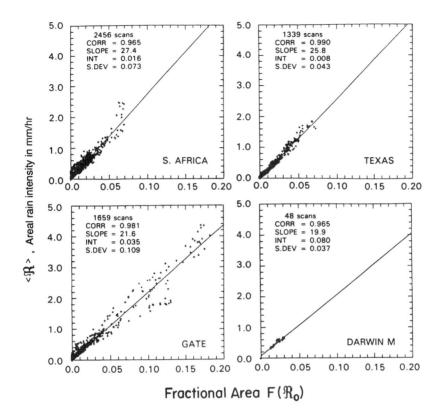

Figure 8.11 The average rain rate–fractional area relationships for convection over South Africa, Texas, GATE, and Darwin. Each point is one radar scan and the rain rate is an average over the area between the 31 and 90 km range markers. The fractional area is the area covered by rain intensities exceeding a threshold of 6 mmhr^{-1} (Rosenfeld et al., 1990).

of 0.07 mmhr^{-1} based on their analyses of the data collected from storms measured over Texas.

Four example scatter plots of $< \Re >$ as a function of F_0 are shown in Fig. 8.11 for radar data collected in South Africa, Texas, during GATE, and at Darwin. According to (8.27a) the rainfall rate varies linearly with F_0 for a given value S_0. The scatter plots shown in Fig. 8.11 apply to a single threshold $\Re_0 = 6$ mmhr^{-1}. The fact that the data support (8.27) suggests that all rain systems within a given region are characterized by a single pdf.

In estimating the volume of rainfall from (8.26), it is important to stress again that the slope parameter S_0 depends on the rainfall

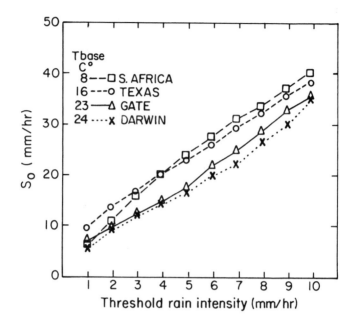

Figure 8.12 The dependence of the slope of the $< \Re >$ versus F_0 scatter-grams on the rain rate threshold $\Re_0$ for the data shown in Fig. 8.11 (Rosenfeld et al., 1990).

threshold chosen. Figure 8.12 further emphasizes this point showing the slope S_0 as a function of the threshold $\Re_0$ for all four convective regimes represented by the data displayed in Fig. 8.11. The coefficient increases as $\Re_0$ increases. The reason for this is simple to understand. To maintain the observed volumetric rain accumulated throughout the life cycle of a storm, defined as the product of S_0 and A_0, the slope parameter S_0 must increase when A_0 decreases. In turn, the latter, decreases as the threshold $\Re_0$ increases.

The analysis described earlier applies to an instantaneous, areawide view of the rainfall. It is a simple matter to obtain the total volume of rainfall integrated throughout the duration of the storm, namely

$$V = \int_{\text{time}} V(t)dt$$

where $V(t)$ is the instantaneous volume of rainfall as defined by (8.26). Substituting (8.26) into this integral leads to

$$V = S_0 \int_{\text{time}} A_0(t) dt$$

where the integral factor is referred to as the area–time–integral (ATI). We write this equation as

$$V = S_0 \text{ATI} \tag{8.27c}$$

Doneaud et al. (1984) analyzed radar data collected at North Dakota using this approach. With the reflectivity threshold $Z_0 = 25$ dBZ, they deduced that $S_0 = 3.7$. The relationship, (8.27c), and even the numerical value of the coefficient S_0, bear a striking resemblance to the OLR rainfall relationship discussed previously in Chapter 7, a resemblance further explored by Atlas and Bell (1992).

8.3 Multiparameter Radar Measurements of Rainfall

Multiparameter radar methods seek to exploit more than one parameter (such as reflectivity, attenuation, polarization state, etc.) to estimate other information that can be related to rainfall. Two particular techniques that provide an estimate of two precipitation–related variables are described shortly: the dual wavelength method in which Z and k_p are measured, and the dual polarization method which provides Z at two different polarizations. In the context of the Marshall–Palmer distribution, (8.15), the basic approach of multiparameter radar is to use two independent measurements (polarization or wavelength dependence or a mixture of both) to estimate the two size distribution parameters: drop concentration N_o and the slope $\Lambda = 1/D_o$. Rainfall rate then follows by application of this size distribution to the relationship between the terminal velocity of the rain drops and their diameter.

8.3.1 A Dual Wavelength Method

Using dual wavelength measurements, it is possible to estimate both Z and k_p independently and, with some assumption about particle scattering, the parameters of the size distribution expressed by (8.15). The dual wavelength approach uses measurements of backscattered power at two wavelengths, a long wavelength that is largely unattenuated and a short wavelength that is attenuated by rainfall.[5] Let us assume for convenience that the two radar beams are matched in terms of the volume of atmosphere they resolve and that the radar constants are also the same. The power backscattered from the volume at a range R are

$$\mathcal{P}_L(R) = \frac{C \mid K \mid^2 Z}{R^2} \tag{8.28a}$$

$$\mathcal{P}_S(R) = \frac{C \mid K_S \mid^2 Z_e}{R^2} \exp[-2 \int_0^R \sigma_{ext} dr] \tag{8.28b}$$

where the subscripts S and L refer to short and long wavelengths respectively and Z_e is the effective reflectivity of the shorter wavelength echo. From measurements of $\mathcal{P}_S$ at two ranges, say R_1 and R_2, the logarithm of the received power follows from (8.28b) as

$$\ln \left[\frac{\mathcal{P}_S(R_1)}{\mathcal{P}_S(R_2)} \frac{Z_e(R_2)}{Z_e(R_1)} \frac{R_1^2}{R_2^2} \right] = -2 \int_{R_1}^{R_2} \sigma_{ext} dr \tag{8.29}$$

Provided that the rainfall is not too heavy (and the precipitation particles not too large) then $Z_e \approx Z$. Therefore the measurement of the reflectivity at the longer wavelength, related to the measured power $\mathcal{P}_L$ by (8.28a), can be used in (8.29) to obtain

$$\frac{1}{R_2 - R_1} \ln \left[\frac{\mathcal{P}_S(R_1)\mathcal{P}_L(R_2)}{\mathcal{P}_S(R_2)\mathcal{P}_L(R_1)} \right] = \frac{2}{R_2 - R_1} \int_{R_1}^{R_2} \sigma_{ext} dr \approx 2\bar{\sigma}_{ext} \tag{8.30}$$

where $\bar{\sigma}_{ext}$ is the extinction averaged along the path from R_1 to R_2 and follows directly from measurements of the echo powers.

For a drop size distribution given by (8.15), σ_{ext} is defined as

$$\sigma_{ext} = \frac{N_o \pi}{4} \int e^{-\Lambda D} Q_{ext}(D) D^2 dD = \Gamma(n+1) c N_o \Lambda^{-n+1} \tag{8.31}$$

---

[5] In principle the dual wavelength method applies to any two attenuating wavelengths. The case considered here for simplicity assumes one wavelength is unattenuated. This dual wavelength approach is common to a variety of sensing methods discussed in this book. It is the principle of Dobson measurements of ozone and it is also the principle used in the analyses of data from a tunable lidar system to obtain trace gas concentrations. This lidar method is referred to as differential absorption lidar (DIAL).

which follows from substitution of (8.15) into (5.25), where Γ is the gamma function and $\pi Q_{ext}(D)D^2 \approx cD^n$. The reflectivity factor Z also follows from (8.16). From σ_{ext} and Z, and for an assumed value of n, the two unknown precipitation size distribution parameters, namely Λ and N_o, can be determined.

The retrieval of rainfall via this approach is straightforward in principle, but it suffers for a number of reasons in practice. It is generally not possible to match the radar volumes for the two wavelengths under consideration so other factors that cancel in the ratio quantity of (8.29) enter into the analyses. Furthermore, the retrieval approach described is subject to the experimental error in the measurement of Z (which is typically ± 1dB) as well as the assumption that $Z_e \approx Z$ which is only likely to be reasonable for lighter rain rates. As a result, the estimation of $\bar{\sigma}_{ext}$ is inherently inaccurate and it is necessary to average over a few range gates which reduces the spatial resolution of the approach.

8.3.2   Dual Polarization Method

The physical basis underlying the dual polarization method of rainfall retrieval was introduced in Section 5.7. The method relies on the relation between the polarization of backscattered radiation and the asymmetry of drop shape. The association between this asymmetry and the mass of the drop provides the connection to rainfall rate. In still air, raindrops fall with their long axis aligned horizontally. The amplitude of the scattered wave parallel to this long axis is significantly (i.e., measurably) larger than the amplitude of the wave scattered perpendicular to this axis. The result is that the backscattered power associated with the horizontally polarized component of the electromagnetic field exceeds the power of the vertically polarized component of the backscattered field. The difference in the measured polarized reflectivities thus provides a measure of the aspect ratio of the precipitation drop and thus of rainfall.

Polarization–based measurements of radar reflectivity are expressed by a radar equation that accounts for the polarization of both the transmitted and the backscattered beams. We introduce the following vector

$$\tilde{\mathcal{P}}_{r,V/H} = \begin{pmatrix} \mathcal{P}_{H,V/H} \\ \mathcal{P}_{V,V/H} \\ \mathcal{P}_U \\ \mathcal{P}_v \end{pmatrix} \qquad (8.32)$$

to represent the full polarization state of the received electromagnetic wave after scattering of a transmitted pulse that is either vertically or horizontally polarized. The first of the subscripts on the first two elements of this vector corresponds to the polarization of the received beam and the second subscript (i.e., V/H) refers to the polarization state of the transmitted beam. For a transmitted beam that is either horizontally or vertically polarized, the transmitted power follows from (5.71) as

$$\tilde{P}_{t,H} = \mathcal{P}_H \begin{pmatrix} 1 \\ -1 \\ 0 \\ 0 \end{pmatrix} \qquad \text{and} \qquad \tilde{P}_{t,V} = \mathcal{P}_V \begin{pmatrix} 1 \\ 1 \\ 0 \\ 0 \end{pmatrix} \qquad (8.33)$$

respectively.

For the particle geometries and orientations relevant to (5.41), the matrix for backscattering has the form

$$\mathbf{C}_b = \frac{1}{k^2} \begin{pmatrix} S_{11} & S_{12} & 0 & 0 \\ S_{12} & S_{22} & 0 & 0 \\ 0 & 0 & S_{33} & S_{34} \\ 0 & 0 & -S_{34} & S_{44} \end{pmatrix} \qquad (8.34)$$

Since we identify $Z_{V/H,V/H}$ with $\int n(r)C_{b,V/H,V/H}\,dr$ it follows that we may write

$$\begin{pmatrix} \mathcal{P}_{HH} \\ \mathcal{P}_{VV} \end{pmatrix} = \frac{|K|^2}{R^2} \begin{pmatrix} C_V & 0 \\ 0 & C_H \end{pmatrix} \begin{pmatrix} Z_{V,V} & Z_{V,H} \\ Z_{H,V} & Z_{H,H} \end{pmatrix} \qquad (8.35)$$

where $Z_{H,H}$ and $Z_{V,V}$ are the measured radar reflectivities of the horizontal and vertical polarization states as denoted by the first of the subscripts for horizontally and vertically polarized transmission as given by the second subscript. $Z_{V,H}$ and $Z_{H,V}$ are the cross-polarized reflectivities and C_H and C_V are the radar constants that refer to the individual polarizations of the transmitted beams. C_H generally differs from C_V due to different characteristics of the two polarized beams as well as to differences in the transmitted power.

The Z_{DR} is a ratio quantity

$$Z_{DR} = 10\log(\frac{Z_{H,H}}{Z_{V,V}}) = 10\log\left[\frac{\int n(r)(S_{11} - 2S_{12} + S_{22})dr}{\int n(r)(S_{11} + 2S_{12} + S_{22})dr}\right]$$
$$(8.36)$$

expressed in decibels. It is also a quantity that is independent of the number concentration of drops (i.e., N_o), and for matched beams, it is independent of the radar constant. The right–hand side of this expression is taken from (5.73) and it is understood that the intensity functions are defined for the scattering angle $\Theta = 180°$.

In precipitating regions of clouds, Z_{DR} is typically positive and largest for heaviest rain which is associated with the largest and most asymmetrical rain drops. Backscattering by graupel and hail also produces large values of reflectivity (especially for wetted particles) but Z_{DR} for these particles is nearly 0 because the particles are either approximately spherical in shape or they tumble as they fall, effectively scattering as spheres.[6] Table 8.3 summarizes the expected relationship between $Z_{H,H}$ and Z_{DR} for various modes of precipitation.

With the assumption that the Marshall–Palmer raindrop–size distribution represents the actual distribution of raindrops, then Z_{DR} depends on the slope Λ. We see this dependence when (8.15) is substituted in (8.36)

$$Z_{DR} = 10 \log \left[\frac{\int_0^{D_{max}} D^6(S_{11} - 2S_{12} + S_{22})e^{-\Lambda D}dD}{\int_o^{D_{max}} D^6(S_{11} - 2S_{12} + S_{22})e^{-\Lambda D}dD} \right] \quad (8.37)$$

The right–hand side of this expression can be evaluated if a relationship between the scattering amplitudes and the particle size is known. Seliga and Bringi (1976) used a simple scattering theory for oblate spheroids to evaluate (8.37) as a function of Λ and arrived at the following empirical relationship

$$\Lambda = 2.603 Z_{DR}^{-0.63} \quad (8.38)$$

which is shown in Fig. 8.13 for $D_{max} = 8$ mm and $T = 20°C$. Results derived from a more exact scattering theory by Al–Khatib et al. (1979) are also presented on this diagram for comparison.

With all the assumptions inherent to (8.38) in mind, the procedure for estimating rainfall from polarization diversity is simple. Measurements of Z_{DR} lead to an estimate of Λ from (8.38). The number density N_o follows from the measurement of $Z_{H,H}$ via (8.16) once Λ is known. An example of (8.16) is also shown in Fig. 8.13

---

[6] Backscattering by conically shaped graupel can, however, produce negative values of Z_{DR}.

Table 8.3 Characteristics of Z and Z_{DR} at 10 cm

Hydrometeor Type	Z	Z_{DR}	Comments
Rain	High	High	With large oblate drops
Drizzle, cloud or fog	Low	Low	Small spherical drops of water or ice particles
Dry snow flakes	Medium –Low	Medium –Low	Large horizontally oriented low–density aggregates
Sleet, wet snow	High	High	Large oblate horizontally oriented particles
Wet graupel	High	Negative	Large conical vertically oriented particles
Wet hail	High	Variable	Large particles: seldom spheres
Dry hail or other high–density ice particles	Medium	Low	

From Hall et al. (1980).

as a function of Λ. Since measurements of $Z_{H,H}$ and Z_{DR} provide estimates of both N_o and Λ, then the rain rate $\Re$ follows from (8.18). Whereas the retrieval of rain rate using this approach is relatively simple, results depend on the details of the shape of the size distribution and on the assumed value of D_{max} used to derive (8.38).

Figures 8.14a and b are RHI displays of the horizontal reflectivity $Z_{H,H}$ and the differential reflectivity Z_{DR} measured in rain cells. Comparison of these reflectivity parameters at a range of 35 km reveals a column of high reflectivity ($Z_{H,H} \approx 55$ dBZ) but low values of Z_{DR} above about 2 km. This circumstance is most likely to be a result of scattering by randomly oriented ice hydrometeors. A comparison of Z_{DR} and $Z_{H,H}$ in this way provides a useful way of discriminating ice phase from water phase in convective rain clouds. A rapid change in Z_{DR} occurs near the 0°C isotherm marking the transition between ice and water. Melting ice particles are approximately spherical and affect the reflectivity as if they are large water spheres

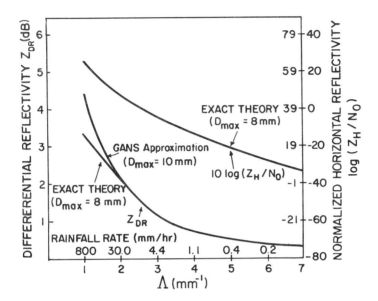

Figure 8.13 Variations of Z_{DR} and the normalized horizontal reflectivity $10\log(Z_{H,H}/N_o)$ as a function of Λ. The rainfall rate and dBZ scales apply to $N_o = 8 \times 10^3$ m^{-3}mm^{-1}. $Z_{H,H}$ is expressed in the units of millimeters to the sixth power per cubic meter (mm^{6}m^{-3}).

of the same size. The reflectivity in the vicinity of the melting layer increases by several orders of magnitude. Because Z depends on the sixth power of the hydrometeor size (assuming Rayleigh scattering), it is likely that the dramatic increase in Z occurs as ice crystals coalesce and melt. The increase in fall speed as the ice crystal melts and the ultimate break up of the particle lead to a decrease in reflectivity below the melting layer. Thus a characteristic of the melting layer is a band of high reflectivity, called the *bright band*. In Fig. 8.14a, the bright band appears at a range of 43 km. This is a region where the Z_{DR} is also enhanced since this is a region where wet hydrometeors are most oblate.

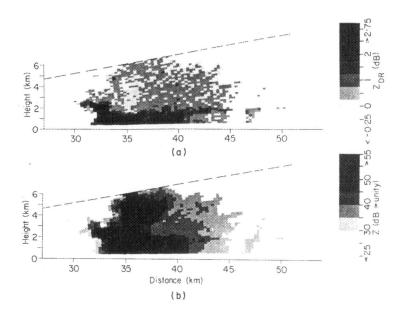

Figure 8.14 An RHI display showing a vertical cross section through rain cells. (a) The horizontal reflectivity $Z_{H,H}$ and (b) the differential reflectivity Z_{DR} (Doviak and Zrnic, 1984).

8.4 General Principles of Lidar Sensing

Lidar systems may be classified according to the particular type of optical interaction sensed by the system. The following are some general categories of lidar systems and the properties they measure:
- backscatter lidar –intensity and polarization
- differential absorption lidar (DIAL) — extinction
- Doppler lidar — phase shift
- fluorescence lidar — intensity at a shifted wavelength
- Raman lidar — intensity at a shifted wavelength

The basic concept of backscatter lidar dates back to the early 1900s when extensive experiments were conducted with large searchlights. Backscatter lidar performs in a manner that is directly analogous to conventional weather radar. A pulse of energy transmitted to the atmosphere and backscattered to the detector is related to some property of that volume. DIAL measures the concentration of a molecular species in the atmosphere by transmitting pulses at two adjacent wavelengths which are characterized by different gas absorption strengths. The difference in the backscattered signal is then related to the absorption by the gas and thus to its concentration. Doppler lidar methods, described in detail at the end of this chapter, use a single wavelength laser together with sophisticated detection techniques to resolve the phase shift in the returned signal that results from the motion of the background Rayleigh and Lorenz–Mie scatterers. The latter two categories of lidar systems, namely fluorescence and Raman lidar, have characteristics that are specialized and will not be discussed any further.

Atmospheric backscatter lidar is perhaps the most common type of lidar in use today. This system usually consists of a nontunable high–power, pulsed laser and employs incoherent detection methods. The lidar is used to sense the backscattering properties of a small volume of atmosphere containing scatterers such as aerosol and cloud hydrometeors. Lidars transmit a narrow beam and a receiver telescope is co–aligned with respect to this beam to collect the backscattered radiation. The field of view of the receiver is generally large enough to encompass the divergence of the laser beam. This particular aspect highlights one of the assets of lidar and a notable difference between radar and lidar; namely, the combination of the short pulse length (of the order of 10^{-8} sec) and the small beam divergence (typically 10^{-3} to 10^{-4} radians) define lidar volumes of only a few cubic meters at ranges of tens of kilometers.

8.4.1 The Lidar Equation

The backscattered radiation detected by a lidar system is described by the lidar equation. This equation follows from precisely the same considerations introduced in relation to the derivation of the radar equation (8.16). The received power $\mathcal{P}_r$ is related to the relative strength of the scattering in a volume by

$$\mathcal{P}_r(R) = \frac{C}{R^2} \frac{h}{2} \frac{\beta}{4\pi} \exp(-2 \int_0^R \sigma_{ext}(r')dr') \qquad (8.39)$$

where C is the "lidar" constant and includes the transmitted power $\mathcal{P}_t$, the receiver cross–section A_r, and other instrument factors. The backscatter factor in the lidar equation, namely $\beta/4\pi$, is the backscattering coefficient expressed in units of $km^{-1}ster.^{-1}$ This factor, in contrast to its cousin, the radar reflectivity Z, cannot be readily derived from scattering theory since backscattering at lidar wavelengths often arises from scattering by a variety of different particles, some of uncertain composition and shape. This aspect of lidar backscattering is clearly exemplified in Fig. 8.15a showing an example of the lidar–measured backscattered power as a function of altitude. This example shows how different sources of backscatter contribute to the measured profiles. Strong aerosol backscatter in the lower portion of the atmosphere decreases systematically to an altitude of 10 km above which the backscatter increases due to the presence of a thin cirrus cloud layer. Above the cirrus layer, backscatter from volcanic ash appears at about 14 km.

Figures 8.15b and c provide further examples of lidar backscatter measured in thick cirrostratus clouds expressed as a function of range (or altitude for the vertically pointing lidar in these examples). The tendency for backscattering to decay in a somewhat exponential manner with altitude suggests attenuation of the lidar pulse as it propagates into the cloud. This attenuation is especially significant in the example of Fig. 8.15c. Attenuation is a more prevalent feature of lidar backscatter than of radar backscatter and is a significant limitation to the application of lidar for probing optically thick clouds.

8.4.2 Inversion of the Lidar Equation: Backscatter–to–Extinction Ratios

Just as in the case of the radar equation which we express in terms of Z and k_p, much of the information about the atmosphere obtained from lidar backscatter measurements is contained in the extinction and backscattering parameters σ_{ext} and β. Retrieving these parameters from (8.39) is a long–standing problem in lidar research. In considering the inversion methods developed for this purpose, we introduce the signal variable,

$$S(R) = \ln\left[R^2 \mathcal{P}_r(R)\right] \tag{8.40}$$

If the signal $S_0 = S(R_0)$ is selected at a reference range R_0, then (8.39) may be re–arranged in the form

$$S - S_0 = \ln(\frac{\beta}{\beta_0}) - 2\int_{R_0}^{R} \sigma_{ext}\, dr \tag{8.41}$$

or in differential form as

$$\frac{dS}{dR} = \frac{1}{\beta}\frac{d\beta}{dR} - 2\sigma_{ext} \tag{8.42}$$

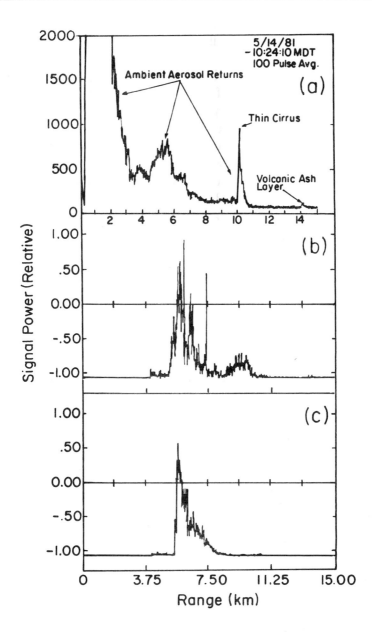

Figure 8.15 (a) A "typical" vertical backscatter profile averaged over 100 pulses. (b) Profiles of received backscatter power as a function of altitude through a deep cirrostratus cloud showing slight attenuation of the pulse. (c) Same as (b) but for strong attenuation (lower two panels from Platt, 1981).

The solution to (8.42) requires that we know or assume a relationship between β and σ_{ext}, the so-called backscattering–to–extinction ratio. For the special circumstance that $d\beta/dR = 0$, then the extinction may be estimated from the slope of the plot of S versus R. This is the so-called slope method of inversion. σ_{ext} is typically determined from a least–squares fit to any portion of the $S - R$ relationship that appears approximately linear.

The condition that $d\beta/dR = 0$ implies that the scatterers are homogeneously distributed with range along the lidar path. This sometimes applies for portions of the path, but generally does not, so the slope inversion method has limited applicability. One method that avoids this homogeneity assumption uses an a priori relationship between β and σ_{ext} typically of the form[7]

$$\beta = b\sigma_{ext}^n \tag{8.43}$$

where b and n are specified constants. Substitution of (8.43) into (8.42) yields

$$\frac{dS}{dR} = \frac{n}{\sigma_{ext}}\frac{d\sigma_{ext}}{dR} - 2\sigma_{ext} \tag{8.44}$$

which has a general solution at range R (Carswell, 1983)

$$\sigma_{ext} = \frac{\exp[(S - S_0)/n]}{\sigma_{ext,0}^{-1} - \frac{2}{n}\int_{R_0}^{R}\exp[(S - S_0)/n]\,dr}, \tag{8.45}$$

where $\sigma_{ext,0}$ is the extinction coefficient at some a priori reference range R_0. Unfortunately, the inversion of (8.39) in the form of (8.45) is limited not only by the validity of the backscattering–to–extinction assumption but also on the stability of (8.45) with respect to $\sigma_{ext,0}$. This instability arises from the decay of signal due to attenuation as R increases beyond R_0. The solution to (8.45) is thus determined as the ratio of two numbers each progressively decreasing with increasing range.

A modification of (8.45) was proposed by Klett (1981) to overcome this problem. Instead of specifying $\sigma_{ext,0}$ at the beginning

---

[7] This is formally analogous to the function introduced to obtain (8.31) for radar. The backscatter of Rayleigh particles varies as D^6 and the power–law relationship introduced for extinction in relation to (8.31) in turn dictates backscatter–to–extinction ratio power law similar to that introduced here for lidar.

of the backscattering profile, he expressed the solution to the lidar equation in terms of the extinction at a predetermined end range R_m so that the solution is generated for $R \leq R_m$ rather than for $R \geq R_0$. The result at R for a constant n is

$$\sigma_{ext} = \frac{\exp[(S - S_m)/n]}{\sigma_{ext,m}^{-1} + \frac{2}{n} \int_R^{R_m} \exp[(S - S_m)/n] \, dr} \tag{8.46}$$

This seemingly trivial change to (8.45) makes a significant difference to the behavior of the solution (which is explored further in Problem 8.8). In the Klett method, the extinction coefficient σ_{ext} is determined as the ratio of two numbers that become progressively larger as the range decreases from R_m.

Figure 8.16a provides an example of aerosol backscatter derived from lidar measurements of the El Chichon volcanic cloud. The data, composited and presented as seasonal averages, show the volcanic cloud high in the stratosphere during the fall of 1982. The subsidence of the aerosol into the troposphere occured in the winter and spring of 1983, producing increased levels of tropospheric aerosol. Figure 8.16b presents data on stratospheric aerosol taken at a fixed site at 37N and 76W. This time series of the aerosol backscattering integrated vertically from the tropopause to 30 km at a wavelength 0.6943 μm demonstrates the longevity of volcanic aerosol in the stratosphere.

8.4.3   High Spectral Resolution Lidar

In the method previously described, inversion of the lidar equation to obtain profiles of both extinction and backscattering coefficients also requires information or an assumption about how each is related to the other. The method described above introduces this information via the assumption expressed in (8.43). Unfortunately, this is a major limitation as the extinction and backscattering properties of irregularly shaped cloud and aerosol particles are, in principle, poorly known.

In general, the backscattered lidar pulse, say at a visible wavelength, arises from the backscattering by molecules which we describe in terms of Rayleigh scattering and from backscattering by aerosol which we commonly approximate using Lorenz–Mie theory. The backscattering by molecules, in turn, is determined by the known atmospheric density and this provides a known calibration target at each range. The problem is, how do we separate the extinction and

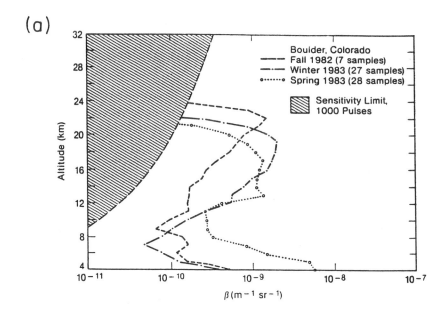

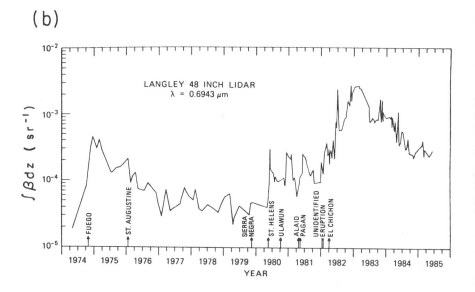

Figure 8.16 (a) Vertical profiles of lidar backscatter β averaged over the three seasonal periods indicated. The aerosol loading of the stratosphere due to the El Chichon volcanic eruption appears in the fall 1982 season and falls into the troposphere in the following seasons (Post, 1986). (b) Long–term variation of integrated stratospheric aerosol backscattering with the times of various volcanic eruptions indicated.

backscattering by the particles from the extinction and backscattering by the N_2 and O_2 molecules? If we can do this, then we will have a retrieval of the properties of the aerosol and avoid assumptions like (8.43). A solution to this problem is provided by a high spectral resolution lidar (HSRL).

Up to this point, our discussion of active systems has not touched upon the issue of the spectral width of the received pulse. In fact, we have tacitly assumed that both the transmitted source and received pulses are monochromatic. Actually, the incident pulse has a characteristic spectral width and the received pulse is broadened as it returns to the instrument. This broadening arises from the random motions of the scatters in the same way as absorption lines are broadened by Doppler effects. Figure 8.17a presents a typical spectrum of a hypothetical backscattered pulse. Different spectral signatures of molecules and aerosol appear in this high–resolution view of the backscatter. The scattering by both the molecules and aerosol is broadened by their thermal motions. The broadening of the heavier, more slowly moving aerosol particles, however, produces a backscattering that is spectrally much narrower than is the broadening by the lighter molecules. The aerosol backscatter appears as a narrow spike in Fig. 8.17a superimposed on the broader molecular spectrum that is typically about 2 GHz in width.

A combination of a narrowband, tunable laser, and a very narrow spectral filter provides a way of disentangling the contribution to the backscattering by molecules from aerosol and this is the principle of the HSRL. This basic idea is conveyed in Fig. 8.17a showing a typical response function of an atomic filter superimposed on the backscattering spectrum. When the laser is tuned to a frequency off the resonance of the atomic filter (i.e., off the minimum of transmission), the returned lidar signal contains both a Rayleigh and aerosol scattering component. An example of the backscattering measured by such a system is shown in Fig. 8.17b indicating the presence of a cloud just above 5 km. When the laser is tuned off the resonance, the measured filtered backscattering contains contributions from both particles and molecules and the cloud appears in the profile as shown. When the laser is tuned to the resonance line, only a molecular scattering contribution is detected eliminating any contribution from the cloud (Fig. 8.17c).

Because the HSRL separately measures both molecular and particulate backscatter, two lidar equations describe these measure-

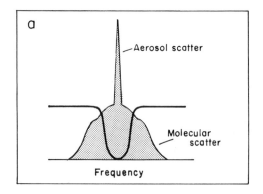

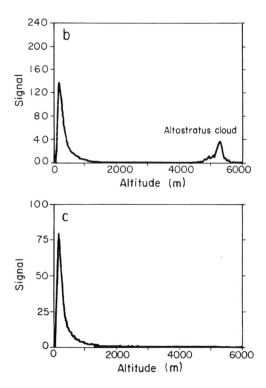

Figure 8.17 (a) Schematic of a Rayleigh plus Lorenz–Mie backscattering spectrum with an approximate width of 2 GHz along with a transmission function of an atomic resonance filter (heavy solid line). The return signals from a lidar after passing through a barium filter are shown in (b) and (c). In (b), the laser source is tuned off the resonance of the barium filter so the aerosol plus Rayleigh backscattering is observed (indicated as an altostratus cloud). In (c) the transmitter is tuned on the resonance of the filter and the Lorenz–Mie signal is removed (from She, 1990).

ments, namely

$$\mathcal{P}_{Ray}(R)R^2 = C\frac{\beta_{Ray}}{4\pi}\exp[-2\int_0^R \sigma_{ext}(r')dr'] \qquad (8.47a)$$

$$\mathcal{P}_a(R)R^2 = C\frac{\beta_a}{4\pi}\exp[-2\int_0^R \sigma_{ext}(r')dr']. \qquad (8.47b)$$

The backscattering coefficients for molecules and aerosol are β_{Ray} and β_a, respectively. We can readily solve these equations for β_a

$$\beta_a(R) = \beta_{Ray}(R)\frac{\mathcal{P}_a(R)}{\mathcal{P}_{Ray}(R)} \qquad (8.49)$$

in terms of factors that are either known (as in the case of β_{Ray}) or measured ($\mathcal{P}_a, \mathcal{P}_{Ray}$). Direct inversion of (8.47a) leads to the profile of the extinction via

$$\sigma_{ext} = -\frac{1}{2}\left\{\frac{d\ln[\mathcal{P}_{Ray}(R)R^2]}{dR} - \frac{d\ln[\beta_{Ray}]}{dR}\right\}, \qquad (8.50)$$

which is the equivalent of (8.42) except that $\mathcal{P}_{ray}$ is now measured and β_{Ray} is calculated from the atmospheric density profile. Thus the extinction profile may be derived using the slope procedure described earlier.

8.4.4 Multiple Scattering Effects

The lidar equation, (8.39), accounts for single scattering events only. In a medium like cloud, however, and for observations at typical lidar wavelengths, multiple scattering can also contribute to the observed signal.

The conceptual problems that multiple scattering introduce are illustrated in Fig. 8.18. The volume of the cloud that contributes to the backscatter alters considerably as the beam enters the cloud. Forward scattering acts to broaden the beam and more of the receiver field of view is now illuminated (region B). Region A represents the continuation of the original beam and is the volume in which single scattering predominates. The lidar detected backscatter signal from regions B and C only comes from higher order scatterings.

The relative contributions of regions A, B, and C to backscatter thus depend on the divergence of the transmitted beam and the receiver field of view. The multiple scattering contributions therefore

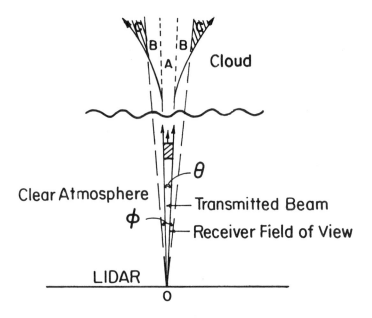

Figure 8.18 Schematic of the beam behavior in a turbid medium. Returns from area A are primarily from single scatter whereas returns from areas B and C arise from the multiple scattering of laser light (from Carswell, 1983).

depend on the distance of the cloud to the lidar. As a result of such factors, the simple lidar equation requires some type of modification for application to multiple scattering situations. For example, the simple R^{-2} dependence cannot always be assumed, the backscatter parameter β is not just simply related to number density and the attenuation of the pulse must account for the fact that photons lost from the original beam re–enter through multiple scattering. It is generally not easy to correct for multiple scattering, and a number of ad hoc methods have been proposed (a few of these are mentioned in the comments at the end of this chapter).

8.5   The Differential Absorption Lidar

The **differential absorption lidar** (DIAL) approach has moved to the forefront of lidar analysis methods. DIAL exploits the strong wavelength dependence of gaseous absorption in the atmosphere and requires either a tunable laser or a two–wavelength laser transmitter, where one wavelength is tuned to the peak (maximum) of an absorption line of the gas of interest and a second wavelength is tuned to a region of low absorption. Comparison of the backscatter at the absorbed wavelength to the backscatter at the weakly absorbed wavelength provides a direct measure of the concentration of the species. To illustrate the approach, consider the extinction as the following combination

$$\sigma_{ext} = \sigma_{ext,a} + Nk \tag{8.51}$$

One component is due to extinction by aerosol ($\sigma_{ext,a}$), and another is due to the molecular absorption. The volume absorption coefficient is expressed here as the product of the number concentration N of the absorbing molecule and an absorption coefficient k.

In principle, the concentration of the absorber gas may be obtained from the inversion of (8.39) using, for example, Klett's retrieval approach applied to measurements obtained at a single wavelength. However, this requires that $\beta, \sigma_{ext,a}$ and k be known accurately among the other factors in the lidar equation. The DIAL technique eliminates the need to specify these parameters by making measurements at two wavelengths, say λ_1 and λ_2, that are close enough that the differences in particle optical properties, namely $\sigma_{ext,a}$ and β, are small. Under this assumption, a ratio of two lidar equations produces

$$\ln\left(\frac{\mathcal{P}_1(R)}{\mathcal{P}_2(R)}\right) \approx -2\int_0^R N\{k_1 - k_2\}dr \tag{8.52}$$

where $\mathcal{P}_1(R)$ is the backscattered signal power received at range R normalized to the transmitted power (i.e., $\mathcal{P} = \mathcal{P}_r/\mathcal{P}_t$) at λ_1 and $\mathcal{P}_2$ is the equivalent ratio at λ_2.

DIAL can be applied either to CW systems to provide a measure of range integrated concentrations or to range resolved systems to obtain the concentration as a function of range. These concentrations $\bar{N}(R)$, averaged over the lidar volume extending from R to $R + \Delta R$, are

$$\bar{N}(R) = \frac{\ln\left(\frac{\mathcal{P}_1(R+\Delta R)}{\mathcal{P}_2(R+\Delta R)}\right) - \ln\left(\frac{\mathcal{P}_1(R)}{\mathcal{P}_2(R)}\right)}{2\Delta k\Delta R} \tag{8.53}$$

where $\Delta k = k_1 - k_2$.

There are some limitations of the DIAL approach that need to be considered in any given experiment. Most are related to how well the absorption coefficient is known for the particular species of interest. The width of the absorption line is also an issue, especially since the laser source has to be accurately tuned to the absorption line. These limitations are summarized in the review article of Grant (1991).

During the past 25 years, a number of DIAL systems have been developed primarily to measure water vapor, and a number of references to this general topic are included in the notes at the end of this chapter. An example of DIAL measurements of water vapor profiles is given in the upper panel of Plate 5 (refer to the front of the book). This diagram shows daytime profiles of DIAL water vapor measurements on an aircraft as it crossed the coastline of Virginia. The deeper boundary layer over land is reflected in a deep moist layer (to the left of the coast indicator at about 1445 LST) followed by a shallower moist boundary layer less than about 1-km deep over the water.

A topical application of DIAL is given in the work of Browell (1989), who uses DIAL to measure vertical profiles of ozone. In that study, Browell derives profiles of O_3 between approximately 10 and 20 km obtained from an upward–pointing airborne DIAL system which is configured to transmit simultaneously at 0.301, 0.311, 0.6 and 1.064 μm. These profiles were obtained as a function of the flight path inside and outside the Antarctic polar vortex. These DIAL measurements provide a unique map of ozone concentration with a vertical and horizontal resolution of 500 m and 60 km, respectively. The lower panel of plate 5 (refer to the front of the book) is an example of the derived ozone data obtained from a DC–8 flight over Antarctica on September 26, 1987. The reduction of ozone between 15 and 22 km poleward of about 62° S is clearly evident and compares well to correlative data obtained from TOMS. These DIAL data provided the first information on the large–scale vertical variability of ozone during the formation of the ozone hole. Browell also provides an analysis of the errors in these ozone retrievals which includes the effects of wavelength variations of both backscattering and extinction by aerosol and by Rayleigh scattering. Aspects of his error analysis are explored in Problem 8.12.

8.6   Doppler Wind Measurements

Perhaps the most common way of remotely sensing winds in the atmosphere is by measuring the motion of some sort of tracer. This tracer may be a natural component of the atmosphere. It may be as large as a cloud or as small as micron–sized aerosol or cloud particles. It may also be the size of an insect, or it may be meters in size as in the case of atmospheric refractive index fluctuations. Tracers may also be launched into the atmosphere, as in the example of chaff or balloons, to provide the necessary scattering target.

Another approach to the measurement of winds is the Doppler method that relies on the measurement of the shift in phase of an electromagnetic pulse scattered by a moving target. The principle of this approach was introduced in Chapter 2. We will now examine these principles in relation to three different Doppler systems: meteorological radar, lidar, and longer wavelength radars operating at UHF/VHF frequencies.

Meteorological Doppler radars, operating at wavelengths of 3–10 cm, typically observe the motion of hydrometeors, such as rain and snow. The operation of these radars in clear air is often marginal and generally requires the presence of other targets lifted by the wind, such as pollens, seeds, or insects. Radars operating at a variety of longer wavelengths between 30 cm and 6 m detect signals from refractive index fluctuations that arise from turbulent mixing of temperature and moisture variations in the clear air (known as Bragg scattering). We refer to these radars as UHF/VHF radars. Their long operating wavelengths require large antennas so it is difficult both to scan these instruments and measure wind fields near the surface because of significant side lobe effects. Doppler lidars complement both radar systems by operating in clear air like the UHF/VHF radars. The highly collimated nature of the lidar beam means that the instrument can be used to scan close to the ground and around complex terrain features, which offers advantages for observing fine structures in the wind field.

Excursus: Beating the Signal — Heterodyne Detection

The Doppler shifts associated with scattering from moving targets in the atmosphere are so small that they require special methods of detection. The method adopted by most systems is *heterodyne* detection. The technique of heterodyning is well known in radio detection and is referred to as "optical beating" when used to detect shifts at optical frequencies. The approach is to mix the weak returned signal with a stronger signal associated with another os-

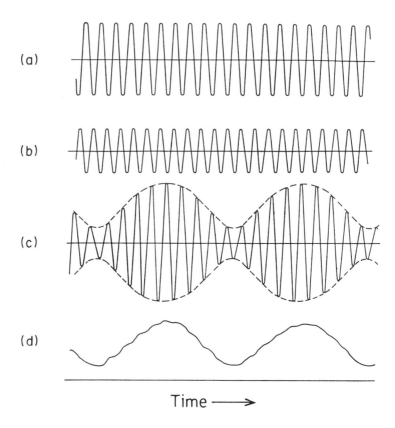

Figure 8.19 The heterodyning of two frequencies. The addition of (a) and (b) yields the waveform (c) which is rectified to form the beat signal (d) at a much lower frequency than either that of (a) or (b).

cillating source known as a local oscillator. The frequency of the latter is chosen to be close to that of the returned signal so that the frequency difference between the two is low enough that it can readily be separated from the frequencies of the original oscillators using commercially available low pass filters.

The general idea of heterodyne detection is illustrated in Fig. 8.19. Signals at the two input frequencies ν_1 and ν_2 shown in Figs. 8.19a and b are added to give the result in Fig. 8.19c which is a waveform amplitude modulated at a frequency $(\nu_1 - \nu_2)$ due to the alternating constructive and destructive interference (or beating) of the two signals. The more slowly varying part of this signal yields the output (Fig. 8.19d) which contains a nonvarying contribution (a D.C. level) and an output varying at the beat frequency $(\nu_1 - \nu_2)$.

This principle applies either to optical frequencies of lasers or to microwave frequencies of radars. We may observe the beating of the two mixed sources by measuring an output that is proportional to the intensity of the incident light. At time t these fields have the form

$$\mathcal{E} = \mathcal{E}_1 \cos(2\pi\nu_1 t + \phi_1)$$
$$\mathcal{E}' = \mathcal{E}_2 \cos(2\pi\nu_2 t + \phi_2)$$

where ν_1 and ν_2 are the frequencies of the oscillators and ϕ_1 and ϕ_2 are arbitrary phases. Since the output of the detector is proportional to the square of the total electric field, then

$$i(t) = B[\mathcal{E}_1 \cos(2\pi\nu_1 t + \phi_1) + \mathcal{E}_2 \cos(2\pi\nu_2 t + \phi_2)]^2$$

is the signal at the sensor where B is a constant. Simple expansion leads to

$$i(t) = B[\frac{1}{2}\left(\mathcal{E}_1^2 + \mathcal{E}_2^2\right) + \mathcal{E}_1\mathcal{E}_2 \cos(2\pi(\nu_1 - \nu_2)t + (\phi_1 - \phi_2))]$$

The first term in parenthesis is the D.C. term proportional to the total intensity and the second term contains the beat frequency $(\nu_1 - \nu_2)$ which is proportional to the product of the amplitudes $\mathcal{E}_1$ and $\mathcal{E}_2$ or $\sqrt{I_1 I_2}$, where I_1 and I_2 are the intensities of the two beams. The contributions to the detected signal by components at the original frequencies ν_1 and ν_2 are filtered from the detector and these have been omitted from the previous expression.

The beat frequency component of the detector is dependent on the product of the intensities of each of the beams. Thus, if we wish to detect a very weak Doppler shifted signal of intensity I_1, say, it is clearly advantageous to supply a "reference" beam I_2 of comparatively large amplitude. This has the advantage of providing a way of amplifying a weak Doppler signal.

8.6.1 General Principles of Doppler Wind Measurement

It follows from (2.22) that the radial velocity is related to the Doppler shifted frequency

$$\Delta\nu_D = \frac{2v_r}{\lambda} \tag{8.54}$$

where the sign[8] of the phase shift determines the direction of the particle motion relative to the receiver and where v_r is the radial

---

[8] The convention that $v_r > 0$ is used throughout for particles moving away from the observer. This is the convention most commonly used in Doppler radar and lidar literature, although the reverse convention is also often used.

velocity. The Doppler shifts in frequency are extremely small compared with the carrier frequency corresponding to the wavelength of the system (see Problem 2.8) and heterodyning detection methods are required to measure them. Whereas the Doppler lidar system has many similarities to the Doppler radar system, especially with regard to techniques for processing the signals, there are significant differences between the two. One of these differences is that radar measures the Doppler shift by observing the phase of the signal over a sequence of many pulses, whereas lidar measures the Doppler shift from a single pulse.

Estimating the frequency $\Delta\nu_D$ of an oscillating phenomenon requires measurement at two points in the cycle and thus at a frequency of at least $2\Delta\nu_D$. For example, if the oscillation was purely sinusoidal, and observations were made at the same frequency of this sinusoid, then all the observations of the signal would be of the same amplitude. Sampling at a rate $2\Delta\nu_D$ reveals alternating high and low values of the signal and thus reveals the existence of a frequency $\Delta\nu_D$. The rate at which pulses are transmitted — the PRF—defines the maximum possible frequency shift that can unambiguously be assigned to the velocity of the scatterer. Based on these considerations, the maximum Doppler shift must be

$$\Delta\nu_{D,max} = \frac{PRF}{2}$$

which corresponds to a maximum Doppler velocity of

$$v_{max} = PRF\frac{\lambda}{4} \tag{8.55a}$$

according to (8.54). The maximum detectable velocity is thus proportional to both the PRF and the wavelength of the system. If it is desirable to measure large velocities, then long wavelengths and high PRFs are required. For example, for $\lambda = 10$ cm and PRF $= 8000$ s^{-1}, it follows that $v_{max} = 200$ ms^{-1}. One way to interpret (8.55a) is that a Doppler system cannot measure the radial velocity unambiguously when the scatterer moves at high speed and more than the distance of one wavelength between pulses. Under these circumstances, the magnitude of the true velocity exceeds v_{max} and the velocity is said to be folded.

Velocity data can occasionally be unfolded but this is generally undesirable and can be avoided by selecting a high enough value of the PRF. Unfortunately, this creates other problems since the maximum Doppler shift and velocity are proportional to the PRF, it also follows that they are uniquely related to the maximum unambiguous range, R_{max}. This range simply follows as

$$R_{max} = \frac{1}{2}(c/PRF)$$

where c is the speed of propagation and substitution into (8.55a) leads to

$$v_{max} = \frac{\lambda}{8} \frac{c}{R_{max}} \tag{8.55b}$$

If a Doppler radar operates at 10 cm and v_{max} is 200 ms^{-1}, then the maximum range would be 18.7 km. Ideally, the PRF should be set low enough so that R_{max} can be set to the desired value. If the PRF is too large, then range folding occurs. Thus we are faced with a compromise — the PRF must be low enough to avoid range folding yet large enough to avoid velocity folding. This is sometimes known as the *Doppler dilemma*.

8.6.2 The Doppler Spectrum

The signal returned to the instrument contains information about the scattering cross–sections and radial velocities of the ensemble of particles in the volume. In processing the returned signal, it is convenient to present the backscattered power as a function of the Doppler shift frequency. Such a function is termed the *Doppler spectrum* and may be defined in the following way

$$\mathcal{P}_r = \int_{-\infty}^{\infty} S(\Delta\nu_D)d\Delta\nu_D$$

where $S(\Delta\nu_D)d\Delta\nu_D$ is the power contained in the interval of Doppler frequency between $\Delta\nu_D$ and $\Delta\nu_D + d\Delta\nu_D$. Because a simple transformation between $\Delta\nu_D$ and v_r follows from (8.54), we can also write

$$\mathcal{P}_r = \int_{-\infty}^{\infty} S(v_r)dv_r$$

Figure 8.20a provides a sketch of a typical Doppler spectrum and the various parameters used to characterize this spectrum. Two of these are the average Doppler shift $< \Delta\nu_D >$ and the spectral width w_D.

The width of the Doppler spectrum depends on a number of factors, some meteorological as described shortly, and others specific to the system under consideration. In the case of weather radar, the spread of the Doppler spectrum depends on the range of terminal velocities of raindrops, which in turn is determined by the spread of the raindrop sizes. This is graphically illustrated in Figs. 8.20b and c which respectively show the broadened spectra of rain and the narrower spectra associated with falling snow. The width of the Doppler spectrum is also determined by variations of wind across the backscattering volume, including the effects of turbulence and wind shear, and studies that attempts to correlate the width of the Doppler spectrum directly to wind shear and turbulence parameters are referenced at the end of this chapter.

8.7   Analyses of Doppler Measurements

8.7.1 Vertical Incidence

A vertically pointing Doppler radar or lidar observes the vertical component of the movement of the scatterers. This movement has two components. The first is due to the vertical movement of air, w, and the second is the terminal velocity v_t of the falling scatterer, so that

$$v_f = w + v_t$$

where the radial velocity is v_f is the fall velocity. Thus if v_t is known or inferred in some way, then w can be obtained. For example, if the measured Doppler velocity is 5 ms^{-1} for raindrops, and the terminal velocity of these drops is -5 ms^{-1}, then these drops exist in an updraft of 10 ms^{-1}. One method used in analyses of radar data is to employ a relationship between v_t and reflectivity Z of the form

$$< v_t >= aZ^b(\frac{p_o}{p})^{0.4}$$

where the pressure factor in parenthesis accounts for the increase in particle fall speed with height resulting from decreased drag. In this expression $< v_t >$ is a "reflectivity weighted terminal velocity"

$$< v_t >= \frac{\int_0^\infty v_t(D)n(D)D^6 dD}{\int_0^\infty n(D)D^6 dD}$$

Therefore if we estimate the terminal velocity from the reflectivity, then the atmospheric vertical velocity w follows as the difference of the measured velocity v_f and this estimated velocity v_t.

(a)

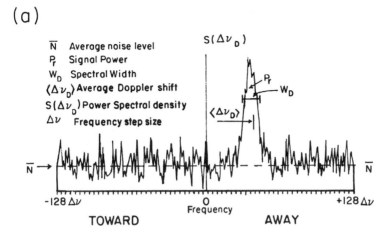

(b)

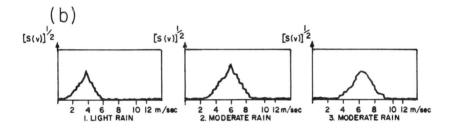

(c)

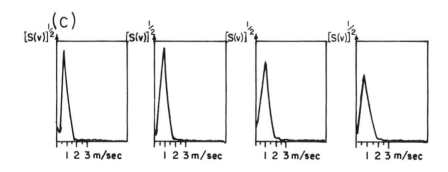

Figure 8.20 (a) A schematic of a Doppler spectrum showing power as a function of Doppler shift and Doppler velocity. The convention adopted here is such that a positive Doppler shift and positive Doppler velocities are produced when scatterers move away from the system. (b) The Doppler spectrum for rain obtained from a radar. (c) The Doppler spectrum of various snow situations [(b) and (c)] (from Battan, 1973).

In regions of stratiform precipitation, however, $|w| \ll |v_t|$, so $v_f \approx v_t$. In these regions, the measured Doppler spectrum then represents the spectrum $S(v_t)$. Since v_t is a function of drop diameter, this spectrum may be related to a spectrum $S(D)$ for drops in the size range D to $D + dD$. Assuming Rayleigh scattering for rain, then $S(D)$ is proportional to $D^6 n(D)$, and the measured velocity spectrum can be inverted, in principle, to obtain the drop distribution $n(D)$. However, small errors in the portion of the measured Doppler spectrum corresponding to small drop diameters leads to large errors in the retrieved size distribution because of the division by D^6 in the retrieval.

8.7.2 Scanning methods for Single Doppler systems

Doppler systems provide a measure of the velocity along a single direction, the radial direction. Measuring the full three–dimensional wind field, in principle, requires measurements of the same volume by three systems. This is generally not possible or practical, and most Doppler measurements are obtained from a single system or occasionally from two systems. We are thus forced to make some simplifying assumptions to derive the full wind field from these incomplete measurements. In using a single system, the usual approach is to scan the instrument in some way; a common mode of scanning is shown in Fig. 8.21a which is referred to as the Velocity–Azimuth–Display (VAD). The three–dimensional wind field may be obtained from such a scan provided some assumption about the wind field within the volume of the scan is made.

VAD Method–Uniform Wind Field

Perhaps the simplest approach is to assume that the horizontal component of the wind field is uniform throughout the area defined by a VAD scan at a given range. We also assume that the distribution of scatterers (precipitation for radar and aerosol for lidar) is uniform across this area. Under these conditions, the measured velocity is

$$v_r = v_h \cos(\phi - \phi_o) \cos \epsilon + v_f \sin \epsilon \qquad (8.56)$$

where ϕ is the azimuthal angle of the scan relative to a reference azimuth (say to the east or to the north), ϕ_o is the azimuth defining the direction of the wind, ϵ is the elevation angle, and v_h is the horizontal wind. We note that the fall velocity v_f of the particles is negative according to the convention used here unless the scatterer

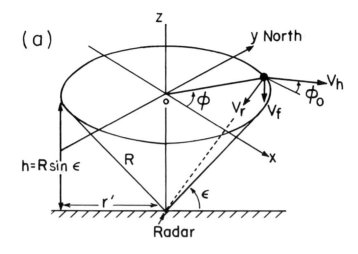

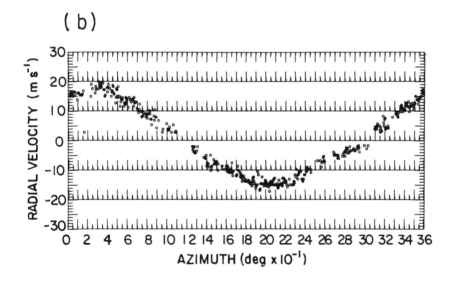

Figure 8.21 (a) The geometry of the VAD scanning mode. (b) An example of an actual VAD scan obtained by a radar at an elevation angle $\epsilon = 0.5°$ (Doviak and Zrnic, 1984).

is being lifted in an updraft. The quantity v_r is actually represented by a spectrum of velocities as discussed previously and the average of this spectrum is used in the VAD analyses. If the wind is uniform over the region being observed, then this mean Doppler velocity will vary sinusoidally with a maximum occurring upwind (i.e., $\phi - \phi_o = 0$) and the minimum occurring downwind (i.e., when $\phi - \phi_o = \pi$). The Doppler velocities at these extreme positions are

$$v_u = v_f \sin \epsilon + v_h \cos \epsilon$$

and

$$v_d = v_f \sin \epsilon - v_h \cos \epsilon$$

respectively. It follows that

$$v_h = \frac{v_u - v_d}{2 \cos \epsilon} \quad \text{and} \quad v_f = \frac{v_u + v_d}{2 \sin \epsilon}$$

When plotted in the VAD form of Fig. 8.21b, v_r has a constant, non-varying component equal to $v_f \sin \epsilon$ and the position of its extrema is defined by a phase shift ϕ_o relative to ϕ.

VAD Method–Linear Wind Field

A slightly more general technique for analyzing VAD data was introduced by Browning and Wexler (1968) for wind fields that vary almost linearly across the scan volume. In this case, the velocity components may be approximated by a Taylor series expansion of the form

$$u \approx u_o + \frac{\partial u}{\partial x} x + \frac{\partial u}{\partial y} y$$

$$v \approx v_o + \frac{\partial v}{\partial x} x + \frac{\partial v}{\partial y} y \qquad (8.57)$$

$$v_f \approx v_{fo} + \frac{\partial v_f}{\partial x} x + \frac{\partial v_f}{\partial y} y$$

where the subscript "o" denotes the value at the center of the circle being scanned. Since the Doppler velocity v_r is periodic in ϕ, it is thus appropriate to represent this velocity in the form of a Fourier series

$$v_r = \frac{1}{2} a_0 + \sum_{n=1}^{\infty} a_n \cos n\phi + b_n \sin n\phi \qquad (8.58)$$

We can establish the relation between the Fourier coefficients a_n and b_n and the wind field in the following way. First we expand (8.56) to

$$v_r = u \cos \phi \cos \epsilon + v \sin \phi \cos \epsilon + v_f \sin \epsilon \qquad (8.59)$$

where $u = v_h \cos \phi_o$ and $v = v_h \sin \phi_o$, and then substitute (8.57) into (8.59) for u and v using $x = R \cos \epsilon \cos \phi, y = R \cos \epsilon \sin \phi$. The final step is to equate each term of this expression to the Fourier series expansion, (8.58). Matching the first three Fourier coefficients, for example, gives

$$a_o = R \cos \epsilon \left(\frac{\partial u}{\partial x} + \frac{\partial v}{\partial y} \right) + v_{fo} \sin \epsilon$$

$$a_1 = u_o \cos \epsilon$$

$$b_1 = v_o \cos \epsilon \qquad (8.60)$$

$$a_2 = \frac{1}{2} R \cos^2 \epsilon \left(\frac{\partial u}{\partial x} - \frac{\partial v}{\partial y} \right)$$

$$b_2 = \frac{1}{2} R \cos^2 \epsilon \left(\frac{\partial u}{\partial y} + \frac{\partial v}{\partial x} \right)$$

The term in parenthesis in the expression for a_o is the two–dimensional divergence of the wind field, whereas those for a_2 and b_2 are the stretching deformation and the shearing deformation, respectively. Thus, the kinematic properties of the wind follow directly from the Fourier coefficients. The wind direction also follows as

$$\phi_o = \frac{2 \pm \pi}{2} - \frac{1}{2} \tan^{-1} \frac{a_1}{b_1}$$

where the $\pm$ is defined by the sign of b_1. In practice, the Fourier expansion coefficients are computed from measurements of v_r at set azimuth intervals (say at every 10 degrees) according to

$$a_0 = \frac{1}{18} \sum_{i=1}^{36} v_{r,i}$$

$$a_n = \frac{1}{18} \sum_{i=1}^{36} v_{r,i} \cos n \phi_i$$

$$b_n = \frac{1}{18} \sum_{i=1}^{36} v_{r,i} \sin n \phi_i$$

The Extended VAD Method

The VAD methods described earlier assume a fixed elevation angle ϵ. If the instrument carries out VAD scans through a series of elevation angles, then it is possible to obtain a more complete three–dimensional sample of the atmosphere. The extended VAD (EVAD) introduced by Srivastava et al. (1986) is such a method. The idea follows from the first of the coefficients given by (8.60) which is re-arranged in the form

$$\frac{2a_0}{r' \cos \epsilon} = DIV - 2v_{fo}\frac{h}{r'^2}$$

where h is the height of the scan point corresponding to a range R and elevation ϵ, r' is the radius of the scan at this height, and DIV denotes the term in parenthesis. When the divergence DIV and velocity v_{fo} are horizontally uniform and functions only of height, then plots of $2a_0/r' \cos \epsilon$ versus h/r'^2 for VAD scans at different elevations form straight lines. If this occurs with real data, then it is likely that these assumptions are reasonable and that both DIV and v_{fo} follow from the intercept and slope of this straight–line relationship, respectively. These conditions perhaps apply more readily to regions of stratiform precipitation where it is reasonable to assume that the air velocity is negligible compared to the terminal velocity of the falling particles. The quantities DIV and v_{fo} are typically determined for layers composed of contiguous thin–altitude slices. Vertical profiles of DIV and v_{fo} are then constructed by applying the EVAD approach to several of these layers.

8.7.3 PPI Signatures of Horizontally Uniform Winds

Vertical variations of both the wind speed and wind direction, even when these properties of the wind field are horizontally uniform, produce complex PPI displays of radial velocities. Some idea of these effects is given in Figs. 8.22a and b showing the VAD scan geometry and a PPI associated with this scan, respectively. As the radial distance increases from the scanning system, the height of the radar volume increases. Vertical shear of the wind field (either directional or speed shear or both) thus complicates the Doppler velocity pattern shown on the PPI. As an example, consider the wind field and its associated Doppler pattern shown in Figs. 8.22 c and d respectively. As the elevated beam passes upward through the atmosphere, the wind direction changes in this example linearly through

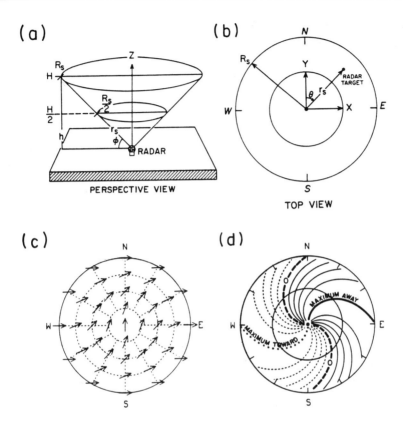

Figure 8.22 (a) The VAD geometry and (b) the associated PPI. (c) Plan view of wind field and (d) the corresponding Doppler velocity along an inverted conical surface defined by a constant elevation angle of the scan (Wood and Brown, 1986).

south–southwesterly to westerly at the edge of the display. The corresponding Doppler velocity measurements are shown in Fig. 8.22d. The solid lines are velocities away from the system, and the dashed lines indicate movement toward the system. The heavy dashed line is where the Doppler velocity is 0 (i.e., where the wind blows in a direction perpendicular to the beam of the radar or lidar).

Further examples of hypothetical PPI displays of Doppler velocity are given in Figs. 8.23 a–c. In Fig. 8.23a, the speed and direction of the wind are uniform at all heights as indicated by both the conventional wind symbols to the left and by the lower and upper boxes to the right of these symbols. In this example, the wind is from the southwest (225 degrees) at a constant speed. In Fig.

8.23b, the wind exhibits a backing with height up to the maximum height H associated with the maximum range of the scan. In this example, the zero Doppler velocity bisects the display in the form of an inverted "S." Since this zero contour corresponds to the wind flow perpendicular to the viewing direction of the instrument, this example indicates southerly flow at the surface backing with height to southeasterly flow at the edge of the display. Figure 8.23c is an example of a wind field of fixed direction (westerly) but with a wind speed that doubles from the ground to the maximum height at $H/2$. In this case, the contours of the Doppler velocity form a concentric oval–like pair.

8.7.4 Multiple Doppler Systems

A single Doppler radar or lidar measures the radial component of the moving scatterer but cannot give its complete three–dimensional velocity structure without the kind of assumptions mentioned earlier. If we employ Doppler systems at three different locations, then the three–dimensional wind field can, at least in principle, be determined. Over the years a number of schemes have been developed to combine velocity data from two or more Doppler radars to produce the wind fields surrounding convective storms.

The most common application is the dual–Doppler approach viewing the scattering volume from different directions. The precision of the measurement of velocity obtained from two radars depends on the angle formed by the direction of the two beams. This angle must not be too small. The minimum error occurs for beam crossing angles of 90 degrees and degrades beyond acceptability at approximately 30 degrees. The domain of measurements common to two radars is thus restricted to two approximately circular areas situated on each side of the line joining the radars (the *baseline*). Increasing the spatial coverage by multiple systems, therefore, is best achieved by forming an array of dual–Doppler pairs.

A dual–Doppler analysis may also be synthesized using data from only one radar and employing some type of time–to–space transformation assuming that the wind field is not evolving significantly. An example of this is found in the application of airborne Doppler radar to study atmospheric motion [e.g., Hildebrand and Mueller (1985); Ray et al. (1985) and others].

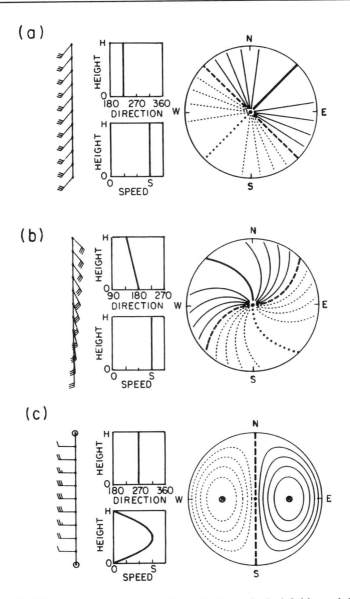

Figure 8.23 Vertical profiles of uniform horizontal wind fields as defined by conventional wind symbols (left) and by the wind speed and direction profiles (center boxes). The corresponding PPI of the Doppler winds are given to the right. (a) The wind blows uniformly at all heights from 225 degrees, (b) the wind speed is uniform with height but the direction backs from the south at the surface to southeast at H, and (c) the wind field has a uniform direction (from the west) but increase in speed to a maximum at height $H/2$ (Wood and Brown, 1986).

Figure 8.24 A typical beam configuration of a dipole array radar used for wind profiling. The most common configuration is one with three beams: one vertical and two tilted 15 degrees from the zenith (to the east and north). Some systems, such as that at Colorado State University, include two additional beams say to the south and west.

8.8   UHF/VHF Radars

Ultrahigh frequency (UHF) and very high frequency (VHF) radars receive echoes caused by scattering from regions of variable refractive index that, in turn, arise from variations in humidity and temperature in both the clear and cloudy atmosphere. These radars currently form a network over the United States to provide routine observations of winds in the atmosphere. The UHF and VHF radars typically operate from the low VHF band (30–300 MHz) to the UHF band (300–3000 MHz). This wide range of possible operating frequencies allows for a variety of different antenna configurations, from steerable dishes at higher frequencies to dipole arrays at lower frequencies — an example of the latter is shown schematically in Fig. 8.24.

8.8.1 Echoing Mechanisms

At the frequencies of UHF/VHF radars, the refractive index m of air in the troposphere and lower stratosphere depends on the pressure p, the temperature T and the partial pressure of water vapor e according to

$$(m - 1)10^6 \approx \frac{77.6}{T}(p + 4810\frac{e}{T})$$

where variations in both e and T produce variations in m which scatter radiation at the frequencies of UHF and VHF radars.

Scattering at the frequencies of UHF/VHF radar principally arises from two different mechanisms. The dominant mechanism responsible for these echoes in the lower atmosphere is turbulent scattering. Turbulent scatter arises from randomly distributed irregularities of refractive index which have dimensions comparable to half the wavelength of the radar. Thus, a typical 50 MHz VHF radar is sensitive to 3 m scale irregularities, a 400 MHz UHF profiler is sensitive to 38 cm scale irregularities, and a 3 GHZ microwave Doppler radar is sensitive to 5 cm scale irregularities. An important consideration for probing the free atmosphere is that the radar half–wavelength needs to be somewhat greater than the dissipation scale of turbulence (Tatarskii, 1971), which is typically a few centimeters in the free atmosphere. Consequently, radars operating at wavelengths greater than about 20 cm are needed to observe rfefractive index irregularities in the troposphere routinely.

The second mechanism that produces echoes at these frequencies is associated with Fresnel reflections that occur, for example, in stable layers or when the refractive index variations perpendicular to the beam are coherent in structure. These latter echoes are evident primarily at wavelengths longer than about 1m and, for backscattering, are largely confined to radar beams at vertical incidence. Figure 8.25 is a sketch of the pertinent refractive index structure as a function of range. In each panel two vertical profiles of radio refractive index encountered by two radar beams are shown. In panel A each beam illuminates a volume of turbulence and the structure of the radio refractive index is random both along the beam and across the beam. We deduce the latter from the fact that the two profiles of refractive index appear to be uncorrelated. In panel C a much different structure is illustrated; isolated sharp gradients that are transversely coherent are shown. These are associated with layers that cause a partial reflection of an incident radar pulse. In B

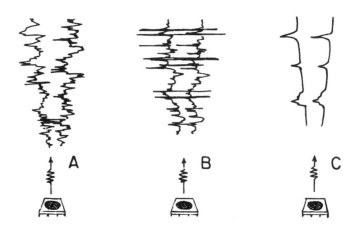

Figure 8.25 Depiction of the primary scattering mechanisms producing echoes at VHF and UHF. Description of the different panels is given in the text (Gage and Balsley, 1980).

an intermediate, and likely more realistic, example of atmospheric structure is illustrated. In this case the profile illustrates random-ness in the vertical but maintains some transverse coherence. This structure is thought to be pertinent to a multiple partial reflection process that is sometimes referred to as "Fresnel scattering".

The quasi–specular echoes associated with structures B and C are very persistent and are closely associated with stable regions of the free atmosphere. They can usually be seen most clearly, for example, just above the tropopause in the lower stratosphere. They can be identified by comparing the signal received from a vertically pointing VHF radar to one received when the antenna is pointed just off the vertical. Figure 8.26 compares the vertical and oblique power profiles from the Sunset radar located near Boulder, Colorado. Vertical profiles of the temperature and the stability derived from co-incident radiosonde data are also shown for comparison. The power profile at vertical incidence correlates with the stability index $\Delta\theta/\Delta z$ where θ is the potential temperature. A practical application of this enhanced specular reflection is given in Fig. 8.26b where the heights of these specular reflections are used to diagnose the height of the tropopause.

The schematic view of atmospheric structure contained in Fig. 8.26 is obviously an idealized abstraction of the refractivity structure of the real atmosphere. The complexities of the real atmosphere lead to many effects that are a challenge to understand theoretically. For example, partially reflecting layers will evolve with time and possess

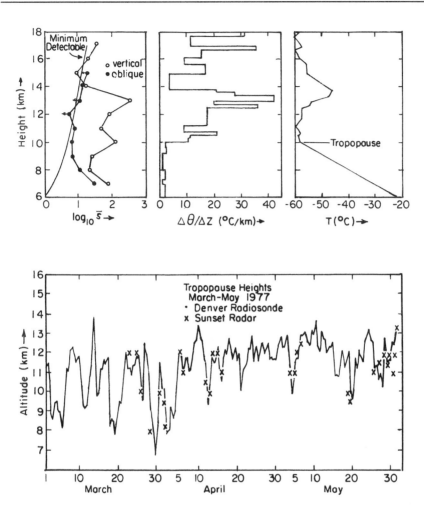

Figure 8.26 (a) Vertical profiles of the range squared normalized signal par-mater S observed by the Sunset radar near 0000 UT, March 26, 1977, (left), the vertical profile of potential temperature gradient (center) and the vertical profile of temperature (right) from coincident radiosonde data. (b) Tropopause heights near Denver, Colorado, for the period from March to May, 1977. The solid line is the height deduced from routine radiosonde data and the crosses are the heights deduced from the sunset radar (Gage and Green, 1978).

varying degrees of spatial (transverse) coherence. They will generally be tilted by internal wave motions causing the quasi–specular echoes to fade. Furthermore, the echoing medium appears different when probed by radio waves of different wavelengths. As a result of these and many other complications, a statistical approach is usually needed to account for the echoes observed by UHF/VHF radars.

8.8.2 Wind Measurements

UHF/VHF radars have found their greatest application in the meteorological community as profilers of wind (hence the term *wind profilers*). The method, in principle, is the same as that used in VAD analyses although a beam is not scanned. Wind profilers transmit multiple beams at fixed directions (such as illustrated in Fig. 8.25). At least three beams are needed to measure the three components of the wind field: typically one directed vertically, and two inclined at 15 degrees off the vertical directed to the east and to the north. According to the analyses of Clark et al. (1986) five beams are needed if additional information about the divergence is required. One of the major advantages of UHF/VHF radars as wind profilers is that they provide measurements of the wind field near continuously in time.

8.8.3 Precipitation Echoes

Even the long wavelength VHF radars are capable of detecting precipitation echoes. In early thunderstorm studies, Green et al. (1978) reported bimodal echoes seen by the Sunset radar that contained precipitation echoes as well as clear–air echoes. Since then, considerable research has been done on this topic (e.g., Fukao et al., 1985). It appears that information on the drop–size distributions as well as precipitation rates may also be contained in echoes from wind profiling radars (Wakasugi et al., 1987). An example of simultaneous clear–air and precipitation echoes seen at 400 MHz is shown in Fig. 8.27 obtained during the Microburst and Severe Thunderstorm (MIST) experiment by Prof J. Forbes and colleagues at Penn State during an episode of steady light rain.

8.9   Notes and Comments

A large body of literature already exists on the topic of the radar remote sensing of the atmosphere. The texts of Battan (1973) and Ulaby et al. (1982) are notable. A recent compendium, providing

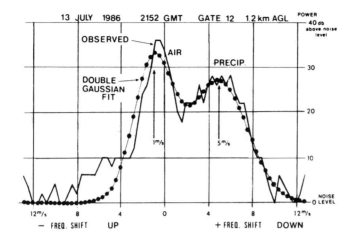

Figure 8.27 The Doppler spectrum showing the clear air and precipitation echoes during light rain. The data were obtained by the Penn State 400 MHz profiler (Gage, 1990). Note the different sign convention used in this Doppler spectrum display.

various historical perspectives of the topic as well extensive overviews of various aspects of conventional radar as well as UHF/VHF radar, is edited by Atlas (1990).

Lidar literature is not nearly as extensive as radar literature. A general text on laser sensing of the atmosphere is that of Measures (1984). A number of review articles discuss some general lidar capablities (e.g., Bilbro, 1980), or certain specific applications, such as Grant (1991) and Browell (1993) on the DIAL application, Sassen (1991) on polarization diversity measurements, Menzies and Hardesty (1989) on Doppler lidar applications, and She (1990) on the HSRL technique.

8.4. Equation (8.40) has an elementary structure, namely that of the Bernoulli or homogeneous Ricatti equation. The solution, in the form of (8.41), first appears in the context of the remote sensing of rain from radar at attenuating wavelengths (Hitchfield and Bordan, 1954).

In a series of papers, Platt (1979, 1981), and Platt and Dilley (1979, 1981) introduced a method for deriving the optical properties of clouds using a combination of lidar and infrared radiometer measurements (this method is referred to as the LIRAD method). The idea is to derive the infrared optical depth from the radiometer measurements, and to relate this to the optical depth at the wavelength of the ruby lidar. With these combinations of measurements, Platt (1979) claims to obtain the lidar backscatter profile and optical depth of cirrus clouds provided attenuation of the lidar beam is light. This procedure was introduced in an effort to derive the backscatter to extinction ratio from independent observations.

There is presently no simple radiative transfer model to account for the effects of multiple scattering in such a way as to provide a viable alternative to (8.35). A number of more approximate approaches have been employed as well as complex modeling efforts using Monte Carlo methods to aid in the interpretation of lidar backscatter (e.g., Platt, 1981). Despite the complexities introduced by multiple scattering, lidar has demonstrated real value in multiple scattering situations; it even provides unique information about the multiple scattered laser light. For example, the lidar system can be simply configured to measure only backscattered radiation that has undergone more than a single scatter event by using a focal plane stop to block detection of the backscatter from region A in Fig. 8.18. Detailed information about the nature of the multiply scattered returned radiation can be obtained by changing the configuration of these stops to block out differing amounts of the backscattered radiation (e.g., Ryan et al.,1979).

The HSRL system developed at the University of Wisconsin is described by Grund and Eloranta (1991) and has been applied to measure the optical depth of cirrus clouds in the study reported by Grund and Eloranta (1990).

8.5. The DIAL approach was first proposed by Schotland (1964) in an attempt to measure water vapor. Since then it has been used to measure the concentrations of NO_2, SO_2, and O_3. Citations to literature describing these measurements may be found in Browell et al. (1979) and Grant (1991).

8.6. Most of the early Doppler lidar measurements of wind have used $10\mu m$ CO_2 lasers since these offer high power and stable–frequency operation (e.g., Huffacker et al., 1970). A comprehensive review of work in this area up to 1980 can be found in Bilbro (1980). A later review of the topic is given by Menzies and Hardesty (1989).

8.8. To distinguish the UHF/VHF radar from the more conventional weather radar technique, the term clear air radar is used despite the fact that useful measurements can be made in both clouds and precipitation. This terminology evolved to distinguish the longer wavelength radars from the microwave radars that primarily observe hydrometeors. The distinction is artificial since microwave radars are also often capable of observing turbulence in the boundary layer. Even VHF radars can observe precipitation echoes on occasion. Moreover, UHF/VHF radars can observe the cloudy atmosphere without any difficulty at all. Nevertheless, the term clear–air radar has been widely used for several decades and is a more inclusive term than wind profiler, which focuses attention on only one measurement capability. Other terms in common usage by researchers include stratosphere–troposphere (ST) radar. Clearly, no single term is sufficient to capture all aspects of these radars, so they are referred to in this book simply in terms of their operating frequencies, namely as UHF/VHF radars.

Hardy and Gage (1990) provide an historical survey of clear–air radar measurements of the atmosphere and a brief discussion of the clear–air scattering mechanisms including turbulence and the effects of stable layers is provided by Gossard (1990) and Gage (1990).

8.10   Problems

8.1 Explain or interpret the following:
 a. Light particles spectrally broaden backscattering more than heavy particles.
 b. Comparison of Z and Z_{DR} is a useful way of identifying hail in precipitating systems.
 c. Convective cloud systems that possess deep regions of high radar reflectivity extending well above the melting level are often highly electrified.
 d. The average rain rate above some threshold is a well–defined function of the fractional area of rain exceeding this threshold.
 e. Doppler spectra of snow are typically narrower than spectra of rain.
 f. The DIAL technique works better when looking down from an aircraft than when pointed upward.

8.2. Given the characteristics of the CHILL radar listed in the following table and the following properties of a lidar system (wavelength=10.6 μm, pulse duration= 3 μsec, a beam divergence of 100 μrad, and a width of 20 cm), calculate the lidar and radar volumes at 30 km stating any assumptions.

8.3. A ground–based meteorological radar has the following characteristics: peak transmitted power $\mathcal{P}_t = 10^6$ W, pulse duration $t = 1$ μs, effective area=10 m^2, wavelength= 10 cm, and noise figure =10 dB. Determine the received power for a target cross–section [i.e., βdV in (8.8)] of 5 m^2 at distances of 50 and 200 km.

8.4. The WSR–57 and CSU–CHILL radars have characteristics described in the following table. Assume that a bumblebee can be approximated by a 2cm diameter sphere when in flight and that the dielectric constant of the bee is that of water at these wavelengths (i.e., $|K| = 0.93$). If a bumblebee scatters as a Rayleigh particle,

	WSR-57	**CHILL**
Band	S	S
Wavelength	10.3 cm	11 cm
Peak Power	500 kW	800 kW
PRF	658	658
Pulse Divergence	0.5 μ sec	0.5, 1.0 μ sec
Beamwidth	2°	0.96°
Beam type	Conical	Conical
Ant. Diameter	12 ft	28 ft
Reflector	Paraboloid	Paraboloid
Min. Signal	1.3×10^{-14}W	6×10^{-15}W

a. Estimate the radar cross–section of the bumblebee.

b. How far away can the WSR–57 and the CHILL detect a single bumblebee?

c. Suppose a swarm of bumblebees are distributed in the atmosphere at random, spaced 100 m apart on average. What is the radar reflectivity of the distribution?

d. Find the equivalent radar reflectivity Z_e for (c) for the two wavelengths.

8.5. Calculate the equivalent radar reflectivity factor Z_e for a target assuming the following distributions (for simplicity assume the radar volume is 1 m^{-3}):

a. A monodisperse raindrop distribution for raindrops with a diameter of 300 μm and with a concentration of 1 L^{-1}.

b. A Marshall–Palmer raindrop distribution

$$n(D)dD = \frac{N_o}{D_o}\exp(-D/D_o)dD$$

with $N_o = 1.4 \times 10^4$ m^{-3}, $D_o = 300$ μm.

c. A gamma raindrop distribution

$$n(D)dD = \frac{N_o}{\Gamma(3)}\left(\frac{D}{D_o}\right)^2 \frac{1}{D_o}\exp(-D/D_o)dD$$

with $N_o = 3.8 \times 10^4$ m^{-3}, $D_o = 100$ μm.

d. A monodisperse ice aggregate distribution with a diameter of 300 μm and a concentration of 1 L^{-1}.

8.6. Discuss the range–versus–Doppler ambiguity for the following cases:

a. A ship radar, 50 km range and 80 kmhr^{-1} maximum speed.

b. An airborne radar, 100 km range and 1000 kmhr^{-1} maximum speed.

c. A weather radar, 200 km range and a maximum speed specified by you.

8.7. A radar operating at a frequency of 1.2 GHz is used to track an aircraft to a distance of 300 km. If the maximum speed of the aircraft is 300 ms^{-1}, what is the unambiguous range? What is the maximum speed that can be detected for this range before folding occurs.

8.8. If the probability density function of radar reflectivity Z is determined to be

$$p(Z)dZ = \frac{\beta^{-\alpha}}{\Gamma(\alpha)}Z^{\alpha-1}\exp[-Z/\beta]dZ$$

a. Determine the expected rain rate $\Re$ given the following $Z - \Re$ relationship

$$\Re = aZ^b$$

b. Derive the slope S_0 between rain rate and fractional area as a function of rain rate threshold $\Re_0$.

8.9. The integral of (8.44) may be expressed as

$$S(R) - S(R_0) = n\ln\frac{\sigma_{ext}}{\sigma_{ext,o}} - 2\int_{R_0}^{R}\sigma_{ext}(r')dr'$$

Suppose that $S(R_0) = 1$ at $R_0 = 100$ m and that $n = 1$, then for the following extinction profile:

R (m)	10	20	30	40	50	60	70	80	90	100
σ_{ext} km^{-1}	10	10	10	15	20	20	20	15	10	10

calculate the profile of the signal function $S(R)$ assuming $S(R = 10\text{m}) = 1$. From this profile solve for the extinction σ_{ext} using (8.45) assuming $\sigma_{ext,o} = 9.9$ and 10.1 km^{-1}. Repeat this exercise using (8.46) together with $\sigma_{ext,m} = 15$ and 5 km^{-1} at $R_m = 100$ m.

8.10. Show that inversion of two lidar equations, one on (subscript 1) and one off (subscript 2) an ozone absorption line, yields

$$N_{O_3} = \frac{1}{2\Delta k} \left[-\frac{d[S_1(R) - S_2(R)]}{dr} + \frac{d\ln\beta_1/\beta_2}{dr} \right] - \frac{1}{\Delta k}\Delta\sigma_{ext}$$

for ozone concentration N_{O_3}. If the aerosol extinction has the form $\sigma_o\lambda^{-\alpha}$, estimate the uncertainty in ozone concentration as a function of σ_o due to the assumption that $\Delta\sigma_{ext} = 0$. Suppose that $\lambda_1 = 0.286$ nm, $\lambda_2 = 0.300$ nm, $\Delta k = 4.27 \times 10^{-3}$ ppbv–km^{-1}, $\alpha = 4$, and $\sigma_o = 0.12$ km^{-1}.

8.11. Using the data presented in Fig. 8.21b, deduce (1) the direction of the wind field, (2) the horizontal wind speed, and (3) the fall velocity of the scatterer, stating any assumption you make.

8.12. Consider Fig. 8.28;

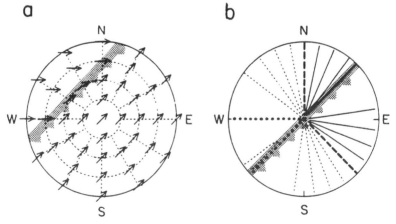

Figure 8.28 Wind vector fields in the vicinity of a surface front (from Wood and Brown, 1986).

a. The portion of this figure labeled (a) represents the horizontal wind vector fields in the vicinity of a surface front (shown as a PPI format). The location of the front is represented by the stippled area. Draw the corresponding Doppler velocity pattern at zero elevation angle.

b. From the PPI display of the Doppler velocity shown in Fig. 8.28 and labeled as b, qualitatively sketch the vertical profiles of wind speed and direction assuming that the wind field at any given level is uniform.

Appendix 1: Frames of Reference

A1.1 Necessary Bookkeeping: Coordinate Systems

Any meaningful discussion of multiple scattering needs to be set on a specific coordinate system. This is usually done using a cartesian coordinate system specified by three orthogonal axes, x, y, and z, and their corresponding unit vectors, $\hat{i}, \hat{j}$, and $\hat{k}$, respectively. Examples of two frames of reference commonly used are shown in Fig. A1.1. We refer to either one as a *terrestrial frame of reference*. These two systems are distinguished from each other by the way the x and z axes are anchored relative to each other. In both cases, the z axis is parallel to the local vertical but in one case z increases in a direction opposite to the direction of $\hat{k}$. Another special property of the two reference frames shown is that the x axis is aligned so the $x - z$ plane contains the sun. This is a special situation and such a frame is known as a *sun–based frame of reference*.

A general reference point within a cartesian frame of reference may be indicted by the position vector $\vec{r}$ such that

$$\vec{r} = (x, y, z)$$

where (x, y, z) defines the coordinates of the tip of this vector. We can also define a direction vector in terms of a unit position vector $\vec{\xi}$ which has its base at the origin and tip at the point (a, b, c) where this point lies on the unit sphere that surrounds the origin. In this case $\sqrt{a^2 + b^2 + c^2} = 1$. The unit direction vector may also be defined in terms of a general point (x, y, z) by

$$\vec{\xi} = \vec{r}/\mid r \mid = (x, y, z) / \left(x^2 + y^2 + z^2\right)^{\frac{1}{2}}$$

A more trigonometrical interpretation of the direction vector follows by considering Fig. A1.2a. For a point (a, b, c) on the unit sphere, it follows that

$$a = \vec{r} \cdot \hat{i} = \cos \phi \sin \theta$$
$$b = \vec{r} \cdot \hat{j} = \sin \phi \sin \theta$$
$$c = \vec{r} \cdot \hat{k} = \cos \theta = \mu$$

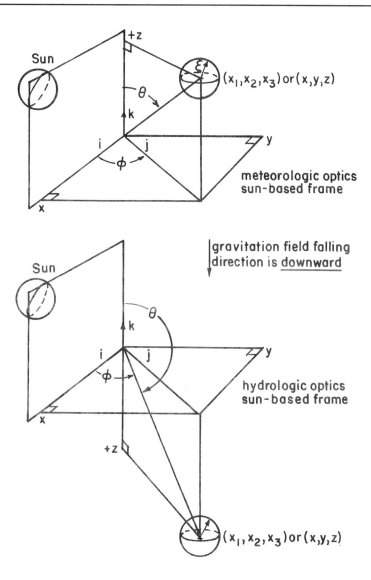

Figure A1.1 Two sun–based terrestrial frames of reference commonly used in studies of the scattering of radiation in the Earth's atmosphere.

where θ is the zenith angle and ϕ is the azimuth angle. The latter, in this case, is measured positive counter clockwise from the x axis. Since $\vec{\xi} = (a, b, c)$, then

$$\vec{\xi} = (\cos\phi\sin\theta, \sin\phi\sin\theta, \cos\theta) \qquad (A1.1)$$

are the three components of the direction vector. We often find it convenient to replace $\vec{\xi}$ with the angle pair (θ, ϕ) where the latter means (A1.1). We will also use $\mu = \cos\theta$ throughout the book.

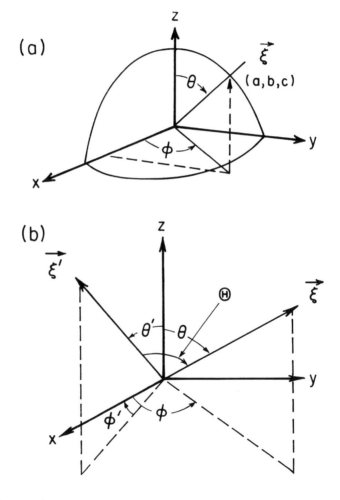

Figure A1.2 (a) Angle and direction definitions defined with respect to a unit sphere. (b) Scattering geometry and the scattering angle on the unit sphere.

A1.2   Scattering Angle

The scattering angle Θ is the angle between the direction of incident radiation and the direction of scattering. If the former direction is $\vec{\xi}$ and the scattering direction is $\vec{\xi}'$, then

$$\cos\Theta = \vec{\xi} \cdot \vec{\xi}' \qquad (A1.2)$$

We can schematically represent Θ and the two directions in question on a unit sphere (Fig. A1.2b). It follows from (A1.1) and (A1.2) that Θ can be stated in terms of the two pairs of angles (μ', ϕ') and (μ, ϕ)

$$\cos\Theta = \mu\mu' + \left(1 - \mu^2\right)^{\frac{1}{2}} \left(1 - \mu'^2\right)^{\frac{1}{2}} \cos(\phi' - \phi) \qquad (A1.3)$$

It is occasionally convenient to replace $\cos\Theta$ with the notation (μ, μ', ϕ, ϕ').

A1.3   Solid Angle and Hemispheric Integrals

Many radiation problems, particularly those dealing with scattering, require some type of integral over solid angle. A simple and convenient way to think about the solid angle is to imagine that a point source of light is located at the center of our unit sphere and that a small hole of area A exists on its surface allowing light to flow through it. This light is contained in a small cone of directions which is represented by the solid angle element

$$\Omega = \quad \text{area of opening on unit sphere} \ \Xi$$

that is

$$\Omega = \frac{A}{r^2}$$

where Ξ symbollically represents the unit sphere and where r is the radius of this unit sphere (i.e., $r = 1$). The area of the opening is then

$$r^2\Omega = r^2 \int_A d\Omega = r^2 \int\int_A da\,db = r^2 \int\int_A \sin\theta\,d\theta\,d\phi$$

Since $r = 1$, the solid angle element $d\Omega$, which has units of *steradian*, is related to θ and ϕ according to

$$d\Omega = \sin\theta\,d\theta\,d\phi$$

For example, the solid angle associated with all directions around a sphere has the same value in steradians as the surface area of a unit sphere, namely 4π.

Suppose we wish to integrate some function, like the intensity, over a complete hemisphere of directions. To fix ideas, consider the intensity flowing to some point on a horizontal surface from the hemisphere above it. The hemispheric integral of this intensity is then

$$h = \int_0^{2\pi} d\phi \int_{90°}^{0°} I(\theta, \phi) \sin\theta d\theta = \int_0^{2\pi} d\phi \int_0^1 I(\mu, \phi) d\mu$$

An important quantity in radiation studies is the hemispheric flux F defined as

$$F = \int_0^{2\pi} d\phi \int_0^1 I(\mu, \phi) \mu d\mu$$

Note how this quantity differs from h through the appearance of the factor μ in the integrand. The hemispheric flux defined in this way is a measure of the energy flowing through a horizontal surface per unit area and per unit time. Recall from Chapter 2 that the intensity is a measure of the energy flowing though a surface normal to the flow per unit area, per unit time, and per unit solid angle. The cosine factor, therefore, accounts for the projection onto a horizontal surface of the area normal to the flow of photons.

Appendix 2: Class Projects

The data and programs required to carryout the projects identified in the table are available via anonymous ftp on herschel.atmos.colostate .edu under directory class. Each project has its own directory and all programs and data are located in the directory. The following table summarizes the contents of these directories.

Directory	Files
project1	atmos.dat, cavitybb.dat, clearspc.dat, cloudspc.dat, coldbb1.dat, hotbb1.dat, febsond.dat, jansond.dat, planck.f
project2	mlo78jan.84, mlo78jan.87, pmrayl.f
project3	btemp90.jan, emiss.f, sst90.jan read.f, coef.f
project4	bismark.dat, hirslaba.f, hirslab.f hirst.out, hirtcol1.dat, mcatmmls.dat mcatmmlw.dat, mcatmtrp.dat
project5	vad.dat

A2.1 Interferometer Measurements of Atmospheric Emission

The CSU interferometer was used to measure the infrared spectra of the atmosphere. The project seeks to familiarize the student both with the workings of an interferometer and how it is calibrated. The project also seeks to make the student aware of general properties of the emission spectrum of the atmosphere. A project similar to the one now described is discussed in Stephens et al. (1993) in more detail than given here.

A2.1.1 Procedures

The CSU interferometer spectrometer is a commercial laboratory spectrometer manufactured by Bomen, Inc., that has been modified for measuring atmospheric emission spectra. The major modification concerns setting up a calibration procedure developed for these measurements. This involves a rotating mirror to select a cold blackbody source, a hot blackbody source, or the observation window. All spectra are actually measured relative to a separate reference hot blackbody (i.e., the interferograms are formed as the difference between the reference source and the target source). The instrument outputs interferograms (intensity vs. mirror distance) which are converted to raw spectra through a Fourier transform procedure. The raw spectra are then calibrated using corresponding cold and hot blackbody spectra.

Interferograms of (1) the cold blackbody, (2) the hot blackbody, (3) an intermediate temperature cavity blackbody, and (4) at least one atmospheric scan are provided for this project as well as sounding data of the atmosphere. The interferograms were converted to raw spectra and provided on the floppy disk. The raw spectra are given as functions of wavenumber extending from 200 cm^{-1} to 2500 cm^{-1}, and examples are shown in Fig. A2.1.

A2.1.2 Tasks

1. Explain the basic operation of a interferometer spectrometer using a simple diagram. Include enough detail to show why a Fourier transform is required to convert the interferograms to spectra. Describe the reciprocal relationship between resolution and range in the interferogram and the spectrogram.

2. Assuming a linear (slope/offset) intensity response of the instrument determine the calibration formula in terms of the cold and hot blackbody spectra. [*Hint*: Treat each wavenumber separately. The calibrated value of intensity at wavenumber $\tilde{\nu}$ is

$$I_{\tilde{\nu}} = \frac{C_{\tilde{\nu}}}{r_{\tilde{\nu}}} - I_{\tilde{\nu}}^o$$

where $r_{\tilde{\nu}}$ is the responsivity (inverse of slope), $C_{\tilde{\nu}}$ is the uncalibrated spectrum and $I_{\tilde{\nu}}^o$ is the offset emission of the intstrument.]

3. Plot the response curve (i.e., the slope of the calibration) as a function of wavenumber. Discuss the shape and characteristics of the spectral response curve. Explain the effects of detector sensitivity on the error level of the spectra.

4. Calibrate the cavity blackbody spectra and plot it along with the appropriate Planck function curve. Explain why the cavity

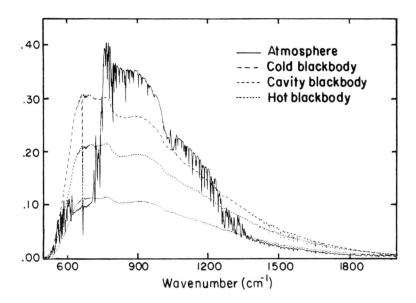

Figure A2.1 Raw, uncalibrated spectra obtained from the interferometer. The data displayed are contained in files **atmos.dat, cavitybb.dat, coldbb1.dat, hotbb1.dat**

blackbody spectrum is not completely smooth after it has been calibrated.

5. Calibrate and plot the atmospheric spectra. While the previous plots should be in radiance units $[\mathrm{Wm}^{-2}(\mathrm{cm}^{-1})^{-1}\mathrm{ster}^{-1}]$, plot the atmospheric spectra in both radiance units and brightness temperature. Indentify the major spectral features of the infrared spectra. Discuss the differences between the various atmospheric spectra (those you calibrated and those provided in files **cloudspc.dat** and **clearspc.dat**) in terms of possible different atmospheric conditions. Focus on the concepts of radiative transfer in your explanations of the spectral differences.

A2.1.3 Computing

All files are ASCII text with the numbers in columns separated by spaces. They may be easily read by a FORTRAN program using free format:

READ (1,*) X, Y

The first column of the spectra is wavenumbers (per centimeter); the second is the intensity at that wavenumber. The intensity values of the raw spectra are arbitrary (that is why you are calibrating them). The values of the comparison spectrum are in $[Wm^{-2}(cm^{-1})^{-1} ster^{-1}]$. The raw spectra go from 200 to 2500 cm^{-1} with a spacing of about 1.0 cm^{-1}. The comparison spectra range from 500 to 2000 cm^{-1} with a resolution of about 0.5 cm^{-1}. The comparison spectra were taken on January 14; the raw spectra on February 7. On January 14, the ceilometer recorded cloud base at 10000 feet MSL for the cloudy spectrum. The sounding data are self explanatory.

The blackbody temperatures for calibration are: first set; Tcold = -46.2C and Thot = 1.3 C; second set; Tcold = -51.0 C, Thot = -1.3 C. The cavity temperature is -17.9 C. Also assume the blackbodies have an emissivity of 0.99. The following files are located in directory **project1**:

File	Description
COLDBB1.DAT	Raw spectrum of cloud blackbody
HOTBB1.DAT	Raw spectrum of hot body
ATMOS.DAT	Raw clear sky spectrum (Feb)
CAVITYBB.DAT	Raw spectrum of cavity black body
CLOUDSPC.DAT	Calibrated cloudy spectrum (Jan)
CLEARSPC.DAT	Calibrated clear spectrum (Jan)
JANSOND.DAT	Atmospheric sounding for comparison day
FEBSOND.DAT	Atmospheric sounding for measurement day
PLANCK.F	Subroutine to calculate Planck's function

A2.2   Turbidity at Mauna Loa

The purpose of this project is to carry out an analysis of solar extinction data to derive information about atmospheric aerosol.

A2.2.1 Procedure and Tasks

Carry out the following tasks:

1. Plot the 1984 and 1987 data on separate Langley diagrams and determine the slopes and intercepts of these plots.

2. Using the subroutine in file **pmrayl.f**, determine the aerosol optical depth at three specified wavelengths.

3. Calculate the wavelength exponent α for the two periods and discuss any inferences you may make about the relative size of the aerosol measured, bearing in mind the time of the El Chicon eruption. You will note that this value of α depends on which particular pair of wavelengths you choose. Can you give physical account for why this is the case?

A2.2.2 Computing

The directory **project2** has three files, **pmrayl.f, mlo78JAN.84**, and **mlo78Jan.87**. The latter two files contain direct beam solar radiation measurements at 380 nm, 500 nm, and 788 nm for January 1984 and January 1987. The location of these measurements is Mauna Loa, Latitude 19.533N and Longitude 155.578W at an elevation of 3400 m. A sample of the data from file mlo78Jan.84 is:

Year	Day	Time GMT	V1 (380)	V2 (500)	V3 (778)	Air Mass Mass
84	7	17.9333	2.988	3.190	6.396	4.9435
84	7	17.9500	3.071	5.254	6.423	4.8611
84	7	17.9667	3.157	5.315	6.453	4.7808
84	7	17.9833	3.235	5.361	6.465	4.7037
84	7	18.0000	3.320	5.422	6.497	4.6293
84	7	18.0167	3.403	5.483	6.523	4.5566
84	7	18.0333	3.486	5.544	6.553	4.4868
84	7	18.0500	3.564	5.593	6.570	4.4187
84	7	18.0667	3.650	5.652	6.602	4.3533
84	7	18.0833	3.730	5.706	6.624	4.2898
84	7	18.1000	3.809	5.757	6.648	4.2278
84	7	18.1167	3.889	5.808	6.670	4.1681

File **pmrayl.f** contains a FORTRAN subroutine that calculates the sum of Rayleigh and ozone optical depths for the three wavelengths of the measurements given previously. You need to specify a surface pressure to calculate the Rayleigh optical depth. For this

```
      SUBROUTINE PMRAYL(IMYR,IMO,IDAY,CHNO,NCH,SPRES,RAYO3)

C     VERSION: MAY 1, 1990 - P. REDDY

C     THIS SUBROUTINE CALCULATES THE SUM OF THE RAYLEIGH OPTICAL
C     DEPTH AND THE OZONE OPTICAL DEPTH (AT 500 NM) FOR THE MLO
C     PMOD SUNPHOTOMETER.  THE ROUTINE IS CURRENTLY CONFIGURED TO
C     HANDLE 3 CHANNELS (380 NM, 500 NM, AND 778 NM).  THE CENTRAL
C     WAVELENGTHS AT THE NOMINAL VALUES CHANGE WITH TIME.  THE ROUTINE
C     DETERMINES THE SUM BASED ON THE WAVELENGTH VALID ON THE DAY AND
C     YEAR PASSED BY THE CALLING PROGRAM.

C     THE OZONE ABSORPTION VALUE AT 500 NM IS COMPUTED FROM THE
C     MONTHLY MEAN OZONE.
C     THE CORRECTIONS TO REFLECT YOUNG'S (1981) CORRECTION TO
C     ELTERMAN'S RAYLEIGH VALUES ARE +.003 AT 380 NM, -.002 AT 500 NM,
C     AND -.001 AT 778 NM.  THE DEFAULT STATION PRESSURE USED IS THE MEAN FOR
C     THE MLO OBSERVATORY FOR 1982-1988 AND IS 680.64.
C     THE VALUES FOR RAYO3 ERRONEOUSLY USED IN THE PMOD PROCESSING PROGRAM
C     PRIOR TO 2/16/1990 WERE .297, .1066, AND .016 FOR 380 NM, 500 NM,
C     AND 778 NM RESPECTIVELY
C
C*******************************************************************************
C     INTEGER VARIABLES:
C
C     IMYR - YEAR, PASSED BY CALLING PROGRAM.
C     IMO - MONTH PASSED BY CALLING PROGRAM.
C     IDAY - JULIAN DAY, PASSED BY CALLING PROGRAM.
C     JYR - YEAR INDEX FOR OZONE WHERE 1983 = 1, 1990 = 8.
C     NCH - NUMBER OF CHANNELS OR WAVELENGTHS, PASSED BY CALLING
C            PROGRAM.
C     CHNO - CURRENT CHANNEL NUMBER, PASSED BY CALLING PROGRAM.
C
C*******************************************************************************
C     REAL VARIABLES:
C
C     YR - THE DECIMAL YEAR.
C     RAYO3(CHNO) - THE SUM OF RAYL. OPTICAL DEPTH AND O3 ABSORPTION
C            PASSED BY PMRAYL.
C     SPRES - THE MEAN STATION PRESSURE FOR THE 2 HRS OR SO OF THE
C            LANGLEY PLOT.
C     O3 - OZONE OPTICAL DEPTH AT 500 NM.
C     OZONE(JYR,IMO) - MONTHLY MEAN MLO OZONE IN DOBSON UNITS.
C*******************************************************************************
```

Figure A2.2 Listing of the calling sequence and comments of the subroutine contained in file **pmrayl.f**.

purpose, assume the climatological value of surface pressure given in the subroutine. Figure A2.2 includes the comments and calling sequences and provides enough information for the user to run the program.

A2.3   Retrieval of Moisture Parameters Using SSM/I Data

The object of this exercise is to have the student become familiar to the topic of microwave remote sensing of column water vapor. These quantities are derived starting with the radiative transfer equation for a nonscattering atmosphere and arriving at the equations used by Greenwald et al. (1993). Results of this physical retrieval are to be compared to the results found by the statistical method of Alishouse (1990).

The data to be used in this project will be described in more detail shortly and will include SSM/I microwave brightness temperatures. The SSM/I makes observations at four frequencies in the microwave region (19, 22, 37, 85 GHz). Horizontal and vertical polarization measurements are made at 19, 37, and 85 GHz while 22 GHz has only vertical polarization measurements. Only the 19, 22, and 37 GHz data are provided in gridded, average form in the file **btemp90.jan**.

A2.3.1 Procedure and Tasks

We begin by carrying out the following tasks:
1. Use the integral form of the radiative transfer equation to show that the brightness temperature received by a satellite can be approximated by

$$T_b^{V,H} \approx T_s \mathcal{R}^{V,H} \mathcal{T}^2$$

where T_s represents the sea surface temperature (SST), $\mathcal{T}$ is the atmospheric transmission function, and $\mathcal{R}^{V,H}$ is the surface reflectivity with V and H indicating the vertical and horizontal polarization states. Assume, for the moment, that $\mathcal{T} \approx 1$ and that the atmosphere is isothermal.
2. Derive (7.25) given the preceding relation.
3. Using the following values, produce global maps of monthly precipitable water[1] and cloud liquid water along with the zonally averaged values for each quantity. Given knowledge of the basic general circulation of our planet, identify features which show both low and high values of precipitable water and cloud liquid water. How do you think these maps and figures would change for July 1991?
4. Compare the results given by this physical retrieval with the results from the statistical retrieval found in Alishouse (1990). What differences do you see between these two methods and why do you think these differences exist?

A2.3.2 Data Information

Two data sets are required. The first is the 1 degree latitude/longitude resolution data set which contains monthly averaged

---

[1] It is usual to carry out the retrievals on the pixel level and then grid and average the data accordingly. This is clearly impractical for such an exercise. Instead, the brightness temperature data have been gridded and averaged to produce monthly mean fields and the retrieval is then applied to these data.

brightness temperatures for 5 SSM/I channels. These data are contained in the file **btemp90.jan** and are formatted into five columns corresponding to a particular frequency and polarization. The first column is the 19H GHz channel followed by 19V, 22V, 37H, and 37V. Each row represents a grid box area with the first row corresponding to the area from 90S to 89S and 0E to 1E. These are followed by a further 359 rows corresponding to the grid box area defined by the next 1 degree longitude increment going eastward. Each set of 360 rows following the set then proceeds northward by 1 degree until 90N.

The second data set is the 2-degree latitude/longitude resolution data set containing sea surface temperatures in Celsius. The structure of the data is similar to the brightness temperature data with (1,1) representing the area from 90S to 88S and 0E to 2E.

It should be noted that all values of 999.99 in the brightness temperature data are areas of land, ice, or non oceanic areas which should be treated as missing values. The SST data do *not* have corresponding missing values, so the brightness temperature data should be used as a template to remove the areas of "missing" data. Also, areas north of 80N and south of 80S have been removed and replaced with values 999.99 since little information concerning water vapor can be extracted from the microwave in these regions. A sample FORTRAN read program is provided in file **read.f** to help you read both data sets. It is left to each individual to figure out ways of displaying the data since this will depend on the facilities available to the instructor.

A2.3.3 Computing Comments

In addition to the data, two subroutines are supplied to assist in the retrievals. The program **read.f** in directory **project3** is an example of how you can read both **btemp90.jan** and **sst90.jan**. You will also need to retrieve the surface wind speed in order to derive the surface emissitivity and hence surface reflectivity. For this purpose, we employ the statistical model of Goodberlet et al. (1989) which is simple. [The student should consult this reference to see how well the retrieval works compared to actual surface wind data]. This retrieval reduces to

SPEED=1.0969(T19V) - 0.4555(T22V) - 1.76(T37V) +
　　　　0.786(T37H) + 147.9

where SPEED is the wind speed in meters per second. A quality control check to ensure the wind speed is > 0 is recommended.

Once the wind speed is established, the surface emissivity is derived from the subroutine EMISS provided in file **emiss.f**. The water vapor and liquid water absorption coefficients as well as the oxygen transmission are provided in the subroutine COEF in file **coef.f**. This requires the SST as input. The view angle of the SSM/I satellite (F8) is approximately 53.1 degrees.

A2.4 HIRS Temperature Sounding Project

The purpose of this project is to acquaint the student with the general concepts of weighting functions and temperature profile retrievals. This is a sophisticated project but one that is nontheless rewarding. The student will program algorithms and generate and plot the results. A FORTRAN template is provided containing all the subroutine and data needed. The most relevant subroutine is the one that computes the transmittance to each of the 40 standard TOVS pressure levels for a given HIRS channel. The program that does this is used for TOVS radiative transfer calculations as part of the International TOVS Processing Package (ITPP) developed by CIMSS/NESDIS at the University of Wisconsin–Madison.

Table A2.1 provides various kinds of information about the channels of the HIRS. The code that calculates the transmission functions applies to the HIRS channels on NOAA-11 and channels 10 and 17 differ slightly from the wavelengths of these channels that are given in Table A2.1.

A2.4.1 Weighting Functions

The first exercise is to compute weighting functions for the 15 μm temperature sounding channels $i = 1, \ldots, 7$ and the 7 μm water vapor sounding channels $i = 10, 11, 12$. Use log pressure as the vertical coordinate so that

$$W_i(p, 0.1\mathrm{mb}) = \frac{d\mathcal{T}_i(p, 0.1\mathrm{mb})}{d\log p}$$

Normalize the weighting functions so their peaks are equal to one, and plot channels 1–7 on one plot and 10–12 on another. There is no need to plot the weighting functions all the way to the highest TOVS level (0.1 mb). Make one set of plots for the tropical McClatchey atmosphere (data contained in file **mcatmtrp.dat**) and

Table A2.1 Characteristics of the 20 HIRS channels

Channel Number	Channel wavelength (μm)	Principal Absorbing Species	Level of Peak Energy Contribution
1	15.00	CO_2	30 mb
2	14.70	CO_2	60 mb
3	14.50	CO_2	100 mb
4	14.20	CO_2	400 mb
5	14.00	CO_2	600 mb
6	13.70	CO_2/H_2O	800 mb
7	13.40	CO_2/H_2O	900 mb
8	11.10	Window	Surface
9	9.70	O_3	25 mb
10	8.30	H_2O	900 mb
11	7.30	H_2O	700 mb
12	6.70	H_2O	500 mb
13	4.57	N_2O	1000 mb
14	4.52	N_2O	950 mb
15	4.46	CO_2/N_2O	700 mb
16	4.40	CO_2/N_2O	400 mb
17	4.24	CO_2	5 mb
18	4.00	Window	Surface
19	3.70	Window	Surface
20	0.70	Window	Cloud

This table is from Smith et al. (1979); the central frequencies of channels 10 and 17 applicable to calculations relevant to this project are 12.6 μm and 4.14 μm respectively.

one for the midlatitude winter atmosphere (in file **mcatmmlw.dat**). Compute and print out the upwelling brightness temperatures for all 19 HIRS infrared channels for the tropical standard atmosphere. File **hirst.out** provides the output for these calculations obtained from program **hirslaba.f**.

A2.4.2 Direct Inversion

The second exercise is to perform a temperature profile retrieval by a simple direct method. The purpose of this is to demonstrate the complexities and pitfalls of temperature sounding inversion. If we assume that the channels of the radiometer are spaced closely enough so the Planck function can be considered constant, and that the tranmittance is independent of temperature, then the radiative transfer equation may be written as

$$I_i = T_i(z_s, \infty) B_i(z_s) + \int_{z_s}^{\infty} B_i(z) \frac{dT_i(z, \infty)}{dz} dz$$

where $z = -\log(p)$ is the height coordinate, z_s is the surface, $B_i(z)$ is the Planck function relevant to channel i (take this to be the central frequencies listed on Table A2.1), I_i is the "measured" intensity in the ith channel, and $T_i(z, \infty)$ is the transmittance (from 0.1 mb) to level z for the ith channel. This equation is linear in $B_i(z)$, which is the profile solved for and from which the temperature is derived. The approach we use for the direct method is to express the Planck function profile as a weighted sum over some profile expansion functions $P_j(z)$:

$$B_i(z) = \sum_{j=1}^{N} b_j P_j(z)$$

This reduces the problem to finding N unknowns for M channels ($N \leq M$). Combining the two equations gives:

$$I_i = \sum_{j=1}^{N} b_j \left\{ T_i(z_s, \infty) P_j(z_s) + \int_{z_s}^{\infty} P_j(z) \frac{dT_i(z, \infty)}{dz} dz \right\}$$

$$= \sum_{j=1}^{N} A_{ij} b_j \quad i = 1, \ldots, M$$

The A_{ij}s are found by numerically integrating over z. The linear system is solved using a least squares solver for the b_js, and the Planck function profile and hence the temperature profile are computed by summing over the b_js.

The profile expansion functions used here are linear ramps so that the profile is a linear interpolation between N levels. Polynomials or sine functions could also be used but they are less realistic for a small number of functions. You will perform some retrievals of temperature profiles from radiances computed for HIRS channels 1–7 from the McClatchey tropical atmosphere. This problem, of course, is easier than the real-sounding problem because we are assuming that we can compute the transmission functions exactly, and that the water vapor and ozone profiles are known exactly. The template code in files **hirslab.f** and **hirslaba.f** computes the discrete version of the P_j functions; use the $N = 7$ functions specified in the code. You will need to compute the A_{ij}s from the transmittances from the tropical profile. Using the linear least squares solver compute the profile coefficients, and then the temperature profile. From the retrieved temperature profile compute the upwelling brightness temperatures and compare with the the radiances input to the direct method. Do the retrieval with and without noise added to the simulated radiances. If you wish you can try other numbers of profile functions $N < 7$. Make plots (T vs. $\log p$) comparing the "true" standard profile with the retrieved one up to 10 mb.

A2.4.3 Iteractive Method of Inversion

Direct inversion methods are not suitable for retrieving temperature profiles because of instabilities and because actual temperature profiles have more complex structure than can be described by simple low order expansions. Because simple direct methods are ill-conditioned, additional information about real temperature profiles must be introduced. This is often done using statistical methods, such as regression or the statistical eigenvector approach of Smith and Woolf (1976), which we will not consider here. Another way of introducing additional information is through the first guess of an iterative method. One such example of a "physical" temperature sounding method is that of Smith (1970), which is one we will now explore.

1. For the $n + 1$st iteration, compute a new Planck function profile for each channel from the difference between the observed intensity I_i and the computed intensity $I^{(n)}$ of the nth iteration (I^0 is the initial guess),

$$\mathcal{B}_i[T_i^{(n+1)}(p)] = \mathcal{B}_i[T^{(n)}(p)] + [I_i - I_i^{(n)}]$$

2. Associated with each channel, i, is a temperature profile, $T_i(p)$, which is found by the inverting the Planck function on the left-hand side. Compute a new temperature from a weighted average

of the profiles for each channel

$$T^{(n+1)}(p) = \sum_{i=1}^{M} T_i^{(n+1)}(p)w_i(p) / \sum_{i=1}^{M} w_i(p)$$

where the weighting factors can be approximated by

$$w_i(p) = \begin{cases} W_i(p, 0.1), & p < p_s \\ T_i(p, 0.1), & p = p_s \end{cases}$$

3. Using this new profile, compute the new intensities $I_i^{(n+1)}$ and repeat the steps.
4. Continue until the maximum difference between the observed and computed radiances is less than some small value.

So that you will know the "true" temperature profile do as in Section A2.4.1 and simulate the "observed" radiances for channels 1–7. For this use the Bismark Sea sounding (**bismark.dat**) which was made by combining a real sounding below 19 km with a modified tropical McClatchey sounding. For this exercise, however, compute the weighting factor, $w_i(p)$, and initialize the iterations with standard profiles. Used both the tropical (**mcatmtrp.dat**) and midlatitude summer (**mcatmmls.dat**) profiles for the initialization to see the importance of the first guess profile. Make plots of the difference in temperature profiles, (i.e., initial-truth and retrieved-truth).

A2.4.4 Questions

1. What is the advantage of using 4 μm channels for temperature sounding compared to channels at 15 μm? [Hint: Consider properties of the Planck function].
2. Why do the weighting functions peak at different heights?
3. Why do the weighting function peaks of channels 10–12 shift greatly between the tropical and midlatitude winter soundings?
4. Weighting functions used for temperature sounding ideally should vary little for different profiles. How are the tropical and midlatitude winter weighting functions different for channels 1–7? Can you explain why?
5. Interpret the general features of the 19 HIRS upwelling brightness temperatures in terms of the vertical profile of temperature and the gaseous absorption spectrum (note the results for channels 9 and 17 are spurious).

6. For the direct temperature sounding method why are the up-welling brightness temperatures computed from the profile different from those input?

7. For the direct retrieval method, why is the sounding so much worse for the case with added noise?

8. Why would a scheme that retrieved the water vapor profile in addition to the temperature profile probably improve the accuracy of 15 μm temperature sounding?

A2.4.5 Computing Comments

The programs and data are located in directory **project4**. The main FORTRAN program as it could be supplied to the student is contained in file **hirslab.f**. The file **hirslaba.f** is similar but contains additional coding that is required to carry out the entire project. The comments in these program tell how to use the subroutines and give guidance on what to do. Files of the form **mcatm*.dat** contain the McClatchey standard atmosphere data. The file **bismark.dat** contains an actual radiosonde from the Bismark Sea north of New Guinea. The file **hirtco11.dat** contains relevant data to calculate the transmission functions relevant to HIRS on NOAA 11.

A2.4.6 Additional References

The following addtional references can be used to help clarify aspects of this project: Houghton et al. (1984), Chapter 7; Liou (1980), Chapter 7; Smith (1970); and Smith and Woolf (1976).

A2.5 Doppler Velocity Project

The object of this project is for students to learn how to carry out rudimentary analyses of Doppler radar data. The radar data in file **vad.dat** located in directory **project5** were obtained from the NCAR CP-3. An analysis of the convective system in question and a further description of the radar data collected is described by Rutledge et al. (1988). The data for this project were obtained at 0345Z and correspond to observations made in the trailing stratiform region of the mesoscale complex (Fig. 10 of Rutledge et al. provides a plan view of reflectivity at 0334Z).

A2.5.1 Procedures

Doppler radar VAD data have been processed for 18 elevation angles ϵ. For each elevation angle, the azimuthal angle and radial velocity

(v_r) (per millisecond) is recorded (azimuth is measured from the north). These data are available in the file **vad.dat** and a portion of the data is shown below for the elevation angle of 0.20 degrees. Each VAD scan corresponds to a full azimuth scan and the data are processed so that each scan represents a ring of data on a cylinder of radius $R = 20$ km.

A2.5.2 Tasks

The student is asked to derive the wind field (speed, direction, divergence, and vertical wind from the anelastic form of the continuity equation) from data collected from a pulsed Doppler radar which scans the atmosphere using the velocity-azimuth display (VAD) technique. The approach is to employ the EVAD technique described in Chapter 8. The student should consider applying this technique to perhaps three or four broad layers. You should compare your analyzed wind fields and profiles of divergence to the results published by Rutledge et al. (1988, Fig. 9 of that paper).

References

Abromowitz, M., and I. Stegun, 1971: *Handbook of Mathematical Functions.* Dover, New York, 1046 pp.

Adler, R. F., and E. B. Rodgers, 1977: Satellite-observed latent heat release in a tropical cyclone. *Mon. Wea. Rev.*, **105**, 956-963.

Adler, R. F., R. A. Mack, N. Prasad, H–Y. M. Yeh, and I. M. Hakkarinen, 1990: Aircraft microwave observations and simulations of deep convection from 18 to 183 GHz. Part I: Observations. *J. Atmos. Oceanic Tech*, **7**, 377–391.

Adler, R. F., H–Y. M. Yeh N. Prasad, W–K. Tao, and J. Simpson, 1991: Microwave simulations of a tropical rainfall system with a three–dimensional cloud model. *J. Appl. Meteorol.*, **30**, 924–925.

d'Alemeida, G. A., P. Koepke, and E. P. Shettle, 1991: *Atmospheric Aerosols: Global Climatology and Radiative Characteristics.* A. Deepak Publ., Hampton, Va., 561pp.

Alishouse J. C., S. A. Snyder, J. Vongsathorn, and R. R. Ferraro, 1990: Determination of oceanic total precipitable water from the SSM/I. *IEEE Trans. Geosci. Remote Sensing*, **28**, 811–816.

Al–Khatib, H. H., T. A. Seliga, and V. N. Bringi, Differential Reflectivity and its use in the radar measurement of rainfall: Atmos. Sci. Prog. Rep. No. AS–S–106. Ohio State University, Columbus.

Allen, R. J., and C.M.R. Platt, 1977: Lidar for multiple scattering and depolarization observations. *Appl. Opt.*, **16**, 3193–3199.

Andreae, M.O., 1985: Dimethyl sulfide in the water column and sediment pore waters of the Peru upwelling area. *Limnol. Oceanogr.*, **30**, 1208–1218.

Ångström, A. K., 1929: On the atmospheric transmission of Sun radiation and on dust in the air. *Geogr. Ann.*, **11**, 156–166.

Arfken, G., 1985: *Mathematical Methods for Physicists.* Academic Press, Orlando, Fl., 985 pp.

Arkin, P. A., 1979: The relationship between fractional coverage of high cloud and rainfall accumulated during GATE over the B–scale array. *Mon. Wea. Rev.*, **107**, 1382–1387.

Arkin, P. A., and B. N. Meisner, 1987: The relationship between large–scale convective rainfall and cold cloud over the western hemisphere during 1982–1984, *Mon. Wea. Rev.*, **115**, 51–74.

Arkin, P. A., and P. E. Ardanuy, 1989: Estimating climatic–scale precipitation from Space: a review, *J. Climate*, **2**, 1229–1238.

Arking, A., and J. D. Childs, 1985: Retrieval of cloud cover parameters from multispectral satellite measurements. *J. Clim. Appl. Meteorol.*, 24, 322–333.

Atlas, D., (ed), 1990: Radar Meteorology: Battan Memorial and 40th Anniversary Radar Met. Conference, American Meteorological Society, Boston, 806pp.

Atlas, D., and C. W. Ulbrich, 1974: The physical basis for attenuation rainfall relationships and the measurement of rainfall parameters by combined attenuation and radar methods. *Atmos. Research*, **8**, 275–298.

Atlas, D., and T. L. Bell, 1992: The relation of radar to cloud area–time integrals and implications for rain measurements from space, *Mon. Wea. Rev.*, **120**, 1997–2008.

Atlas, D., D. Rosenfield, and D. A. Short, 1990: The estimation of convective rainfall by area integrals, Part I: The empirical basis, *J. Geophys. Res.*, **95**, 2153–2160. (8)

Bacastow, R. B., C. D. Keeling, and T. P. Whorf, 1985: Seasonal amplitude increase in the atmosphere CO_2 concentration at Mauna Loa, Hawaii, 1959–1982. *J. Geophys. Res.*, **90**, 10529–.

Baggot J., 1992: *The Meaning of Quantum Theory*, Oxford University Press, Oxford, 230 pp.

Baker, D. J., 1990: *Planet Earth: The View from Space.*, Harvard University Press, Cambridge, Ma., 191 pp. (2nd Edition.)

Banwell, C. N., 1983: *Fundamentals of Molecular Spectroscopy*, McGraw–Hill, (U.K.), 338 pp.

Barkstrom, B., and G. L. Smith, 1986: The earth radiation budget experiment: science and implementation. *Rev. Geophys.*, **24**, 379–390.

Barton, I. J., and J. C. Scott, 1986: Remote measurements of surface pressure using the oxygen A–band of absorption. *Appl. Opt.,* **25**, 3502–3507.

Basharinov, A. Y., A. S. Gurvich, and S. T. Yegorov, 1969: Determination of geophysical parameters from data on thermally induced radioemission obtained with the Cosmos 243 satellite. *Dokl. Akad. Nauk. SSSR.,* **188**, 1273–1276.

Battan, L. J., 1973: *Radar Observations of the Atmosphere,* University of Chicago Press, Chicago, 324 pp.

Becker, G. E. and Autler, S. H. (1946): Water vapor absorption of electromagnetic radiation in centimeter wavelength range., *Phys. Rev.,* 70, 300–307.

Bilbro, J. W., 1980: Atmospheric laser Doppler velocimetry: an overview. *Opt Engineer.,* **19**, 533–542.

Bohren, C., 1987: *Clouds in a Glass of Beer.* Wiley, New York, 195 pp.

Bohren, C., and D. R. Huffman, 1983: *Absorption and Scattering of Light by Small Particles.* Wiley, New York, 530 pp.

Bohren, C., and T. J. Nevitt, 1983: Absorption by a sphere: a simple approximation. *Appl. Opt.,* 22, 774–775.

Bohren, C., and S. B. Singham, 1991: Backscattering by non-spherical particles: a review of methods and suggested new approaches, *J. Geophys. Res.,* **96**, 5269–5277

Born, M., 1965: *Optik,* 1933: Springer, Berlin, Heidelberg, New York.

Born, M., and E. Wolf, 1964: *Principles of Optics, 2nd Edition,* Macmillan, New York, ...pp.

Box, M. A., and B.H.J. McKellar, 1976: Determination of moments of the size distribution function in scattering by polydispersion. *Appl. Opt.,* **15**, 2610

Brest, C. L., and W. B. Rossow, 1989: Radiometric calibration and monitoring of NOAA AVHRR data for ISCCP. *Int. J. Remote Sensing,* **13**, 235–273.

Bricaud, U. A., and A. Morel, 1986: Light attenuations and scattering by phyplanktonic cells: a theoretical modelling. *Appl. Opt.,* 25(4), 571–580.

Browell, E. V., 1989: Differential Absorption lidar sensing of ozone, *Proceed. IEEE*, **77**, 419–432.

Browell, E. V., 1993: Remote sensing of trace gases from satellite and aircraft. In *Chemistry of the Atmosphere: The Impact on Global Change*, Blackwell Pub.

Browell, E. V., T. D. Wilkerson, and T. J. McIlrath, 1979: Water vapor differential absorption lidar development and evaluation. *Appl. Opt.*, **18**, 3474–3483.

Browning K. A., 1990: Rain, rainclouds and climate. *Q. J. R. Meteorol. Soc.*, **116**, 1025–1051.

Browning, K. A., and R. Wexler, 1968: The determination of kinetic properties of a wind field using doppler radar. *J. Appl. Meteorol.*, **7**, 105–113.

Buettner, K. J. K., 1963: Regenortung vom wettersatelliten mit hilfe von zentimeterwellen (Rain localization from a weather satellite via centimeter waves). *Naturwiss.*, **50**, 591.

Burroughs, W. J., 1979: The water dimer: a meteorologically important molecular species. *Weather*, 34, 233–236.

Carswell, A. J., 1973: Polarization properties of lidar backscatter from clouds. *Appl. Opt.*, **12**, 1530–1535.

Carswell, A. J., 1983: Lidar measurements of the atmosphere. *Can. J. Phys.*, **61**, 378–395

Chandrashekar, S., 1960: *Radiative Transfer*, Dover, 393 pp.

Chen, H. S. 1985: *Space Remote Sensing Systems: An Introduction*, Academic, Orlando, Fl., 257 pp.

Chu, W. P., and M. P. McCormick, 1979: Inversion of stratospheric aerosol and gaseous constituents from spacecraft solar extinction data in the 0.38–1.0 μm wavelength region. *Appl. Opt.*, **18**, 1404–1414.

Chu, W. P., M. P. McCormick, J. Lenoble, C. Broniez, and P. Pruvost, 1989: SAGE II Inversion Algorithm, *J. Geophys. Res.*, **94**, 8339–8351.

Clark, W. L., J. L. Green, and J. M. Warnock, 1986: The use of vertical beam clear-air Doppler radar to measure horizontal divergence of the wind field. Preprints *23rd Radar Met. Conf.*

and Cloud Physics Conf., Snowmass, Co, Amer. Met. Soc., 38–40.

Clough S. A., F. X. Kneizys, R. Davis, R. Gamache, and R. Tipping, 1980: Theoretical line shape for H_2O vapor: application to the continuum. In *Atmospheric Water Vapor*, A. Deepak, T. D. Wilkerson, and L. H. Ruhnke (eds.), Academic, New York, pp. 22–46.

Conrath, B. J., 1969: On the estimation of relative humidity profiles from medium resolution infrared specta obtained from a satellite. *J. Geophys. Res.*, **74**, 3347–3361.

Coakley, J. A., 1991: Reflectives of uniform and broken layered clouds, *Tellus*, **43B**, 420–433.

Coakley, J. A., and F. P. Bretherton, 1983: Cloud cover from high–resolution scanner data: detecting and allowing for partially filled fields of view, *J. Geophys. Res.*, **87**, 4917–4932.

Coulson, K. L., 1975: *Solar and Terrestrial Radiation*, Academic, New York, 322 pp.

Coulson, K. L., 1983: Effects of El Chicon volcanic cloud in the stratosphere on the polarization of light from the sky, *Appl. Opt.*, **22**, 1036–1050.

Coulson, K. L., 1988: *Polarization and Intensity of Light in the Atmosphere,* A. Deepak Publ., Hampton, Va., 596 pp.

Cox, C. S., and W. H. Munk, 1955: Some problems in optical oceanography, *J. Marine Res.*, **4**, 63–78.

Cutler, C. C., A. P. King, and W. E. Kock, 1947: Microwave antenna measurements, *Proc. IRE*, **35**, 1462–1471.

Dave, J. V., 1968: Subroutines for computing the parameters of the electromagnetic radiation scattered by a sphere. Report 320–3237, IBM Scientific Center, Palo Alto, CA

Dave, J. V., and C. L. Mateer, 1964: Measuring of successive iteration of the auxiliary equation in the theory of radiative transfer. *Astrophys. J.*, **140**, 1292–1303

Dave, J. V., and C. L. Mateer, 1967: A preliminary study on the possibility of estimating total ozone from satellite measurements. *J. Atmos. Sci.*, **24**, 414–427.

Debye, P. 1929: *Polar Molecules*, Chemical Catalog Co., New York, Chapter 5.

Decker, M. T., E. R. Westwater, and F. O. Guiraud, 1978: Experimental evaluation of ground–based microwave radiometric sensing of atmospheric temperature and water vapor profiles. *J. Appl. Meteorol.*, **17**, 1788–1795.

De Zafra, R. L., M. Jaramillo, J. Barrett, L. K. Emmons, P. M. Solomon, and A. Parrishs, 1989: New observations of a large concentration of CIO in the springtime lower stratosphere over Antarctica and its implications for ozone–depleting chemistry. *J. Geophys. Res.*, **94**, 11423–11428.

Dobson, G.M.B. 1957: *Observers' Handbook for the Ozone Spectrometer*

Doneaud, A. A., S. I. Niscos, D. L. Priegnitz, and P. L. Smith, 1984: The area–time integral as an indicator for convective rain volumes. *J. Clim. Appl. Meteorol.*, **23**, 555–561.

Doviak, R. J. and D. S. Zrnic, 1984: Doppler Radar and Weather Observations, A.P., Orlando Fa, 458 pp.

Drain, L. E., 19??: *The Laser Doppler Technique*, Wiley, New York, 241 pp.

Draine, B. T., 1988: The discrete dipole approximation and its application to interstellar graphite grains. *Ap. J.*, **333**, 848–872.

Dungey, C.C.E., and C. F. Bohren, 1992: Backscattering by nonspherical hydometers as calculated by the coupled dipole method : an application in radar meteorology. *J. Geophys. Res.*

Durkee, P. A., F. Pfeil, E. Frost, and R. Shema, 1991: Global analysis of aerosol particle characteristics. *Atmos. Environ.*, **25A**, 2457–2471.

Elachi, C. 1987: *Introduction to the Physics and Techniques of Remote Sensing*, Wiley, New York, 413 pp.

Elgered, G., 1992: Tropospheric radio path delay from ground based microwave radiometry. In *Atmospheric Remote Sensing by Microwave Radiometry*, Janssen (ed.), Wiley, New York.

Evans, K. F., and G. L. Stephens, 1993: Microwave radiative transfer through clouds composed of irregularly shaped ice crystals. Submitted to *J. Atmos. Sci.*.

Evans, K. F., and J. Vivekanandan, 1990: Multiparameter radar and microwave radiative transfer of nonspherical atmospheric ice particles. *IEEE Trans. Geoscience and Remote Sensing*, **28**, 423–437.

Farman, J. C., R. T. Murgatroys, B. G. Gardiner, and J. D. Shanklin, 1985: Large losses of total ozone in Antarctica reveal seasonal ClO_x/NO_x interaction. *Nature*, **315**, 207.

Feynman, R., R. B. Leighton, and M. Sands, 1977: *Feynman Lectures on Physics*, Addison–Wesley, Boston, Vols I and II (sixth printing).

Fischer, J., W. Cordes, A. Schmitz–Peiffer, W. Renger, and P. Morel, 1991: Detection of cloud–top height from backscattered radiances within the oxygen–A band. Part 2: measurements. *J. Appl. Met.*, **30(9)**, 1260–1267.

Fischer, J., and H. Grassl, 1991: Detection of cloud–top height from backscattered radiances within the oxygen–A band. Part I: theoretical study. *J. Appl. Met.*, **30(9)**, 1245–1259.

Flatau P. J., 1992: Scattering by irregular particles in anomalous diffraction and discrete dipole approximations. Ph.D. Thesis, Dept. Atmos. Sci., Colorado State Uni., Ft. Collins, Co.

Flatau P. J., G. L. Stephens, and B. T. Draine, 1990: Light scattering in the discrete dipole approximation: Exploiting the block Toeplitz structure. *J. Opt. Soc. Amer. A*, **7**, 593–600.

Foukal, P. V., 1990: The variable sun. *Scientific Am.*, Feb., 34–41.

Fukao, S., T. Sato, T. Tsuda, I. Kimura, N. Takeuchi, M. Matsuo, and S. Kato, 1985: Simultaneous observation of precipitating atmosphere by VHF band and C/Ku band radars. *Radio Sci.*, **20**, 622–630.

Fuller, K. A., and G. W. Kattawar, 1987: Consumate solution to the problem of classical electromagnetic scattering by ensembles of spheres. II: clusters of arbitrary configuration. *Opt. Lett.*, **13**, 1063–1065.

Fymat, A. L. 1978: Analytical Inversions in remote sensory of particle size distributions. 1: Multispectral extinctions in the anomalous diffraction approximation. *Appl. Opt.*, **17**, 1675–1676.

Fymat, A. L., and C. B. Smith, 1979: Analytical Inversions in remote sensory of particle size distributions. 4: comparisons of

Fymat and Box–McKellar solutions in the anomalous diffraction approximation. *Appl. Opt.*, **18**, 3595–3598.

Gage, K., 1990: Radar observations of the free atmosphere: structure and dynamics. In *Radar in Meteorology: Battan Memorial and 40th Radar Met. Conf.*, Atlas (ed.), Amer. Met. Soc., 534–565.

Gage, K., and J. L. Green, 1978: Evidence for specular reflection from monostatic VHF radar observations of the stratosphere. *Radio Sci.*, **13**, 991–1001.

Gage, K., and B. B. Balsley, 1980: On the scattering and reflection mechanisms contributing to clear air radar echoes from the troposphere, stratosphere, and mesosphere. *Radio Sci.*, **15**, 243–257.

Garratt, J. R., 1992: *The Atmospheric Boundary Layer*, Cambridge University Press, Cambridge, 316 pp.

Gille, J. C., and F. B. House, 1971: On the inversion of limb radiance measurements. 1: temperature and thickness. *J.Atmos.Sci.*, **29**, 1427–1442.

Goedecke, G. H., and S. G. O'Brian, 1988: Scattering by irregular inhomogeneous particles via the digitized Green's function algorithm. *Appl. Opt.*, **27**, 2431–2438.

Goodberlet, M. A., and C. T. Swift, 1989: Remote sensing of ocean surface winds with special sensor microwave/images, *J Geophys Res.*, **94**, 14,547–14,555.

Goodman, J. J., B. T. Draine, and P. J. Flatau, 1991: Application of fast–Fourier transform techniques to the discrete–dipole approximation. *Opt. Letters*, **16**, 1198–1200.

Goody, R. M., and Y. L. Yung, 1989: *Atmospheric Radiation: Theoretical Basis*, Oxford University Press, New York, 519 pp.

Gordon A. L., and J. C. Comiso, 1988: Polynas in the southern ocean, *Sci. Amer.*, June 1988, 90–97.

Gossard, E. E., 1990: Radar research on the atmospheric boundary layer. In *Radar in Meteorology: Battan Memorial and 40th Radar Met. Conf.*, Atlas (ed.), Amer. Met. Soc., 477–527.

Grant, W. B., 1990: Water vapor absorption coefficients in the 8 − −13 μm spectral region: a critical review. *Appl. Optics*, **29**, 451–461.

Grant, W. B., 1991: Differential absorption and Raman lidar for water vapor profile measurements: A review. *Opt. Eng.*, **30**, 40–48.

Green, J. L., R. H. Winkler, J. M. Warnock, W. L. Clark, K. S. Gage, and T.E. VanZandt, 1978: Observations of enhanced clear air reflectivity associated with convective clouds, preprints *18th Radar met. Conf.*, Atlanta, Amer. Met. Soc., 88–93.

Greenwald, T. J., G. L. Stephens, T. H. Vonder Haar, and D. L. Jackson, 1993: A physical retrieval method of liquid water over the global oceans using SSM/I observations. *J. Geophys. Res.* (to appear).

Grody, N. C., A. Gruber, and W. S. Shen, 1980: Atmospheric water content over the tropical Pacific derived from Nimbus-6 scanning microwave spectrometer. *J. Appl. Meteor.*, **19**, 986–996.

Grund, C. J., and E. W. Eloranta, 1990: The 27–28 October 1986 FIRE IFO cirrus case study: cloud optical properties determined by high spectral resolution lidar, *Mon. Wea. Rev.*, **118**, 2344–2355.

Grund C. J., and E. W. Eloranta, 1991: University of Wisconsin high spectral resolution lidar, *Opt. Eng.*, **30**, 6–12.

Gunn, K.L.S., and T.W.R. East, 1954: The microwave properties of precipitation particle. *Quart. J. Roy. Meteor. Soc.*, **80**, 522–545.

Hall, M. P., S. M. Cherry, J.W.F. Goddard, and G.R. Kennedy, 1980: Raindrop sizes and rainfall rate measured by dual–polarization radar, *Nature*, **285**, 195–198.

Hanel, R. A., 1961: Determinations of cloud altitudes from a satellite, *J. Geophys. Res.*, **66**, 1300.

Hanel, R. A. 1983: Planetary exploration with spaceborne Michelson interferometers in the thermal infrared. In *Spectroscope Techniques*, Vol. III, G. A. Vanasse (ed.), Academic, New York, pp. 43–135.

Hansen, J. E., and Hovenier, J. W., 1974: Interpretation of the polarization of Venus. *J. Atmos. Sci.*, **31**, 1137–1160.

Hansen J. E., and J. B. Pollack, 1970: Near–Infrared light scattering by terrestrial clouds. *J. Atmos. Sci.*, **27**, 265–281.

Hansen, J. E., and Travis, L. D., 1974: Light scattering in planetary atmospheres. *Space Sci. Rev.*, **16**, 527–610.

Hardy, K. R., and K. S. Gage, 1990: The history of radar studies of the clear atmosphere. In *Radar in Meteorology: Battan Memorial and 40th Radar Met. Conf.*, Atlas (ed.), Amer. Met. Soc., 130–142.

Hariharan, P., 1992: *Basics of Interferometry.* Academic, San Diego, 213 pp.

Harries, J. E. 1990: *Earth Watch: The Climate from Space*, Horwood, Chichester, West Sussex, UK, 216 pp.

Henderson–Sellers, A., 1984: *Satellite Sensing of a Cloudy Atmosphere: Observing the Third Planet*, Taylor and Francis, London, 340 pp.

Herzberg, G., 1945: *Molecular Spectra and Molecular Structure*, Van Nostrand Reinhold, Princeton, NJ.

Heymsfield A. J., and C.M.R. Platt, 1984: A parameterization of the particle size spectrum of ice clouds in terms of the ambient temperature and the ice water content, *J. Atmos. Sci.*, **41**, 846–855

Hildebrand, P. H., and C. K. Meuller, 1985: Evaluation of meteorological radar. Part I: dual doppler analyses of air motions. *J. Atmos. Oceanic Tech.*, **2**, 362–380.

Hitchfield, W., and J. Bordan, 1954: Errors inherent in the radar measurement of rainfall at attenuating wavelengths. *J.Meteor.*, **11**, 58–67.

Hobbs, P. V. 1974: *Ice Physics.* Oxford University Press, London, 835 pp.

Hollinger, J. P., 1971: Passive microwave measurements of sea surface roughness. *IEEE Trans. Geosci, Electr.*, **GE–09**, No. 3, 165–169.

Houghton J. T., 1961: Meteorological significance of remote measurement of infrared emission from atmospheric carbon dioxide. *Q.J.R. Meteorol. Soc.*, **87**, 102–104.

Houghton J. T. and A. C. Lee, 1972: Atmospheric transmission in the $10 - -12\ \mu m$ window. *Nature*, 238, 117–118.

Houghton J. T., F. W. Taylor, and C. D. Rogers, 1984: *Remote Sounding of Atmosphere*, Cambridge University Press, Cambridge, 343 pp.

Houston, J. D., and A. J. Carswell, 1978: Four–component polarization measurement of lidar atmospheric scattering. *Appl. Opt.*, **17**, 614–120.

Huffacker, R. M., A. Jelalian, and J. A. Thompson, 1970: Laser Doppler system for detection of aircraft trailing vortices, *Proc. IEEE*, **58**, 322–326

Husson, N., B. Bonnet, N. A. Scott, and A. Chedin, 1992: Management and study of spectroscopic information: the GEISA program. *J. Quart. Spectrosc. Rad. Transfer.*

Inoue, T., 1989: Features of Clouds over the tropical pacific during northern hemispheric winter derived from split window measurements. *J. Meteorol. Soc. Japan*, **67**, 621–637.

Ishamari, A., 1978: *Wave Propagation and Scattering in Random Media (Vol I).*, Academic, New York, 255 pp.

Jackson, D. L., and G. L. Stephens, 1993: A climatological study of SSM/I precipitable water over the globe. *J. Climate*, submitted.

Jackson, T. J., T. J. Schmugge, and J. R. Wang, 1982: Passive microwave sensing of soil moisture under vegetation canopies. *Water Resour. Res.*, **18**, 1137–1142.

Jameson, A. R., and K. V. Beard, 1982: Raindrop axial ratios, *J. Appl. Meteorol.*, **21**, 257–259.

Junge, C., 1955: The size distribution and aging of natural aerosol as determined from electrical and optical data on the atmosphere. *J. Meteorol.*, **12**, 13–25.

Kalnay, E., R. Atlas, W. Bakes, and J. Susskind, 1985: GLAS experiments on the impact of FGGE satellite data on numerical weather prediction. In *Proceed. First National Workshop on the Global Weather Experiment*, National Academy Press, Washington, D.C.

Kaplan, L. D., 1959: Inferences of atmospheric structures from satellite remote radiation measurements. *J. Opt. Soc. Am.*, **49**, 1004.

Kaufman, Y., R. S. Fraser, and R. A. Ferrare, 1990: Satellite Measurements of large–scale air pollution methods. *J. Geophys. Res.*, **95**, 9895–9909.

Keeling, R. F., 1988: Measuring correlations between atmospheric oxygen and carbon dioxide mole fractions: a preliminary study of urban air. *J. Atmos. Chem.*, **7**, 153–176.

Kerker, M., 1969: *The Scattering of Light and Other Electromagnetic Radiation.* Academic, New York.

Kidder, S. and T. H. Vonder Haar, 1993: *Satellite Meteorology: An Introduction,* Academic, In Press: Academic Press, Inc.

King, J.I.F., 1958: The radiative heat transfer of planet earth. In *Scientific Uses of Earth Satellites*, 2nd Rev. Edn., University of Michigan Press, Ann Arbor.

King, M. D., 1981: A method for determining the single scattering albedo of clouds through observation of the internal scattering radiation field, *J. Atmos. Sci.*, **38**, 2031–2044.

King, M. D., and Harshvardan, 1986: Comparative accuracy of selected multiple scattering approximations. *J. Atmos. Sci.*, **43**, 784–801.

King M. D., L. F. Radke, and P. V. Hobbs, 1990: Determination of the spectral absorption of solar radiation by marine stratocumulus clouds from airborne measurements within clouds. *J. Atmos. Sci.*, **47**, 894–907.

King, M. D., Y. J. Kaufman, W. P. Menzel, and D. Tanr, 1992: Remote sensing of cloud, aerosol and water vapor properties from Moderate Resolution Imaging Spectrometer (MODIS). *IEEE Trans. Geoscience and Remote Sensing*, **30**, 2–27.

Kirdiashev, K. P., A. A. Chukhlantsev, and A. M. Shutko, 1979: Microwave radiation of the earth's surface in the presence of vegetation cover. *Radio Eng. Electron. Phys.* (Eng. Transl.), **24**, 256–264.

Kittel, C., 1971: *Introduction to Solid State Physics* (4th Ed), Wiley, New York, 766 pp.

Klenk, K. E., P. K. Leonard, A. J. Fleig, V. G. Kaveeschwar, R. D. McPeters, and P. M. Smith, 1982: Total ozone determination from backscattered ultraviolet, (BUV) experiment. *J. Appl. Meteorol.*, **21**, 1672–1684

Klett, J. D., 1981: Stable analytical inversion for processing lidar returns, *Appl. Opt.*, **20**, 211–220.

Klett, J. D., 1984: Anomalous diffraction model for inversion of multispherical extinction data including absorption effects. *Appl. Opt.*, 23, 4499–4508.

Kliger, D. S., J. W. Lewis, and C. E. Randall, 1990: *Polarized Light in Optics and Spectroscopy*, HP., San Diego, 304 pp.

Kobayashi, T., 1961: *Philos. Mag.*, **6**, 1361.

Komhyr, W. D., R. D. Grass, and R. K. Leonard, 1989a: Dobson Spectrometer 83: A Standard for Total Ozone Measurements, 1962–1987.

Komhyr, W. D., R. D. Grass, F. J. Reitelbach, S. E. Kuester, P. R. Franchois, and M. L. Flanning, 1989b: Total ozone, ozone vertical distributions and stratospheric temperatures at south pole, Antarctica, in 1986 and 1987. *J. Geophys. Res.*, **94**, 11429–11436.

Kragh, H., 1991: Ludwig Lorentz: his contributions to optical theory and light scattering by spheres. In *Proceeding of the 2nd Intl. Congress on Optical Particle Sizing*, March 5–8, Tempe, Arizona, 1–6.

Kreiss, W. T., 1969: The influence of clouds on microwave brightness temperature viewing downward over open seas. *Proc. of IEEE*, **57**, 440-446.

Kuik, F., P. Stammes, and J. Hovenier, 1991: Experimental determination of scattering matrices of water droplets and quartz particles. *Appl. Opt.*, **30**, 4872–4881.

Kummerow, C. D., R. A. Mack, and I. M. Hakkarinen, 1989: A self–consistency approach to improve microwave rainfall estimates from space. *J. Appl. Meteorol.*, **28**, 869–884.

Lenoble, J. (ed)., 1985: *Radiative Transfer in Scattering and Absorbing Atmospheres: Standard Computational Procedures.* A. Deepak, Hampton, Va., 300 pp.

Liebe, H. J., 1981: Modeling attenuation and phase of radio waves in air at frequencies below 1000 GHz. *Radio Science*, **16**, 1183–1199.

Liebe H. J., G. A. Hufford, and T. Manabe, 1991: A model for the complex permittivity of water at frequencies below 1 THz, *J. Infra Millim. Waves*, **12(7)** 659–682.

Linke, F., and K. Boda, 1922: *Meteor. Zeitr.*, **39**, 161–166.

Liou, K–N., 1980: *An Introduction to Atmospheric Radiation*. Academic, New York, International Geophys. Series 26, 392 pp.

Liu, W. T., 1986: Statistical relation between monthly mean precipitable water and surface–level humidity over global oceans, *Mon. Wea. Rev.*, **114**, 1591–1602.

Liu, W. T., and P. P. Niiler, 1984: Determination of monthly mean humidity in the atmospheric surface layer over the oceans from satellite data. *J. Phys. Oceanogr.*, **14**, 1451–1457.

Livingston, J. M., and P.B. Russel, 1989: Retrieved of aerosol size distribution moments from multiwavelength particle extinction measurements, *J. Geophys. Res.*, **94**, 8425–8433.

Logan, N. A., 1990: Survey of some early studies of the scattering of plane waves by a sphere. *Proceed of the 2nd Intl. Congress on Optical Particle Sizing*, March 5–8, Tempe, Arizona, 7–15.

Lopez, R. E., J. Thomas, D. O. Blanchard, and R. L. Holle, 1983: Estimation of rainfall over an extended region using only measurements of the area covered by radar echoes. Presented at 21st conference on Radar Meteorology, Am. Meteorol. Soc., Edmonton, Alberta, Canada, 19–23.

Ma, Q., and R. H. Tipping, 1992a: A far wing line shape theory and its application to the water vibrational bands. *J. Chem. Phys.*, **96**, 8655–8663.

Ma, Q., and R. H. Tipping, 1992b: A far wing line shape theory and its application to the foreign–broadened water continuum absorption. *J. Chem. Phys.*, **97**, 818–828.

Maggraf, W. A., and M. Griggs, 1969: Aircraft measurements and calculations of the total downward flux of solar radiation as a function of altitude. *J. Atmos. Sci.*, **26**, 468–477.

Mason, B. J., 1971: *The Physics of Clouds*, Oxford University Press, Oxford, 671 pp.

Marshall, J. S., and W. Palmer, 1948: The distribution of raindrops with size. *J. Meteorol.*, **5**, 165–166.

Mateer, C. L., D. F. Heath, and A. J. Kreuger, 1971: Estimation of total ozone from satellite measurements of backscattered ultraviolet earth radiance. *J. Atmos. Sci.*, **28**, 1307–1311.

McClain, E. P., W. G. Pichel, and C. C. Walton, 1985: Comparative performance of AVHRR–based multichannel sea surface temperatures. *J. Geophys. Res.*, **90**, 11587–11601.

McCleese, D. J. and L. S. Wilson, 1976: Cloud top heights from temperature sounding instruments. *Quart. J. R. Met. Soc.*, **102**, 781-790.

McCormick, M. P., and C. R. Trepte, 1987: SAM II Measurements of antarctic PSCs and aerosols. *Geophys. Res. Lett.*, **13**, 1276–1279.

McCormick, M. P., and R. E. Veiga, 1992: SAGE II measurements of early pinatubo aerosols. *Geophys. Res. Letters*, **19**, 155–158.

McCormick, M. P., P. Hamill, T. J. Pepin, W. P. Chu, T. J. Swissler, and L.R. McMaster, 1979: Satellite studies of the stratospheric aerosol. *Bull. Am. Meteor. Soc.*, **60**, 1038–1046.

McGuffie, K., and A. Henderson–Sellers, 1986: Climatology from space: data sets for climate and climate modelling. In *Remote Sensing Applications In Meteorology and Climatology*, Vaughan (ed.), NATO ASI Series C, Vol. 201, 375–389.

McMillan, L. M., and C. Dean, 1982: Evolution of a new operational technique for producing clear radiances, *J. Appl. Meteorol.*, **20**, 1005–1014.

McPeters, R. D. and W. D. Komhyr, 1991: Long–term changes in the total ozone mapping spectrometer relative to world primary standard Dobson spectrometer 83. *J. Geophys. Res.*, **96**, 2987–2993.

Meador W. E., and W. R. Weaver, 1980: Two stream approximation to radiative transfer in planetary atmospheres: a unified description of existing methods and a new improvement. *J. Atmos. Sci.* , **37**, 630–643.

Measures, R. M., 1984: *Laser Remote Sensing*, Wiley, New York, 510 pp.

Menke, W., 1989: *Geophysical Data Analysis: Discrete Inverse Theory, Revised Edition*, Academic, New York, 289 pp.

Menzies, R. T., and R. M. Hardesty, 1989: Coherent doppler radar lidar for measurement of wind fields. *JEEE*, 77(3), 449–462.

Metejka, T., and R. C. Srivastava, 1991: An improved version of the extended velocity–azimuth display analysis of single Doppler radar data. *J. Atmos. Oceanic. Tech.*, **8**, 453–466.

Mie G., 1908: Beitr: age zur Optik tr: uber medien, speziell kolloidaler metall: osungen. *Ann. Physik*, **25** 377–445.

Michalsky J. J., E. W. Pearson, and B. A. LeBaron, 1990: An assessment of the impact of volcanic eruptions on the Northern Hemispheric Aerosol Burden During the Last Decade. *J. Geophys. Res.*, **95**, 5677–5688.

Minneart, M., 1954: *The Nature of Light and Colour on the Open Air*, Dover, New York, 362 pp.

Mitchell, R. M., and D. M. O'Brien, 1987: Error estimate for passive satellite measurements of surface pressure using absorption in the A band of oxygen. *J. Atmos. Sci.*, **44**, 1981–1991.

Mitchell, R. M., and D. M. O'Brien, 1993: Correction of AVHRR shortwave channels for effects of atmospheric scattering and absorption. *Remote Sens. Envir.*, to appear.

Mo, T., B. J. Choudhury, T. J. Schmugge, J. R. Wang, and T. J. Jackson, 1982: A model for microwave emission from vegetation–covered fields. *J. Geophys. Res.*, **87**, 11 229–11 237.

Mugnai, A., H. J. Cooper, E. A. Smith, and G. J. Tripoli, 1990: Simulation of microwave brightness temperature of an evolving hailstorm at SSM/I frequencies. *Bull. Amer. Met. Soc. Cor.*, **71**, 2–13.

Muinonen, K., and K. Lumme, 1991: Light scattering by solar system dust: the opposition effect and the reversal of polarization. IAU Colloquim 126, *Origin and Evolution of Dust in the Solar System*, Kyoto, Japan (Kluwer Academic Press), 159–162

Nakajima, T., M. D. King, J. Spinhirne, and L. R. Radke, 1991: Determination of the optical thickness and effective radius of clouds from reflected solar radiation measurements. Part II: marine stratocumulus observations. *J. Atmos. Sci.*, **48**, 728–750

Nakajima, T., and M. D. King, 1990: Determination of the optical thickness and effective radius of clouds from reflected solar radiation measurements. Part I: theory. *J. Atmos. Sci.*, **47**, 1878–1893.

Neale, C.M.U., M. J. McFarland, and K. Chang, 1990: Land–surface–type classification using brightness temperature from the special sensor microwave/imager. *IEEE Trans. Geosc. Remote Sensing*, **28**, 829–839.

Nordberg, W., J. Conaway, D. B. Ross, and T. Wilheit, 1971: Measurements of microwave emission from a foam–covered wind driven sea. *J. Atmos. Sci.*, **38**, 497–506.

O'Brien, D. M., and R. M. Mitchell, 1991: Error estimates for retrieval of cloud top pressure using absorption in the A–band of oxygen. *J. Appl. Met.*, **31**, 1179–1192.

Owens, J. C., 1967: Optical refractive index of air: dependance on pressure, temperature and composition, *Appl. Opt.*, **6(1)** 51–59

Pal, S. R., and A. J. Carswell, 1977: The polarization characteristics of lidar scattering from snow and ice crystals in the atmosphere. *J. Appl. Meteor.*, **16**, 70–80.

Parol, F., J. C. Buriez, G. Brogniez, and Y. Fouquart, 1991: Information content of AVHRR channels 4 and 5 with respect to the effective radius of cirrus cloud particles. *J. Appl. Meteorol.*, **30**, 973–984.

Petty, G. W., 1990: On the response of the SSM/I to the marine environment–implications for atmospheric parameter retrievals, PhD Dissertation, Depart. of Atmos. Sci. AK–40, Univ. of Washington, Seattle, WA 98195.

Petty, G. W., and K. B. Katsaros, 1990: New geophysical algorithm for the special sensor microwave imager. In 5th Conf. on Satellite Meteorol. and Oceanog., Sept 3–7, London.

Piexoto, J. P., and A. H. Oort, 1991: *Physics of Climate.* Amer. Inst. Phys., New York, 520 pp.

Pilewskie, P., and S. Twomey, 1987: Discrimination of ice from water in clouds by optical remote sensing, *Atmos. Res.*, **21** 113–122

Post, M. J., 1981: Atmospheric Purging of El Chicon Debris, *J. Geophys. Res.*, **91(4)**, 5222–5228.

Platt, C.M.R., 1979: Remote sounding of high clouds: I. Calculations of visible and infrared optical properties from lidar and radiometer measurements. *J. Appl. Meteor.*, **18**, 1130–1143.

Platt, C.M.R., 1981: Remote sounding of high clouds: III. Monte Carlo calculations of multiply scattered lidar returns. *J. Atmos. Sci.*, **38**, 156–167.

Platt, C.M.R., 1981: Transmission and reflectivity of ice clouds by active probing. In *Clouds, Their Formation, Optical Properties and Effects.* P. V. Hobbs and A. Deepak (Eds), Academic, New York, 407–436.

Platt, C.M.R., and A. C. Dilley, 1979: Remote sounding of high clouds: II. Infrared emissivity of cirrostratus. *J. Appl. Meteor.*, **18**, 1144–1150.

Platt, C.M.R., and A. C. Dilley, 1981: Remote sounding of high clouds: II Observed temperature variations in cirrus optical properties. *J. Atmos. Sci.*, **38**, 1069–1082.

Platt, C.M.R. and A. C. Dilley, 1984: Determination of the cirrus particle single–scattering phase function from lidar and solar radiometric data. *Appl. Opt.*, **23**, 380–386.

Prabhakara, C., G. Dalu, and V. G. Kunde, 1974: Estimation of sea surface temperature from remote sensing in 11 to 13 μm region. *J. Geophys. Res.*, **79**, 5039–5044.

Prabhakara, C., H. D. Chang, and A.T.C. Chang, 1982: Remote sensing of precipitable water over oceans from Nimbus–7 microwave measurements. *J. Appl. Meteorol.*, **21**, 59–68.

Prabhakara, C., G. Dalu, R.C. Lo, and N. R. Nath, 1979: Remote sensing of seasonal distributions of precipitable water vapor over the oceans and inference of boundary layer structure. *Mon. Wea. Rev.*, **107**, 1388–1401.

Prabhakara, C., R. S. Fraser, G. Dalu, M.L.C. Wu, R. J. Curran, and T. Styles, 1988: Thin cirrus clouds: seasonal distribution over oceans deduced from Nimbus–4 IRIS. *J. Appl. Meteorol.*, **27**, 379–399.

Prata, A. J., 1993: Land surface temperatures derived from the AVHRR. I: theory. Submitted to *J. Remote Sensing.*

Preisendorfer, R. W. 1976: *Hydrological Optics.* Vols I&II, NOAA ERL and PMEL.

Prosperso, J. M. 1982: Dust from the Sahara. *Natural History*, **46**, 55–61.

Pruppacher, H. R., and K. V. Beard, 1970: A wind tunnel investigation of internal circulation and shape of water drops falling at terminal velocity in air. *Q. J. R. Meteorol. Soc.*, **96**, 247–256.

Pruppacher, H., and J. D. Klett, 1980: *Microphysics of Clouds and Precipitation*, D. Reidal, Dordrecht, 714 pp.

Purcell, E. M., and C. R. Pennypacker, 1973: Scattering and absorption of light by nonspherical dielectric grains. *Ap. J.*, **186**, 705–714.

Radke L. F., J. A. Coakley, Jr., and M. D. King, 1989: Direct and remote sensing observations of the effects of ships on clouds, *Science*, **246**, 1146–1149.

Ray, P. S., D. P. Jorgensen, and S.L. Wang, 1985: Airborne doppler radar observations of convective storm. *J. Clim. and Appl. Meteorol.*, **24**, 687–698.

Reagan, J. A., and B. M. Herman, 1972: Optical methods for remotely measuring aerosol size distributions. *AIAA Journ.*, **10**, 1401–1407.

Rees, W., 1990:*Physical principles of remote sensing.* Cambridge University Press, Cambridge, 247 pp.

Reynolds, R. W., 1988: A Real–Time Global Sea Surface Temperature Analysis. *J. Climate*, **1**, 75–86.

Rind, D., E-W. Chou, W. Chu, S. Oltmans, J. Lerner, J. Larson, M.P. McCormick, and L. McMasters, 1991: Satellite validation of water vapor feedback in GCM climate change experiments. *Nature*, **349**, 500–503.

Rizzi, R. R. Guzzi, and R. Legnani, 1982: Aerosol size spectra from special extinction data: the use of a linear inversion method. *Appl. Opt.*, **21(9)**, 1578–1587.

Rodgers, C. D., 1976: Retrieval of atmospheric temperature and composition from remote measurements of thermal radiation. *Rev. Geophys. Space Phys.*, **14**, 609– ??? .

Rodgers, C. D., 1990: Characterization and error analysis of profiles retrieved from remote measurements. *J. Geophys. Res.*, **95(05)**, 5587–5595.

Rosenfeld, D., D. Atlas, and D. A. Short, 1990: The estimation of convective rainfall by area integrals. 2. The height–area rainfall threshold (HART) method. *J. Geophys. Res.*, **95**, 2161–2176.

Rossow W. B. 1989: Measuring cloud properties from space: A review. *J. Climate*, **2**, 201–213.

Rossow, W. B., and R. A. Schiffer, 1991: ISCCP data Products. *Bull. Amer. Meteorol. Soc.*, **72**, 2–20.

Rossow, W. B., and L. C. Gardner, 1992: Cloud detection using satellite measurements of infrared and visible radiances for IS-CCP. Submitted to *J. Climate.*

Roth, N., K. Anders, and A. Frohng, 1991: Refractive–index measurements for the correction of particle sizing methods, *Appl. Opt.*, *30*, 4960–4965.

Rothman, L. S., R. R. Gamache, A. Goldman, L. R. Brown, T. A. Toth, H. M. Pickett, R. L. Poynter, J.–M. Flaud, C. Camy-Peyret, A. Barbe, N. Husson, C. P. Rinsland, and M.A.H. Smith, 1987: The HITRAN database: 1986 edition, *J. Appl. Meteor.*, **26**, 4058–4097. (3)

Ruf, C. S., and C. T. Swift, 1988: Atmospheric Profiling of water vapor density with a 20.5–23.5 GHz autocorrelation radiometer. *J. Atmos. Oceanic Tech.*, **5**, 539–546.

Rutledge, S. A., E. R. Williams, and T. D. Keenan, 1992: The down under doppler and electricity experiment (DUNDEE): overview and preliminary results, *Bull. Amer. Meteorol. Soc.*, **73**, 3–16.

Rutledge, S. A., R. A. Houze, Jr., M. I. Biggerstaff, and T. Matejka, 1988: The Oklahoma–Kansas mesoscale convective system of 10–11 June 1985: precipitation structure and single–doppler radar analysis. *Mon. Wea. Rev.*, **116**, 1409–1430.

Ryan, J. S., S. R. Pal, and A. I. Carswell, 1979: Laser backscattering from dense water–droplet clouds. *J. Opt. Soc. Amer.*, **69**, 60–67.

Sabins Jr, F. F. 1982: *Remote Sensing—Principles and Interpretation* (2nd Ed), Freeman, New York, 449 pp.

Sassen, K., 1991: The polarization lidar technique for cloud research: a review and current assessment. *Bull. Amer. Meteorol. Soc.*, **72**, 1848–1865.

Sassen, K., K–N Liou, S. Kinne, and M. Griffin, 1985: Highly supercooler cirrus cloud water: confirmation and climate implications. *Science*, **227**, 411–413.

Schluessel, P., and W. J. Emery, 1990: Atmospheric water vapor over oceans from SSM/I measurements. *Int. J. Remote Sens.*, **11**, 753–766.

Schmidt, E. O., R. F. Arduini, B. A. Wielicki, and R. S. Stone, 1993: Considerations for modeling thin cirrus effects on brightness temperature differences. *J. Appl. Meteorol.*, submitted.

Schotland, R., 1964: The determination of the vertical profile of atmospheric gases by means of a ground based optical radar. In *Proceed. 3rd Symposium on Remote Sens. of Envir.*, Oct. 1964, University of Michigan, Ann Arbor.

Schuster, A. 1905: Radiation through a foggy atmosphere. In *Selected Papers on the Transfer of Radiation,* D. Menzel (ed.), Dover, New York.

Seliga, T. A., and V. N. Bringi, 1976: Potential use of radar differential reflectivity measurements at orthogonal polarizations for measuring precipitation. *J. Appl. Meteorol.*, **15**, 69–76.

Shapiro, P. R., 1975: Interstellar polarization: magnetite dust. *Ap. J.*, **201**, 151–164.

She, C. Y., 1990: Remote measurements of atmospheric parameters: new applications of physics with lasers. *Contemp. Phys.*, **31**, 247–260.

Shifrin, K. S., and A. Y. Perelman, 1966: Determination of particle spectrum of atmospheric aerosol by light scattering. *Tellus* XVIII, 566–572.

Short, N. 1982: *The Landsat Tutorial Workbook: Basics of Satellite Remote Sensing.* NASA Ref. publ. 1078, U.S. Govt. Printing Office, Washington, D.C.

Short, D. A., T. Kozu, K. Nakamura, and T. D. Keenan, 1992: On stratiform rain in the tropics, *J. Meteorol. Soc. Japan,* submitted.

Simpson J., R. F. Adler, and G. R. North, 1988: A proposed tropical rainfall measuring mission (TRMM) satellite. *Bull. Amer. Meteorol. Soc.*, **69(3)**, 278–295.

Singer, S. F., and G. F. Williams, Jr., 1968a: Microwave detection of precipitation over the surface of the ocean. *J. Geophys. Res.*, **73**, 3324-3327.

Singham S. B., and C. F. Bohren, 1988: Light scattering in an arbitrary particle: the scattering–order formulation of the coupled–dipole method. *J. Opt. Soc. Am.*, A, **5**, 1867–1872.

Smith, C. B., 1982: Inversion of the anomalous diffraction approximation for variable complex index of refraction near unity, *Appl. Opt.*, 21, 3363–3366.

Smith W. L., 1970: Iterative solution of the radiative transfer equation for temperature and absorbing gas profiles of an atmosphere. *Applied Optics,* **9,** 1993–1999.

Smith, W. L., 1972: Satellite techniques for observing the temperature structure of the atmosphere, *Bull. Am. Meteor. Soc.,* **53**, 1074– 1082.

Smith, W. L., 1991: Atmospheric soundings from satellites—false expectation or the key to improved weather prediction. *Q.J.R. Meteorol. Soc.*, **117**, 267–297.

Smith, W. L., and H. M. Woolf, 1976: The use of eigenvectors of statistical covariance matrices for interpreting satellite sounding radiometer observations, *J. Atmos. Sci.,* **33,** 1127–1140.

Smith, W. L., and C.M.R. Platt, 1979: Comparison of satellite-deduced cloud heights with indications from radiosonde and carrel–based laser measurements, *J. Appl. Meteorol..*, **17**, 1796–1802

Smith, W. L., and R. Frey, 1990: On cloud altitude determinations from high resolution interferometer sounder (HIS) observations. *J. Appl. Met.*, 29, 658–662.

Smith, W. L., H. M. Woolf, and H. E. Fleming, 1970: A regression method for obtaining real–time temperature and geopotential height profiles from satellite spectrometer measurements and its application to Nimbus–3 SIRS observations. *Mon. Wea. Rev.*, **98**, 582–603.

Smith, W. L., W. M. Woolf, C. M. Hayden, D. Q. Wark, and L. M. McMillan, 1979: The TIROS–N operational vertical sounder, *Bull. Amer. Meteorol. Soc.*, **60**, 1177–1187.

Sobel, M. I., 1987: *Light*, University of Chicago Press, Chicago, 263 pp.

Solomon, S., 1988: The mystery of the Antarctic ozone hole. *Rev. Geophys.*, **26**, 131–148.

Spencer, R. W., 1993: Global Oceanic Precipitation from the MSU during 1979-91 and Comparisons to Other Climatologies. *J. Climate*, **6**, 1301-1326.

Spencer, R. W., B. B. Hinton, and W. S. Olson, 1983a: Nimbus–7 37 GHz radiances correlated with radar rain rates over the Gulf of Mexico. *J. Climate Appl. Meteor.*, **22**, 2095–2099.

Spencer, R. W., W. S. Olson, Wu Rongzhang, D. W. Martin, J. A. Weinman, and D. A. Santek, 1983b: Heavy thunderstorms observed over land by the Nimbus–5 scanning multichannel microwave radiometer. *J. Climate Appl. Meteor.*, **22**, 1041–1046.

Spencer, R. W., J. R. Christy, and N. C. Grody, 1990: Global Atmospheric temperature monitoring with satellite microwave measurements: method and results 1979–1984. *J. Climate*, **3**, 1111–1128.

Srivastava, R. C., T. J. Matejka, and T. J. Lorello, 1986: Doppler radar study of the trailing anvil region associated with a squall line, *J. Atmos. Sci.*, **43**, 356–377.

Stackhouse, P. W., Jr, and G. L. Stephens, 1991: A theoretical and observational study of the radiative properties of cirrus: Results from FIRE 1986, *J. Atmos. Sci.*, **48**, 2044–2059.

Staelin, D. H, K. F. Kunz, R.K.L. Poon, R. W. Wilcox, and J. W. Waters, 1976: Remote sensing of atmospheric water vapor and liquid water with nimbus–5 microwave spectrometer. *J. Appl. Meteor.*, **15**, 1204–1214.

Stephens, G. L., 1990: On the relationship between water vapor over the oceans and sea surface temperature. *J. Climate*, **3**, 644–645.

Stephens, G. L., and S. C. Tsay, 1990: On the cloud absorption anomaly. *Quart. J. Roy. Meteor. Soc.*, **116**, 671–704.

Stephens, G. L., S. K Cox, P. W. Stackhouse, Jr., J. Davis, and the 1992 AT622 class 1993: FIRE in the classroom, Submitted to *Bull. Amer. Met. Soc.*

Stolarski, R., R. Boikov, L. Bishop, C. Zerefos, J. Staehelin, and Z. Zawadny, 1992: Measured trends in stratospheric ozone. *Science*, **256**, 342–349.

Stone, R. S, G. L. Stephens, C.M.R. Platt, and S. Banks, 1990: The remote sensing of thin cirrus clouds using satellites, lidar and radiative transfer theory. *J. Appl. Meteorol.*, **29**, 353–366.

Strivastava, R. C., T. J. Matejka, and T. J. Lorello, 1988: Doppler radar study of the trailing anvil region associated with a squall line. *J. Atmos. Sci.*,43, 356–377.

Suttles, J. T., R. N. Green, P. Minnis, G. L. Smith, W. F. Staylor, B. A. Wielicki, I. J. Walker, D. F. Young, V. R. Taylor, and L. L. Stowe, 1988: Angular radiation models for earth–atmosphere system. NASA Ref. Publ. 1184, July 1988, 144 pp.

Takano, Y., and K. N. Liou, 1989: Solar radiative transfer in cirrus clouds. Part I: single–scattering and optical properties of hexagonal ice crystals. *J. Atmos. Sci.*, **46**, 20–36

Tang, L., J. A. Kong, and R. T. Shin, 1985: *Theory of Microwave Remote Sensing*, Wiley, New York, 613 pp.

Tatarskii, V. I., 1971: *The Effects the Turbulent Atmosphere on Wave Propagation.*, U.S. Dept Commerce, 74–76. [Available from NTIS, Springfield, VA 22151.]

Taylor, F., and J. R. Eyre, 1989: Future Satellite Missions. *Weather*, 298–302.

Tjemkes, S. A., G. L. Stephens, and D. L. Jackson, 1991: Spaceborne observation of columnar water vapor: SSMI observations and algorithm. *J. Geophys. Res.*, **96**, 10941–10954.

Tooma, S. G., R. A. Mennella, J. P. Hollinger, and R.D. Ketchum, Jr., 1975: Comparison of sea–ice type identification between airborne dual–frequency passive microwave radiometry and standard laser/infrared techniques. *J. Glaciol.*, **15**, 225–239.

Toon, O. B., C. P. McKay, and T. P. Ackerman, 1989: Rapid calculation of radiative heating rates and photodissociation rates in inhomogeneous multiple scattering atmospheres. *J. Geophys. Res.*, **94**(D13), 16287–16301.

Townes, C. H., and A. L. Schalow, 1955: *Microwave Spectroscopy*, Dover, New York, 698 pp.

Tucker, C. J., 1978: A comparison of sensor bands for vegetation monitoring. *Phot. Prog. Remote Sens.*, **44**, 1369–1380.

Tucker, C. J., 1979: Red and photographic infrared linear combinations for monitoring vegetation. *Rem. Sens. Env.*, **8**, 127–150.

Twomey, S., 1971: Radiative Transfer: Terrestrial Clouds. *J. Quant. Spectrosc. Radiat. Transfer*, **11**, 779-783.

Twomey, S., 1977: *Introduction to the Mathematics of Inversion in Remote Sensing and Direct Measurement*, Elsevier, New York, 243 pp.

Twomey, S., 1991: Aerosols, clouds and climate, *Atmos. Env.*, **25A**, 2435–2442.

Twomey, S., and K. J. Seton, 1980: Inferences of grass microphysical properties of clouds from spectral reflectance measurements, *J. Atmos. Sci.*, **37**, 1065–1069.

Ulaby, F. T., R. K. Moore, and A. K. Fung, 1981(I), 1982(II), 1986(III): *Microwave Remote Sensing* (Vols. I, II & III), Addison–Wesley, Boston.

Uyeda, H., and K. Kituchi, 1979: Observations of the Three Dimensional Configuration of Snow Crystals of Combination of Bullet Type. *J. Meterol. Soc. of Japan*, **57**, 488-489.

Valley, S. L. (ed.), 1965: *Handbook of Geophysics and Space Environment*. Airforce Cambridge Res. Lab., Cambridge, Mass.

van de Hulst, H. C., 1957: *Light Scattering by Small Particles*. Dover, New York, 470 pp.

van de Hulst, H. C., 1980: *Multiple Light Scattering, Tables, Formulas and Applications* (Vol. I & II), Academic, New York.

Viera, G., and M. A. Box, 1985: Information content analysis of aerosol remote–sensing experiments using an analytic eigenfunction theory: anomalous diffraction approximation, *Appl. Opt.*, **24**, 4525–4533. (6)

Volz, F., 1959: Photometer mit selen–photoelement zur spektralen messung der sonnen strahlung ans zur bestimmung der wallenlangeabbhangigkeit der dunsttrubung, *Arch. Meteorol. Geophys. Bioklimatol*, **10**, 100–131.

von Hippel, A. R., 1954: *Dielectrics and Waves*, Wiley, New York, 284 pp.

Voss, K. J., and E. S. Fry, 1984: Measurement of the Mueller matrix for ocean water, *Appl. Opt.*, 23, 4427–4439.

Wakasugi, K., B. B. Balsely, and T. L. Clark, 1987: The VHF Doppler radar as a tool for cloud and precipitation studies, *J. Atmos. Oceanic Technol.*, **4**, 273–280.

Wakimoto, R. M., and V. N. Bringi, 1988: Dual polarization observations of microbursts associated with intense convection: the 2D July storm during the MIST Project, *Mon. Wea. Rev.,* **116**, 1521–1539. (5)

Wallace, J. M., and P. V. Hobbs, 1977: *Atmospheric Science: An Introductory Survey,* Academic, New York, 467 pp.

Walter, S. J., 1992a: The tropospheric microwave water vapor spectrum: uncertainties for remote sensing. In *Proceedings of Microwave Radiometry and Remote Sensing,* Boulder, Jan. 14–16.

Walter, S. J., 1992b: The slant path atmospheric refraction calibrator: an instrument to measure the microwave propagation delays induced by atmospheric water vapor. *IEE Trans. Geosci. and Remote Sens.,* **30(3)**, 462–471.

Wan, Z–M., and J. Dozier, 1989: Land surface temperature measurement from space: physical principles and inverse modeling, *IEEE Trans. Geosci. and Remote Sensing,* **27**, 268–278.

Warren, S. G., 1984: Optical Constants of ice from the ultraviolet to the microwave. *Appl. Opt.,* **23**, 1206–1225.

Waters, J. W., 1976: Absorption and emission by atmospheric gases. In *Methods of Experimental Physics,* Vol. 12, Part B, Academic, New York, 142–176.

Wayne, R. P., 1985: *Chemistry of Atmosphere,* Oxford University Press, Oxford, 361 pp.

Weinman, J.A., 1988; Derivation of atmospheric extinction profiles and wind speed over the ocean from satellite–borne lidar. *Appl. Opt.,* **27**, 3994–4001.

Wendling, P., R. Wendling, and H. K. Weickman, 1982: Scattering of solar radiation by hexagonal ice crystals. *Appl. Opt.,* **18**, 2663–2671.

Wielicki, B. A., and J. A. Coakley, 1981: Cloud retrieval using infrared sounder data: error analysis, *J. Appl. Meteorol.,* **20**, 157–169

Wielicki, B. A., and L. Parker, 1992: On the determination of cloud cover from satellite sensors: the effect of sensor spatial resolution, *J. Geophys. Res.,* **97**(D12), 12799–12824.

Wienberg, S., 1977: *The First Three Minutes*, Basic, New York, 188 pp.

Wilheit, T. T., 1979: A model for the microwave emissivity of the ocean's surface as a function of wind speed, *IEEE Trans. Geosci. Electronics*, **GE–17(4)**, 244–249.

Wilheit, T., A.T.C. Chang, M.S.V. Rao, E. B. Rodgers, and J. S. Theon, 1977: A satellite technique for quantitatively mapping rainfall rates over the oceans. *J. Appl. Meteor.,* **16**, 551–560.

Williams, E. R., S. A. Rutledge, S. G. Geotis, N. Renno, E. Rasmussen, and T. Rickenbach, 1992: A radar and electrical study of tropical "hot towers," *J. Atmos. Sci.*, **49**, 1380–1395.

Wiscombe, W. J., 1980: Improved Mie scattering algorithms, *Appl. Opt.*, **19**, 1505–1509.

Wood, J. T., and R. A. Brown, 1986: Single doppler velocity signature interpretation of nondivergent environmental winds. *J. Atmos. Oceanic Tech.*, **3**, 144–128.

Wu, M–L.C., 1985: Remote sensing of cloud–top pressure using reflected solar radiation in the oxygen A–band. *J. Clim. Appl. Meteorol.*, **24**, 539–546.

Wu Z., R. E. Newell, and J. Hsuing, 1990: Possible factors controlling global marine temperature variations over the past century, *J. Geophys. Res.*, **95**, 11799–11810.

Wylie, D. P., and W. P. Menzel, 1989: Two years of cloud cover statistics using VAS, *J. Climate*, **2**, 380–392.

Yasim, K., and R. L.Armstrong, 1990: Theoretical modeling of microwave absorption by water vapor. *Appl. Optics*, 29, 1879–1883.

Yamamoto, G. A., and D. Q. Wark, 1961: Discussion of the letter by R. A. Hanel, "Determination of cloud altitude from a satellite." *J. Geophys. Res.*, **66**, 3596.

Subject Index